MULTIDIMENSIONAL
PALAEOBIOLOGY

Related Pergamon Titles of Interest

BOOKS

HANLEY
Microcomputer Applications in Geology II

KOCH
Geological Problem Solving With Lotus 1-2-3

MARSAL
Statistics for Geoscientists

NIELD
Drawing & Understanding Fossils

NIELD
Palaeontology

JOURNALS

Bulletin of Mathematical Biology

Computers & Geosciences

Computers & Mathematics With Applications

Mathematical and Computer Modelling

Full details of all Pergamon publications/free specimen copy of any Pergamon journal available on request from your nearest Pergamon office

613
R495m
1991

MULTIDIMENSIONAL PALAEOBIOLOGY

by

RICHARD A. REYMENT

DEPARTMENT OF HISTORICAL GEOLOGY AND PALAEONTOLOGY,
UNIVERSITY OF UPPSALA, SWEDEN

U.S. GEOLOGICAL SURVEY
MENLO PARK
APR 29 1993
LIBRARY

PERGAMON PRESS

OXFORD · NEW YORK · SEOUL · TOKYO

UK	Pergamon Press plc, Headington Hill Hall, Oxford OX3 0BW, England
USA	Pergamon Press, Inc, Maxwell House, Fairview Park, Elmsford, New York 10523, USA
KOREA	Pergamon Press Korea, KPO Box 315, Seoul 110-603, Korea
JAPAN	Pergamon Press, 8th Floor, Matsuoka Central Building, 1-7-1 Nishi-Shinjuku, Shinjuku-ku, Tokyo 160, Japan

Copyright © 1991 R. A. Reyment

All Rights Reserved. No part of this publication may be reproduced, stored in a retrieval system or transmitted in any form or by any means: electronic, electrostatic, magnetic tape, mechanical, photocopying, recording or otherwise, without permission in writing from the publisher.

First edition 1991

Library of Congress Cataloging-in-Publication Data
Reyment, R. A.
Multidimensional palaeobiology/by Richard A. Reyment.
p.   cm.
Includes bibliographical references.
1. Paleobiology—Statistical methods.
2. Multivariate analysis.
I. Title.
QE721.2.R49   1991   560'.72—dc20   90-22476

British Library Cataloguing in Publication Data
Reyment, R. A. (Richard Arthur) 1926-
Multidimensional palaeobiology.
1. Palaeobiology
I. Title
560

ISBN 0-08-0 37231-7 Hardcover
ISBN 0-08-0 41001-4 Flexicover

Printed in Great Britain by BPCC Wheatons Ltd, Exeter

Contents

Preface

When the step is taken to write a scientific text, there is a tacit understanding that one has something useful to impart to others—established scientists and students alike. This is, as we all know, not always the case and there is no doubt that much of what has been, and will be, published is of little lasting value. In a text of the present kind, it can be difficult to hit upon the happy middle road of creating the right melange of subjects, statistical, biological and geological. One way of testing the product is to try it out on students. Another way is to badger colleagues into criticizing drafts of the text. I have used both of these methods quite extensively.

Encouraged by the overwhelmingly positive response to an earlier work in somewhat the same vein as the present text (*Introduction to Quantitative Paleoecology*, Elsevier, Amsterdam, 1971), I have decided to make use of the same general format again. Thus, the various topics introduced are taken up as case histories, or examples, complete with theoretically oriented introductions, methodological sections and an interpretation of the results.

Over the last 20 years or so, many stimulating books treating "the modern Palaeontology" have appeared. Thus, the volume by Raup and Stanley (1967) showed that it was possible to find a new way to look at palaeontology and, moreover, to make us aware of the fact that there was a wealth of fascinating problems for us to solve. Many of the newer works have been concerned with what has come to be known as Evolutionary Biology, such as the books by Stanley (1979) and Gould (1977). These, and other authors, have done much towards making palaeontology a worthy partner of the more "accepted" branches of biology in using new techniques and concepts for unravelling the mysteries of speciation.

In general, I think it is not unreasonable to say that the student of Palaeontology of today is given a far more interesting time than I had some 40 years ago, notwithstanding the valiant efforts of my respected mentor, Curt Teichert. Apart from the quantitative methods forming the backbone of the present text, the advent of the scanning electronic microscope has, for many of us, been of such unrivalled importance that one might venture to claim that many areas of classical palaeontology are being worked over again to great effect. Much of the material I use to illustrate the case histories in the ensuing text comes from my main field of micropalaeontology. It is perhaps a reflection of the importance of the SEM that my coworkers and I are frequent contributors to the purely taxonomical publication *Atlas of Ostracod Shells*. It also mirrors the inescapable fact that sound taxonomy must be the basis of all palaeontological work. Without a reliable palaeontological footing, no quantitative analysis, no matter how advanced the methods, can yield respectable results.

I believe I have new and useful information to impart, particularly in the fields of the application of multivariate statistical analysis (recently this type of work has been labelled "traditional multivariate statistics" not without a pejorative undertone) to the study of fossils. Also, the application of the methods of quantitative genetics to palaeontological material is a new sphere and one with which I have maintained a close interest for several years. You will doubtless learn on studying these applications that the subject is new, even for theoretical biologists, and there is still a lot to be learned before a fully developed methodology is available.

Perhaps the most important message I bring is that there has been a revolution in applied multivariate statistics over the last few years. Many of the techniques we know, and with which we have become comfortable, are no longer the safe havens we once thought them to be. What does this mean you might well ask? Are the canned programs we find at the computer centre unreliable? Are the diskette programs we use on our personal computers flawed in some obscure manner? Alas, the truth is cruel and it transpires that in many situations, standard forms of multivariate analysis, as available in general packages, are not suitable for the tasks to which they are assigned. As long as your data do not deviate more than slightly from the multivariate Gaussian state, i.e. they conform with the theoretical requirements of the multivariate method of interest, then little can go wrong with your analysis. If, however, you have atypicalities in your data, compositional data, etc., then a great deal can, and does, go awry. I shall be giving these, and other questions, special attention and bringing to your notice the important statistical contributions of applied workers in the multivariate field.

On first sight, you may be disconcerted by the seeming lack of homogeneity in the subject matter covered. This must be so by the very nature of the theme, the application of multivariate statistical methods to a range of suitable topics in palaeobiology which is, by definition, heterogeneous. One of the principal aspirations of my text is to unite exploratory techniques such as **ordination**, multivariate statistical methods aimed at testing hypotheses about the data, and palaeobiological problems tractable to quantitative analysis.

To all who know their work, it will be obvious how much I have gained from my association with Fred L. Bookstein, Norm A. Campbell, John Aitchison, John Birks, Allan Gordon, F. James Rohlf and Robert E. Blackith. I have also profited greatly from contacts with Russell Lande, Leslie Marcus, Arthur Mourant, and many other colleagues and friends. My students and coworkers in Uppsala need special mention.

John Birks kindly provided much valued criticism of Chapters 1, 2, 3 and 6 and Allan Gordon of Chapters 1, 2 and 3. Russell Lande gave me his opinion of Chapters 5 and 6. Fred Bookstein put much work into his evaluation of the manuscript; without Fred's help the book would have had far less merit than it may have now. Leslie Marcus has worked carefully through the whole book and as a result of this, came to provide the diskette of methods that accompanies the volume. Special thanks go to Jim Rohlf for letting us incorporate his program TPSPLINE into the set of procedures available on the diskette. I am grateful to these colleagues for the time and trouble to which they have gone on my behalf. Naturally, all faults and inadequacies remaining are entirely my own responsibility. I must place on record the debt of gratitude I owe to Pergamon Press, Oxford, one of the fruits of my twenty-year association with Mr Robert Maxwell, and to Peter Henn of the Press, for the encouragement to go ahead with the project. Almost all the SEM work underlying many of the examples and illustrations was carried out by my wife Eva Reyment for which mere words of thanks are inadequate. To Uppsala University I express my indebtedness for providing an ideal research environment and to the Swedish Natural Science Research Council (N.F.R.) for grants covering some of the work.

Uppsala, 30 September 1990

CHAPTER 1

Introduction

Contents

1.1 Form of Presentation

What is being taken up in this book? In general, I consider topics in palaeobiology that can be studied profitably in terms of many variables. In most cases, the appropriate tools for doing this will be obtained from mathematical statistics. The most readily accessible source of quantitative information is obviously from morphological measures. We are most familiar with characters that express distance relationships, lengths, heights and breadths, for example; however, it is also possible to find ways of measuring variability in shape by using coordinates and this is a point we shall be giving special attention. I shall also show that the statistical analyses of size measures, as opposed to shape descriptors, are usually made in a different manner.

My aim is to provide the reader with an intellectual framework for finding analytical methods which are capable of giving an adequately informative representation of a particular situation. These situations will be mostly of a static nature, for example, such as exemplified by studies of size, shape and taxonomical relationships. Dynamic situations can also be considered in which some kind of variability is followed over time in relation to interaction with some ecological factor or factors. Clearly, most of the cases I have chosen to illustrate my text are drawn from my own fields of experience. The reader will almost certainly have other interests and will doubtless wish to experiment with them.

Central to the ideas forming the pith of my text is the concept of *statistical design*, so admirably developed by Sir Ronald Fisher (1925) in his fundamental volume on the subject. All too often, statistical advice is sought after the work has been done—rather like bringing in a medical consultant to a wake. It is absolutely necessary, yet fundamental, to the practice of multidimensional palaeobiology that one identify and characterize the scientific problem; plan the selection of diagnostic measures; select the appropriate methods of statistical analysis; investigate the statistical structure of the observations prior to analysis. The mathematical technique forming the uniting thread throughout the book is the algebra of latent roots and vectors which you will encounter in most of the multidimensional methods invoked in the following pages.

The case histories are usually presented at various levels of complexity in order to cater for the goals and ambitions of a wide range of users. To those who already possess a solid background in biometry, I apologize for the triteness of parts of the text, but to many readers with little or no knowledge of multivariate work, these same parts may seem overly erudite. It will be possible to employ almost all techniques in the book at a simple level. Nonetheless, workers with more specialized needs will be able to find more detailed treatments of topics and references to specialized approaches.

The presentations of the case histories take the following form:

1. An introductory overview of the field.
2. A description of the problem and an account of the data.
3. Methods of analysis proposed and a review of the statistical techniques.
4. The analysis and its interpretation.
5. Discussion.

Recent evolutionary ideas will be looked at in a quantitative environment whenever the opportunity presents itself.

1.2 Layout

The way in which the material has been located in the chapters requires a few words. In the present chapter I discuss the **choice of variables**. In addition to the usual statistical differentiation between continuous, discontinuous, dichotomous and qualitative variables, attention is also directed towards distinguishing between measures of size, which engender some, but rather little, information on variability in shape, and variables designed to capture variation in shape. Homology in the choice of shape variables will be elaborated upon and also the importance of producing adequate graphical displays. Some popular multivariate methods in wide use at the present time are almost exclusively aimed at the graphical appraisal of data.

In Chapter 2 the basics of linear algebra are given. I have given the question quite some thought before deciding to include a section on vectors and matrices. There are numerous good books available—why not refer the reader to one or more of them? The reason is that most users of this book will not be interested in acquiring more than the most necessary tools in order to follow the general trend of the theoretical discussions. It would be an ungenerous act to send them out to buy a book they do not want. A further reason for adding my own section on linear algebra is that the operations required for many of the methods are best presented in one and the same connexion.

Likewise, the rationale for including a section on multivariate statistics (Chapter 3) can be debated. There are a great number of excellent textbooks on the market. I have more tangible reasons for making my own presentation, despite the fact that I wish to make quite clear that this is not a text on applications of multivariate statistical analysis. There are many new techniques available of special significance for the palaeobiologist (and others, of course). These hold rather slight interest for the professional statistician, as they hardly touch upon matters of major theoretical significance and are, therefore, no more than mentioned, if at all, in standard textbooks on multivariate statistical analysis. They are, however, of great importance from the point of view of data-analysis. In Chapter 3, the main features of the multivariate methods used are reviewed. Hence, the well-known method of principal components is examined, including the many special uses to which it can be put in biological work. This is followed by assessments of principal coordinates, canonical variate analysis, the multivariate analysis of variance, and the important duality between Q- and R-mode methods, which leads us to the graphical technique of correspondence analysis. The important questions of robust methods, influence functions, atypical observations and compositional data are taken up in this chapter. Also in Chapter 3, problems pertaining to the analysis of single samples are considered. As a first step, it is advisable to ascertain whether our data are homogeneous in the multivariate sense; this can be approached graphically by means of a Q-mode analysis as well as by statistical tests. The properties of multivariate distributions are important. Usually, I employ Gower's (1966) inverted principal component method, known as principal coordinates, for a preliminary appraisal of the empirical distribution. The question of identifying sexual dimorphism and size-correlated polymorphism is taken up here as well as discrete ontogenetic states. The concept of shape analysis is introduced in respect of the representation in terms of latent roots and vectors.

The analysis of several populations forms a natural development in Chapter 3. If a data set is shown to be markedly heterogeneous, it may be possible, using graphical information, to break the original material into subsets. Once this has been achieved, the method of canonical variate analysis may be found to be useful. This chapter also introduces the concepts of linear and quadratic discriminant functions, Mahalanobis' generalized statistical distance, D^2, and the related Hotelling T^2. Problems and pitfalls in canonical variate analysis and discriminant functions are highlighted with emphasis on the stability of the coefficients. This takes us to the study of atypical values and robust procedures of estimation, along with the question of the co-occurrence of biological homogeneity and statistical heterogeneity.

There has been a certain tendency over the last few years to downgrade the role of tests for statistical significance in applied work, particularly in biological connexions. Examples in several chapters are concerned with situations in which it is necessary to pay attention to this question. The labile nature of means in samples showing even slight deviations from the multivariate Gaussian state is discussed as well as the use of the mode instead of the mean in studies of series in time.

In Chapter 4, the topics of **size and shape** are reviewed in the light of classical concepts and modern theory. Huxley's allometric equation and the principal component and factor-analytical elaborations of it are taken up as well as Mosimann's theorem and the newly developed geometric morphometrics of Bookstein. Here, the analysis of shape considered as an elastic deformation and the application of simple tensor-analytical principles to shape studies are examined. Other methods of characterizing outlines, such as Fourier representations and "eigenshapes" are discussed. Shape trajectories, biorthogonal grids and the use of the theory of thin-plate splines as solutions to the graphical ideas of D'Arcy Wentworth Thompson are considered in the light of evolution in shape over time. The subject of *fractals* is introduced here with a simple example.

One of the most exciting and, at the same time, most difficult fields from the standpoint of repeatability is that of the application of the principles of **quantitative genetics** to palaeontological data. The topic is perforce approached via studies on living animals. In particular, the results of Lande (1976, 1979, 1988) are used to develop the notions involved in Chapter 5. Time series are covered from the multidimensional palaeontological aspect in the last part of the chapter. An important feature is provided by trends in evolutionary sequences and the problem of ascertaining whether a varying sequence is stationary, in the statistical sense, or whether it displays significant trend. This leads to the coin-tossing paradigm of Raup and Gould (1974), stochastic sequences, the method of multiple comparisons for cohorts of species. Some elementary methods for the analysis of time series of relevance in palaeontology are presented.

Polymorphism forms the main theme of Chapter 6, the smallest section of the book. I have toyed with the idea of merging this part with Chapter 5, but the methods and ideas are so different, and the subject of such palaeontological significance, that I finally decided to maintain the separation. Polymorphism in shape and size are discussed as well as the analysis of ornamental polymorphism with and without pleiotropic effects. The analysis of ornamental patterns makes use of geometrically oriented morphometrics. There are various kinds of polymorphism and we shall be obliged to distinguish between polymorphism of evolutionary significance and polymorphism without evolutionary significance—solely under the control of environmental forces.

In Chapter 7 multidimensional aspects of Palaeoecology are taken up. The question of correlating between sets is examined, for example, correlating between, say, a set of morphological measures and a set of ecological factors. The most readily accessible multivariate procedure is canonical correlation analysis, notwithstanding that there can be difficulties attached to understanding what the procedure actually does. There is also the problem of deciding when a canonical correlation is biologically significant. Palaeoecological models come to the fore in this chapter and we shall consider an example in which physical and chemical components of the palaeoenvironment are balanced against morphometric variation. Other palaeoecological examples taken up in this section concern fluctuations in molluscan predation pressure and concomitant variation of the prey. The subject of the comparison of curves is important in borehole logging and evolutionary studies. This topic is examined at length and the usefulness of Gordon's (1973) method of slotting of sequences in time is stressed.

Special topics involving multivariate applications are discussed in Chapter 8. These embrace discussions of the general applicability of the methods presented in the book. Some facets of multivariate analysis not treated in Chapters 3 and 4 are presented in conjunction with the case histories. Inevitably there is some repetition of statistical detail, but I have tried to keep this to a minimum. An attempt is made to probe the reliability of the procedures advocated in the text and it is clearly indicated how far it should be possible to go without inviting spurious interpretations. Directions for future research are also indicated and there is a review of the work of authors whose results I think to be fundamentally important from the palaeobiological point of view.

1.3 Computational Aspects

Statistically impressive analyses can compound a lie. Self-deceit and scientific humbug may lurk uneasily in the background—it must be understood that statistics do not provide an elixir for all conceptual ills. Wrong quantitative results are worse than no measurements at all. The danger is insidious, as may be judged from the effects of outliers and influential observations, mixed distributions and pronounced deviations from normality. It is therefore absolutely necessary to adopt a state of continuous awareness in the interpretation of statistical analyses. If you suspect that you are using "dodgy data", you perform a service to yourself, and your peers, if you reject the material. There is also a critical point of deciding whether a quantitative analysis is really necessary; in some cases, visual inspection of the material tells us all we need to know. Be this as it may, it has been my experience that to the untutored, a quantitative analysis may seem to be quite uninteresting and unnecessary, even artificially constructed, because the theoretical aims of the study are quite lost on him. In such circumstances it becomes a matter of educating people in what mathematical biology can do and, perhaps, is not capable of doing. To paraphrase Oscar Wilde, it becomes more than a moral duty to do so; it becomes a pleasure.

Another matter needing mention concerns the **accuracy** of numerical results. You will no doubt encounter the situation in which your results for a set of data differ in some of the decimal places from those obtained for the same set of data analysed in another machine. This is due to the rounding-off procedures used in various computing devices. Double-precision arithmetic can be used to obtain increased accuracy—this has been a particular problem with some main frame computers.

All of the methods used in this book can be run on a personal computer with a mathematical coprocessor. I use an IBM PS model 50 with a hard disk storage capacity of 60 megabytes. I also use the IBM facility at the Uppsala University Computing Centre (UDAC). The microcomputer can be easily linked to the UDAC facilities via the liaising program KERMIT. Leslie F. Marcus has provided a supplement comprising useful computer programs for the PC. I seldom use canned programs, owing to the special nature of palaeobiological problems and the need for customized treatment of each of them. Canned programs often lead to quite misleading and unfortunate results, as we shall see in the following pages. This is a hapless situation, but if the predicament did not exist, there would perhaps have been a little less incentive for producing this book. I should like to point out that the material presented here represents the accumulated wisdom of 30 years in the field of applying multivariate methods to biological situations. Believe me, I have made all the mistakes there are to be made in this connexion and I flatter myself that I have learnt something from these negative experiences. Wise is he who learns from the errors of others.

Although not so widely known as such systems as the SAS-package, the one used in Marcus' programs, I have found GENSTAT to be of inestimable value in my work, particularly GENSTAT V, release 1.3. This system has been developed by statisticians of a high scientific standing working in conjunction with biologists and agricultural scientists at the Rothamsted Agricultural Station, U.K. This prestigious organization was once the scientific home of Sir Ronald Fisher—need more be said?

Let me reflect for a moment on the value of parallel-programming. My programs have been written in FORTRAN 4 and FORTRAN 77. Some of them have been in action for a long time (cf. Blackith and Reyment, 1971) and have been subjected to frequent phases of updating. The risk of introducing computing errors is by no means insignificant and it is therefore good working policy to safeguard oneself with the following measures:

(a) On a PC, you should back up your program before starting any revisory measures; likewise if you are employing a computer facility—make a copy in disk storage of the program you are about to dissect. Speaking from bitter experience, I must advise you to use diskettes of the best quality available. Disk-crashes can, and do, occur.

(b) When you have the program in its final, revised state, alternatively, an entirely new program has been constructed, it is very useful if you can check the results obtained with a set of trial data with those yielded by running the same data in an entirely separate program. This is one of the uses to which I put GENSTAT. A skeleton program which does roughly the same things as your own production, without the specialties, provides an invaluable means of checking your own programming.

(c) It is always useful to work in close computing collaboration with colleagues at another university. I have such an arrangement with several colleagues around the world. If somebody else is engaged in incorporating your betterments into his own system, then the chance that errors will lurk undetected is minimized. Things that work on one PC may not work so well on another owing to some minor constructional oddity. If you are working with other people, it is usually possible to eliminate such obstacles.

1.4 Some Definitions

What do I mean when I speak of Multidimensional Palaeobiology? And why use the word *palaeobiology* when *palaeontology* ought to suffice? There is a new journal bearing the name *Paleobiology*, but the first organized concept of a subset of palaeontology and called *Paläobiologie* derives from the German palaeontologist O. Abel who in 1912 used the word to designate palaeontological interpretations with a basis in the Paléthologie of L. Dollo, to wit, the study of fossils with the end in view of interpreting their adaptational history and mode of life. There is, therefore, a clear overlap with R. Richter's concept of palaeoecology without perhaps absolute identification. As pointed out by Lehmann (1964), Abel's use of the noun palaeobiology constituted a restriction in the current usage of the time, when it seems to have been interchangeable with palaeontology. Be this as it may, scrutiny of the material published in *Paleobiology* since its inception agree largely with the scope conceived by Abel, albeit suitably modernized. Thus we find no taxonomy, stratigraphy, nor regional palaeontology, but there is a clear preoccupation with function, actuopalaeontology (i.e. extrapolating from what can be deduced from living species, their modes of occurrence and post-mortem history, to fossils), experimental studies and computer modelling.

With the foregoing in mind, we can define *Multidimensional Palaeobiology* as being the application of multivariate statistical methods to mainly non-taxonomical problems in palaeontology. This is, as you see, a broad demarcation and one that turns out to be convenient. Notwithstanding the bounds just stipulated, some of the problems reviewed in the following contain a clear taxonomical component which is as it should be for, where would palaeontology be without its taxonomy? You may find some of the material on data in the form of frequencies to be difficult to reconcile with the main surge of the text, but for some of you, at least, these parts may prove to be the most informative.

1.5 On Variables

Although I expect that all users of this book are familiar with classes of variables, it does no harm to comment on this again. Variables can be conveniently thought of as falling into one of the following categories:

Continuous—that is, the distance measures such as length, height and breadth, weight, area, etc.

Discontinuous—that is variables that are measured in integers, whole numbers, such as the number of marginal spines on an ostracod, colour-banding in snails, the number of segments in a trilobite, etc.

Dichotomous—that is, variables of the presence-absence type. Examples of this class are represented in ostracods by the presence or absence of a posterior spine, the ornamental states smooth or costate, smooth or punctate, regular or irregular reticulations, etc., again among ostracods. Another name for two-state variables is binary variables.

Qualitative—a fourth state which normally cannot be measured in any usual manner but which can be made to enter into the computation of similarities by some kind of coding.

Notwithstanding the fact that one may feel that there is a certain degree of arbitrariness in the above classification (which I have taken from Gower, 1971), the four types of variables can and do occur frequently in problems involving ordination. The fourth category of "qualitatives" is specifically catered for in Gower's (1971) similarity matrix. Continuous variables are also called quantitative variables. Discontinuous variables are also referred to as discrete, and meristic (in genetics). The group I have referred to as "qualitative" (the terminology is from Gower 1966, 1970) can be given various definitions. I think of this class as embracing such poorly definable traits as arbitrarily expressed coarseness of sediments, shifts in colouring and the like. Gordon (1981) has provided a more erudite discussion of the nature of variables in statistical work. Hence, he distinguishes between **numeric variables** (quantitative real numbers), my first category above, and **nominal and ordinal variables** (a set of discrete states), corresponding to my second and third categories above. Nominal variables cannot be provided with an ordering, whereas ordinal variables can be ordered. Another general reference is Graybill (1961), in which many of the basic tools of multivariate analytical theory are reviewed.

The variables we normally choose for describing quantitative variability in plants and animals are measures of size. Bookstein (1986) separates these out as measures of morphological distance, a useful way of thinking about them as that is exactly what they are—expressions of the distance from point A to point B. Bookstein (1978, 1986) has likewise popularized the term **landmark** for specifying points on an organism forming a mensurable reference or fixpoint. This definition may seem provocatively unconditional but a moment's reflection should serve to disclose that this is indeed so. There is no doubt that a mesh of distances (truss, trellis) incorporates information on shape, and this can be demonstrated empirically by multivariate analytical methods. None the less, there is a clear difficulty attached to the extraction of information describing variability in shape from a situation dominated by size. More desirable would be a system of measurement that was concentrated entirely on changes in shape and which is not confounded with size variability. In order to achieve this goal, we require a geometrical measure of shape. A useful approach is provided by the triangulation technique expounded by Bookstein (1986). A landmark of diagnostic interest may be charted, individual by individual, in relation to two landmarks selected to be the baseline of the triangle.

The term *Sample Space* needs defining. For the purposes of most discussion in this book, the sample space is the Euclidean line, the two-dimensional Euclidean plane, and the multidimensional Euclidean hyperplane. A *random variable* is any function defined on a sample space. A *statistic* is defined as any function of observable random variables that does not involve unknown parameters.

I have assembled a number of exercises at the end of every chapter from Chapter 2, onwards, one or two of them intentionally droll. The object of this is not only to allow you to test how well you have grasped the contents of a particular chapter, but also to introduce some special ideas and to bring home some more difficult points not explicitly developed in the main text. Above all, the exercises are meant to stimulate you to think analytically about fossils.

What have I left out? I am very aware of being less than attentive to questions of phylogenetic analysis by hierarchical methods such as dendrograms, and their surrogates, for relationships between higher taxa. There are, consequently, no sections devoted to numerical taxonomy, cladistics, and the like. My defence is that the intent of this book is to concentrate attention on adequate methods for studying variability in fossil organisms and to seek out quantitative relationships between organism and environment. There are no doubt countless other topics that could have been included, and many will decry the dearth of examples drawn from Palaeobotany. None the less, the methods reviewed here are quite general in their applicability and if you can master them in the situations I have presented, you should have little trouble in tailoring them to your own needs.

1.6 On Style

I feel I must expend a few words on the style I have adopted. I am fully aware that the use of the first and second persons in scientific texts can be experienced as offensive by some. I have occasionally been taken to task for this by people who also dislike what they call my "racy style", which is said to be easy to follow but to obtrude too much of my personality on the reader. For my part, I am no friend of stuffy circumlocutions such as "the author opines", "the present writer thinks", "the author of these pages intends", "the reader is referred to", etc. I also must point out that subjunctives occur occasionally in this text, for which I make no apology. (I was condemned by a reviewer some years ago for "the strange use of a pronoun, in the singular, with the plural form of the verb"). Subjunctives may be out of vogue in some quarters, but any book on English grammar will suffice to point the dissident and, or, the uncertain, in the right direction (for example, R. W. Zandvoort: *A Handbook of English Grammar*, Longmans (1969), pp. 86–89).

Exercises

1. How would you classify the following characters?
   a. Segments of a trilobite.
   b. Volume of an ostracod carapace.
   c. Colour-banding in *Cepaea*.
   d. Frequencies of blood groups.
   e. Tables of chemical analyses.
   f. Dimensions of a skull.
   g. Angles between measures.
   h. Counts of species in a stratigraphical column.
   i. Colouring in a sedimentary sequence.
2. Comment on the grammatical mood of the verbs in the following sentences:
   a. If I were King for a day.
   b. Suffice it to say.
   c. Be this as it may.
   d. God be with you.
   e. We suggested she come to us for dinner next week.
   f. The inventor may, if he live in London, or visit that city, search the files of the Patent Office.
   g. As it were.
   h. If Edberg play today, we have a good chance of winning the tournament.
3. Can the following measures be provided with a natural ordering?
   a. Ontogenetic appearance of anterior spines in crustaceans (Reyment and Van Valen, 1969).
   b. Girth circumferences in mammals.
   c. Number of pits in reticular ornament of crustaceans.
   d. Serological determinations.
   e. Binary relationship in the posterior spine of ostracods.

CHAPTER 2

Vectors and Matrices

Contents

2.1 Introduction

The following set of definitions is provided in order to make the reading of the text easier and is not meant to be a primer in linear algebra. There are a great number of books on matrix algebra aimed at the elementary and middle levels and you should make a point of consulting one or more of these, particularly if you are interested in writing your own computer programs. A few useful references are: Bellman (1960), Davis (1965), Fox (1964), Gantmacher (1965), Horst (1963), Murdoch (1957), and Searle (1966). In addition, many of the standard textbooks in multivariate statistical analysis include a chapter on Linear Algebra, for example, the summary in Anderson (1984). Normally, the information contained in these chapters is more than enough for most introductory purposes. If you need to learn the basics of programming matrix operations, a manual such as the one by Monro (1987) gives ample information for FORTRAN 77. GENSTAT 5 is a useful language for multivariate work (see note in Jongman *et al.*, 1987). The GENSTAT manual (1988) contains all the information required for operations in linear algebra. The SAS-system is likewise a useful tool for programming matrix operations.

11

The algebra of matrices and vectors constitutes the lifeblood of multivariate morphometrics and if you are seriously interested in understanding what is going on in an analysis, it is absolutely necessary to be acquainted with the basics of the subject. Jöreskog *et al.* (1976) present the fundamentals of matrices and vectors with ample simple illustrations from the Earth Sciences. The information in Searle (1966) provides a valuable introduction, not the least because of its biological orientation. The last part of the present chapter contains a terse account of transformations and definitions in differential geometry.

2.2 Some Definitions

The *Data Matrix* is a $N \times p$ array with N observational vectors each comprising p variables. The data matrix is usually denoted **X** in this book. Each row of **X** corresponds to a specimen, observation, object, etc., and each column corresponds to N observations on each of the p variables in turn. The data matrix is an example of a *rectangular* matrix.

A *Matrix* can be rectangular or square. A **Square Matrix** can be symmetric or asymmetric. If **A** is the array

$$\begin{bmatrix} a_{11} & a_{12} & \dots & a_{1p} \\ & \dots\dots\dots\dots & & \\ a_{p1} & a_{p2} & \dots & a_{pp} \end{bmatrix}$$

the matrix is symmetric if $a_{12} = a_{21}$. More generally, the condition of symmetry may be written $a_{ij} = a_{ji}$.

You will see that matrices and vectors are printed in semi-bold type, which is standard procedure these days. Also, that I sometimes use Greek letters to denote a statistical quantity. This is a convention that was introduced into Statistics some 50 years ago by Sir Ronald Fisher. Matrices printed in Greek letters refer to the population value. The corresponding sample value, that is, the sample estimate of the population quantity, is expressed in the matching Roman letter.

The *Order* of a matrix is the specification of the number of rows and columns. In citing the order of a matrix, the number of rows is given first, followed by the number of columns, as was done above for the data matrix.

In the course of manipulating matrices, it is often necessary to flip one on to its side. This is known as *transposing* the matrix. Thus, if we transpose matrix **A** (8×4) it becomes matrix $\mathbf{A}^T$ (4×8). The most frequently used sign for denoting the transpose of a matrix is a prime. Primes can be hard to make out in complicated formulae, which is why I use the less common convention of providing the matrix with a superscripted T.

A *Vector* can be thought of as a matrix consisting of *one* column and p rows. Alternatively, if the vector is transposed, that is, made to lie on its side, it is said to have one row and p columns. It is sometimes convenient to think of a matrix as being a sheaf of vectors.

A *Scalar* is an everyday number with which all of us are quite familiar. A scalar has, of course, no subscripts, but it can be thought of as a matrix consisting of one row and one column.

2.3 Operations on Vectors

Two vectors are said to be *equal* if each of their corresponding elements are equal. For the p-element row vectors $\mathbf{a}^T$ and $\mathbf{b}^T$,

$$\mathbf{a}^T = \mathbf{b}^T$$

if $a_k = b_k$ for $k = 1, \ldots, p$.

The *addition* of two vectors is a simple operation carried out by adding corresponding elements to each other.

$$\mathbf{a}^T + \mathbf{b}^T = (\mathbf{a} + \mathbf{b})^T. \tag{2.1}$$

The *subtraction* of two vectors yields a vector the elements of which contain the differences between the corresponding elements of the two vectors. Just as for addition, the two vectors must be of the same order.

There are two ways of *multiplying* vectors. One of these methods is known as the **minor product**, with alternative designations: Scalar Product, Inner Product and Dot Product. This product is made by multiplying a row vector (left vector) by a column vector (right vector). The minor product is a scalar: it is the sum of the product of corresponding elements of the two vectors.

$$\mathbf{a}^T \cdot \mathbf{b} = \sum_{k=1}^{p} a_k b_k \tag{2.2}$$

The second way of multiplying vectors is called the **major product** and is the opposite procedure to that used for making the minor product. When a column vector is post-multiplied by a row vector, the result is a matrix. The ij-th element of this matrix is given by the product of the i-th element of the left vector (column vector) and the j-th element of the right vector (row vector):

$$\mathbf{M} = \mathbf{a} \cdot \mathbf{b}^T. \tag{2.3}$$

Another property of vectors of interest in multivariate analysis is the **length**. The length of a vector $\mathbf{a}$, written $|\mathbf{a}|$ is defined as the square root of the sum of its squared projections on to the coordinate axes. Another term for "length" in this connexion is *magnitude*.

$$|\mathbf{a}| = (\mathbf{a}^T \cdot \mathbf{a})^{1/2}. \tag{2.4}$$

Note that this formula relates to the minor product and is, in effect, the minor product of a vector with itself.

The **angle** between two vectors **a** and **b** is obtained from the inner product of these two vectors, divided by their lengths:

$$\cos \theta = \frac{\mathbf{a}^T \mathbf{b}}{(\mathbf{a}^T \mathbf{a})^{1/2} (\mathbf{b}^T \mathbf{b})^{1/2}}. \tag{2.5}$$

A cosine of 0.0 is obtained if the vectors are at right angles to each other. A cosine of 1.0 indicates that the vectors are collinear, that is, parallel to each other.

The **Euclidean Distance** between two points is found from the square root of the minor product of their vector of differences.

$$d = |\mathbf{a} - \mathbf{b}|$$
$$= \{(\mathbf{a} - \mathbf{b})^T (\mathbf{a} - \mathbf{b})\}^{1/2}. \tag{2.6}$$

2.4 Operations on Matrices

The elements the row and column subscripts of which are the same are said to form the *principal diagonal* of the matrix. The trace or spur (*Spur* is the German word for "trace") is the sum of the elements of the principal diagonal.

A *diagonal matrix* is a square symmetric matrix, the off-diagonal elements of which are all nought. A special case of the diagonal matrix is the *Identity Matrix*, which has ones in the principal diagonal and noughts in the other positions. It is the linear algebraic equivalent of the number 1 in scalar arithmetic.

The *null matrix* has all elements equal to zero. Analogously, the null vector has all elements equal to nought.

Two matrices are said to be equal if all the corresponding elements are equal. It follows that equal matrices are of the same order.

Multiplication: The very important operation of the multiplication of matrices differs in many respects from the scalar version. The basic formula can be written:

$$\mathbf{C} = \mathbf{AB}.$$

Each element is formed by the process

$$c_{ij} = \Sigma a_{ik} b_{kj} \tag{2.7}$$

where i, j and k denote the number of rows, respectively columns, of **C**, **A** and **B**. This is a useful representation of the procedure as it gives the exact notation for writing a computer program for multiplication of matrices. Conformity is the key concept here. The product **AB** can be formed if the number of columns of **A** is equal to the number of rows of **B**. The product **BA** can be formed if the number of columns of **B** equals the number of rows of **A**. You should try your hand at an example to show that the order of multiplication of the two matrices does not necessarily yield the same answer. This means that for the majority of cases,

$$\mathbf{AB} \neq \mathbf{BA}.$$

As an illustration, note that if $\mathbf{A}$ is a 3×2 matrix and $\mathbf{B}$ is a 2×7 matrix, the product $\mathbf{AB}$ can be formed, but not $\mathbf{BA}$.

There are two kinds of multiplication of a matrix by itself that are very important in multivariate morphometrics. The **Major Product Moment** is defined as the product of a matrix post-multiplied by its transpose. That is

$$\mathbf{X}_{(N \times p)} \text{ post-multiplied by } \mathbf{X}_{(p \times N)}.$$

$$\mathbf{C} = \mathbf{XX}^T. \tag{2.8}$$

Matrix $\mathbf{C}$, the major product moment, is a square symmetric matrix of order $N \times N$.

The **Minor Product Moment** is defined as the pre-multiplication of a matrix by its transpose. For an $N \times p$ matrix $\mathbf{X}$, usually the data matrix in biometry, the minor product moment matrix $\mathbf{E}$ is of order $p \times p$

$$\mathbf{E} = \mathbf{X}^T\mathbf{X}. \tag{2.9}$$

What we see in equations (2.8) and (2.9) is the simply expressed *duality* between Q-mode and R-mode representations of a data matrix. The Q-mode expression is given by equation (2.8). The R-mode expression is given by equation (2.9).

Product of a Matrix with a Vector: A matrix pre-multiplied by a conformable row vector yields a row vector. We can write this out as follows.

$$(a_1 \, a_2 \, a_3) \begin{bmatrix} b_{11} & b_{12} \\ b_{21} & b_{22} \\ b_{31} & b_{32} \end{bmatrix} = (a_1 b_{11} + a_2 b_{21} + a_3 b_{31} \, ; \, a_1 b_{12} + a_2 b_{22} \, a_3 b_{32}) \tag{2.10}$$

The dimensionality of the row vector is 2, which is the same as the number of columns in matrix $\mathbf{B}$.

The pre-multiplication of a matrix by a unit row vector yields a row vector of which the elements are the sums of the corresponding columns of the matrix.

A matrix *post-multiplied* by a conformable cclumn vector yields a column vector. If the vector is a *unit vector* (i.e. a vector with all elements equal to one), the product is a column vector containing the row sums of the matrix. To test yourself, prove that the following result for a post-multiplication of a matrix by a vector is correct:

$$\begin{bmatrix} 3 & 1 \\ 4 & 0 \\ 1 & 2 \end{bmatrix} \begin{bmatrix} 1 \\ 3 \end{bmatrix} = \begin{bmatrix} 6 \\ 4 \\ 7 \end{bmatrix}$$

The two forms of multiplication of a matrix and a vector just introduced are used in many connexions in the methods employed in this book. For example, the basic steps for computing the linear discriminant function and the generalized statistical distance are no more complicated than what you have just been shown.

A matrix pre-multiplied by a diagonal matrix: A matrix pre-multiplied by a *diagonal* matrix has the elements of its n-th row multiplied by the n-th element of the diagonal matrix.

A matrix post-multiplied by a diagonal matrix has the elements of the j-th column multiplied by the j-th element of the diagonal matrix. These valuable properties are made use of for scaling operations in multivariate statistics.

2.4.1 *Triangular Matrices*

A variant of square matrices is the *triangular matrix*. If the elements of the triangular matrix are located along the diagonal and in the positions above the diagonal, we speak of an Upper Triangular Matrix. If the elements are located along the diagonal and below it, the matrix is said to be Lower Triangular.

2.4.2 *Determinants*

Another subject in matrix algebra of wide applicability is concerned with the theory of Determinants. A determinant is a scalar quantity derived from operations on a square matrix. The determinant of the square matrix $\mathbf{A}$ is usually denoted $|\mathbf{A}|$. For a 2×2 matrix, the determinant is calculated in the following manner:

$$|\mathbf{A}| = a_{11}a_{22} - a_{12}a_{21}. \tag{2.11}$$

The formulae become rapidly more complicated as the number of dimensions increases. For many practical programming purposes, the determinant of a positive definite square symmetric matrix is easily obtained via the latent roots (see p. 20, (2.23)).

A *minor* is a special kind of determinant. Given a square matrix $\mathbf{A}$, the minor $\mathbf{M}_{ij}$ is defined as the determinant of a matrix formed by deleting the i-th row and j-th column of $\mathbf{A}$. There will therefore be one minor corresponding to each element of $\mathbf{A}$. Special interest attaches to the minors associated with the principal diagonal of square symmetric matrices. A square symmetric matrix is positive definite if all the minors associated with the elements of the principal diagonal are greater than nought. A matrix is said to be *non-singular* if its determinant is not zero.

Consider the *quadratic form* $\mathbf{x}^T\mathbf{A}\mathbf{x}$, where $\mathbf{A}$ is a symmetric matrix. The matrix $\mathbf{A}$ and the quadratic form are called *positive semidefinite* if $\mathbf{x}^T\mathbf{A}\mathbf{x} \geq 0$ for all $\mathbf{x}$. If $\mathbf{x}^T\mathbf{A}\mathbf{x} > 0$ for all $\mathbf{x} \neq \mathbf{0}$, $\mathbf{A}$ and the quadratic form are called *positive definite*. The generalized distance is produced by means of a quadratic form, a scalar. The triple product $\mathbf{1}^T\mathbf{X1}$, that is the data matrix $\mathbf{X}$, pre-multiplied by a unit vector and post-multiplied by a unit vector (a vector composed of ones), is a quadratic form; it gives a scalar which consists of the sum of all the elements of $\mathbf{X}$.

Vector of Differential Operators: A determinant that occurs in the section on quantitative genetics (Chapter 5) is the *Jacobian*, which is formed from vectors of differential operators. The subject is relatively complicated and I refer you to Searle (1966, pp. 203–209) for a biologically oriented account of differential operators and the selection index of Hazel (1943).

2.4.3 *Inverse of a Matrix*

The operation called Division in scalar arithmetic requires a more complicated operation when it comes to matrices. I refer you therefore to any of the standard texts on applied linear algebra for the step-by-step details, for example, the book by Fox (1964). In scalar arithmetic, the reciprocal of a number multiplied by the original number yields 1. In matrix algebra, the reciprocal $\mathbf{C}$ of a square matrix $\mathbf{B}$ satisfies the relationship

$$\mathbf{BC} = \mathbf{CB} = \mathbf{I}. \tag{2.12}$$

The *Identity Matrix* $\mathbf{I}$ is the analogue of the number one in scalar arithmetic. The reciprocal of $\mathbf{B}$ as defined above is called the inverse $\mathbf{C}$ of $\mathbf{B}$. It is written $\mathbf{C}^{-1}$.

The Generalized Inverse: A singular matrix does not have an inverse in the above sense. In order to circumvent the various difficulties attaching to this situation, the concept of a generalized inverse has been developed. The standard reference is the book by Rao and Mitra (1971). The subject is quite new in the statistical literature; the invention of a generalized inverse has been of great value in the development of many branches of applied multivariate analysis. The way of writing the generalized inverse of matrix $\mathbf{A}$ is $\mathbf{A}^-$.

Matrix inversion forms an essential part of many multivariate statistical techniques, from multiple regression to linear discriminant functions.

2.4.4 *Normal and Orthonormal Forms of Matrices and Vectors*

A *normalized vector* (see 2.4) is one that has been made to have unit length. Any vector $\mathbf{x}$ can be normalized by dividing each of its components by the length of that vector. Hence, the normalized form of $\mathbf{x}$ is

$$\mathbf{x}/(\mathbf{x}^T\mathbf{x})^{1/2}. \tag{2.13}$$

This is standard procedure for latent vectors in principal component analysis, canonical variate analysis, etc.

Two vectors are said to be *orthogonal* to each other if their minor product is zero; this indicates that the vectors are at right-angles to each other.

If the minor product moment of matrix **X** is a diagonal matrix, **D**, that is:

$$\mathbf{X}^T\mathbf{X} = \mathbf{D} \tag{2.14}$$

then all possible pairs of column vectors of **X** are orthogonal. This is an important property and one that plays a substantial role in many of the statistical procedures that are based on the algebra of latent roots and vectors of square symmetric matrices.

Orthonormal Matrices: When the minor product moment of a matrix is the identity matrix, that matrix is said to be *orthonormal*. This property signifies that the column vectors of the matrix are normalized vectors and that they are orthogonal to each other. An example of such a matrix is:

$$\mathbf{A} = \begin{bmatrix} \cos\phi & -\sin\phi \\ \sin\phi & \cos\phi \end{bmatrix}$$

You might like to verify that the product of this matrix with its transpose is **I** (i.e. *the matrix with ones down its diagonal and noughts in all other positions*).

This kind of matrix is particularly prominent in principal component factor analysis (Jöreskog *et al.*, 1976) in the form of a square orthonormal matrix, say, **Q**. The two fundamental properties of relevance here are:

$$\mathbf{Q}^T\mathbf{Q} = \mathbf{I}$$

and

$$\mathbf{QQ}^T = \mathbf{I}.$$

A convenient reference for orthonormal matrices is Horst (1963).

Square matrices have square roots and they can be powered. The operation for powering a matrix follows the rules of matrix multiplication:

$$\mathbf{A}^2 = \mathbf{AA}.$$

A matrix is said to be *idempotent* if it is equal to its square, or, for that matter, to any of the powers to which it is raised.

$$\mathbf{A}^2 = \mathbf{A}.$$

If **A** is a square symmetric idempotent matrix, then the rank of **A** is equal to the trace of **A**. Idempotent matrices play an important part in the theory of Burnaby's growth invariant discriminant functions. Graybill (1961) gives a comprehensive account of idempotent matrices.

The *rank* of a matrix is the smallest common order amongst all pairs of matrices whose product is the matrix (Horst, 1963, p. 335). This is one definition of rank; we have already met other ways of saying this earlier on in this chapter. The rank of a matrix cannot be greater than its smallest dimension. Following on to the present definition we have that a vector that is a scalar multiple of another vector is said to be linearly dependent. A vector is linearly independent of a set of vectors if it is neither a scalar multiple, nor a weighted or unweighted sum of any combination of the members in that set.

Finding the rank of the data matrix is a commonly required exercise in applied multivariate work, not the least in the interpretation of some analyses of size and shape (Chapter 4). If the column vectors constituting a square symmetric matrix are linearly independent, that matrix will have a rank equal to the dimensionality of the matrix. If the vectors are linearly dependent, the rank of the matrix will be less than the dimensionality. The topic of rank is taken up at length, with illustrations and examples, in Jöreskog *et al.* (1976, pp. 35–41). The subject of factor analysis depends strongly on the determination of rank.

2.5 Latent Roots and Vectors

Many methods of multivariate analysis lean heavily on the algebra of latent roots and vectors, for example, principal component analysis, principal coordinates, factor analysis, correspondence analysis, canonical variates, and canonical correlations. Thus, latent variables can be said to form the uniting thread of the main part of this book.

For our present purposes, we restrict this overview to the latent roots and vectors of a real symmetric matrix **R**. This could be the major or the minor product moment of a data matrix, the correlation matrix, or the covariance matrix. The following summary connects to Section 3.6.1.

A latent vector of **R** is a vector **u** with components, not all nought, such that

$$\mathbf{Ru} = \mathbf{u}\lambda \qquad (2.15)$$

where λ is an unknown scalar. Another form of (2.15) is

$$\mathbf{RU} - \mathbf{u}\lambda = \mathbf{0} \qquad (2.16)$$

where **0** is the null vector.

Equation (2.16) represents a system of homogeneous equations. The solution for **u** and λ is begun by solving the following determinantal equation

$$|\mathbf{R} - \Lambda\mathbf{I}| = 0. \qquad (2.17)$$

The roots of this equation are the latent roots (synonyms: eigenvalues, proper values, characteristic roots). In general, (2.17) is a polynomial of degree p, the number of variables, and there will be p latent roots. If two or more latent roots are equal, we say that there are one or more multiple roots. If **R** is not positive definite, but positive semidefinite, it will have at least one zero latent root. Such matrices occur in multivariate statistics, for example, where we are dealing with compositional data matrices; i.e. matrices in which the rows sum to a constant.

Equation (2.16) can be written more generally as

$$\mathbf{RU} = \mathbf{U}\Lambda. \tag{2.18}$$

The matrix $\mathbf{U}$ is square orthonormal so that

$$\mathbf{U}^T\mathbf{U} = \mathbf{U}\mathbf{U}^T = \mathbf{I}. \tag{2.19}$$

Some useful facts to have at hand are:

1. Post-multiplication of (2.18) by $\mathbf{U}^T$ gives

$$\mathbf{R} = \mathbf{U}\Lambda\mathbf{U}^T \tag{2.20}$$

which is an equation of great importance in linear algebra and multivariate statistical analysis. It shows that a square symmetric matrix may be expressed in terms of its latent roots and vectors. Another way of writing (2.20) is

$$\mathbf{R} = \lambda_1\mathbf{u}_1\mathbf{u}_1^T + \ldots + \lambda_p\mathbf{u}_p\mathbf{u}_p^T. \tag{2.21}$$

The importance of the representation in (2.21) is that it is a decomposition of the square symmetric matrix $\mathbf{R}$ into a weighted sum of matrices, each of which is of order $p \times p$ and of rank 1. Moreover, each term is orthogonal to all other terms since for $i \neq j$, $\mathbf{u}_i^T\mathbf{u}_j = 0$. The property outlined in equation (2.21) is exploited by several multivariate methods.

2. Pre-multiplication of (2.18) by the transpose of $\mathbf{U}$ yields

$$\Lambda = \mathbf{U}^T\mathbf{R}\mathbf{U} \tag{2.22}$$

which shows that the square symmetric matrix $\mathbf{R}$ is reduced to diagonal form by $\mathbf{U}$. Relationship (2.22) is very useful for computer programming of large-scale multivariate operations. Instead of having to use the rather rambling formulae for computing determinants, it is possible to obtain the determinant of $\mathbf{R}$ (positive definite) by multiplying together its latent roots.

$$|\mathbf{R}| = \lambda_1\lambda_2\ldots\lambda_p. \tag{2.23}$$

The latent roots of a matrix are commonly required in statistical work and it is therefore not associated with any great trouble to obtain the determinant of a square symmetric matrix.

Similarly,

$$\mathrm{trace}(\mathbf{R}) = \lambda_1 + \lambda_2 + \ldots + \lambda_p. \tag{2.24}$$

3. Another useful result is that the *rank* of a matrix is equal to the number of non-zero latent roots possessed by that matrix (see above on p. 18). More precisely, if for a square symmetric matrix $\mathbf{R}$, of order p, there are m zero latent roots, the rank of the matrix is $(p - m)$.

4. The square root of a square symmetric matrix is sometimes needed. This can be a rather complicated procedure but it can be conveniently approached by taking the square root of the elements of the diagonal matrix **D** corresponding to **R**.

$$\mathbf{R}^{1/2} = \mathbf{U}\mathbf{\Lambda}^{1/2}\mathbf{U}^T \tag{2.25}$$

by virtue of equation (2.20), where $\mathbf{\Lambda}^{1/2}$ is a diagonal matrix with the square roots of the latent roots down its diagonal.

Geometrical Significance of Latent Roots and Vectors

The geometrical significance of latent roots and vectors can be worth considering for a moment since the concepts involved are exploited in practical work. The major and $(p-1)$ minor axes of the hyperellipsoid (which subsumes multivariate normality) can be found from the latent vectors of, say, the covariance or correlation matrix:

$$\mathbf{w}^T\mathbf{R}^{-1}\mathbf{w} = c \tag{2.26}$$

where **R** is the correlation or covariance matrix, **w** is a vector of coordinates for the points on the hyperellipsoid, and c is a constant. A fairly detailed discussion is given by Jöreskog *et al.* (1976, pp. 45–47). Here it is shown that

$$\mathbf{R}\mathbf{w} = \lambda\mathbf{w} \tag{2.27}$$

with $\mathbf{w}^T\mathbf{w} = \lambda$ so that **w** must be a latent vector of **R** corresponding to the largest latent root of **R**. This latent vector gives the direction of the major axis of the hyperellipsoid (*hyper* denotes the concept of abstract dimensionality—more than three dimensions). The length of the major axis, with $\mathbf{w}^T\mathbf{w} = 1$, is $\lambda^{1/2}$. Similarly, the second latent vector corresponds to the largest minor axis of the hyperellipsoid, and so on.

The geometrical description of latent roots and vectors leads to some conclusions about the properties of clusters of points in hyperspace. Firstly, the location of latent vectors along the principal axes of an ellipsoid means that these coincide with the directions of *maximum variance* (i.e. of linear combinations of variables with coefficients summing to one when squared). We make use of this in several applications.

Secondly, the location of the latent vectors may be thought of as being a rotation or transformation, whereby the original variables are rotated to new positions. The rows of the matrix of latent vectors are the coefficients that rotate the axes of the variables to new positions along the major and minor axes. The elements of the vectors are cosines of the angles of rotation from the old to a new system of coordinates.

Thirdly, latent vectors are linearly independent vectors that are linear combinations of the original variables. They can therefore be viewed as new variables with the desirable property that they are uncorrelated and, furthermore, are bound to successively decreasing parts of the variance as expressed by the sum of all the latent roots. The square root of the latent root corresponding to a particular latent vector can be used as a standard deviation of the "new" variable represented by that latent vector.

Fourthly, if a latent root is zero, the corresponding minor axis is of zero length. This condition indicates that the dimensionality of the space containing the data points is less than suggested by the number of variables, i.e. $<p$.

2.6 Basic Structure of a Matrix

The basic structure of a matrix, or singular value decomposition of a matrix, as it is perhaps more commonly called, is a vital concept in multivariate analysis, not the least from computational aspects. The Eckart–Young Theorem expresses how a data matrix can be decomposed into two orthonormal matrices and a diagonal matrix; these are the singular values. Details of the theorem can be obtained, for example, from Jöreskog *et al.* (1976, pp. 47–52) and Press *et al.* (1986, pp. 52–64).

Thus, if

$$\mathbf{X} = \mathbf{VGU}^T \tag{2.28}$$

where $\mathbf{V}$ is an $N \times r$ matrix, the columns of which are orthonormal, $\mathbf{U}$ is a $p \times r$ matrix with orthonormal columns, and $\mathbf{G}$ is a diagonal matrix of order $r \times r$; the diagonal elements of $\mathbf{G}$, $g_1, \ldots, g_r$ are the singular values of $\mathbf{X}$, the order of which is $N \times p$ and rank r. The designation "basic structure of a matrix" is closely bound to the theory of factor analysis. The term "singular value decomposition" is more current in standard statistical and general mathematical literature.

An interpretively useful rendition of the decomposition expressed by (2.28) is

$$\mathbf{X} = g_1 \mathbf{v}_1 \mathbf{u}_1^T + \ldots + g_r \mathbf{v}_r \mathbf{u}_r^T \tag{2.29}$$

which shows that $\mathbf{X}$ of rank r is a linear combination of r matrices of rank 1 (see also equation 2.21).

The relationships between $\mathbf{G}$, $\mathbf{U}$, and $\mathbf{V}$ in the basic structure of $\mathbf{X}$ express the link between Q-mode and R-mode multivariate analytical procedures. The principal features of this link are summarized below:

(1) Consider the major product moment $\mathbf{XX}^T$, $N \times N$. It is square symmetric and has r positive latent roots and $(N - r)$ zero latent roots. Let the positive latent roots be

$$g_1^2, \ldots, g_r^2$$

and the corresponding latent vectors

$$\mathbf{v}_1, \ldots, \mathbf{v}_r.$$

The singular values are the square roots of the positive latent roots of $\mathbf{XX}^T$. The columns of $\mathbf{V}$ are the corresponding latent vectors.

(2) The minor product moment $\mathbf{X}^T\mathbf{X}$, which is square symmetric and of order $p \times p$ has r positive latent roots and $(p - r)$ zero latent roots. The positive latent roots are

$$g_1^2, \ldots, g_r^2$$

and the corresponding latent vectors

$$\mathbf{u}_1, \ldots, \mathbf{u}_r.$$

(3) Hence, the latent roots of $\mathbf{XX}^T$ and $\mathbf{X}^T\mathbf{X}$ are the same. Furthermore, if $\mathbf{v}_m$ is a latent vector of $\mathbf{XX}^T$ and $\mathbf{u}_m$ a latent vector of $\mathbf{X}^T\mathbf{X}$, corresponding to the latent root g_m^2, the following relationships between $\mathbf{v}_m$ and $\mathbf{u}_m$ hold for $m = 1, \ldots, r$.

$$\mathbf{v}_m = (1/g_m)\mathbf{Xu}_m \tag{2.30}$$

and

$$\mathbf{u}_m = (1/g_m)\mathbf{X}^T\mathbf{v}_m. \tag{2.31}$$

Providing that the representation of $\mathbf{X}$ is in a form compatible with the foregoing discussion, the singular value decomposition provides a convenient and economical way of obtaining the latent roots and vectors of very large matrices, such as can occur in principal coordinate analysis and correspondence analysis. An example of the singular value decomposition of a matrix is given in Table 2.1.

TABLE 2.1. *Illustration of the singular value decomposition of a matrix*

In order to furnish an example of how the concept of the basic structure of a matrix works, I have taken a matrix from Jöreskog *et al.* (1976). This matrix, which is a data matrix $\mathbf{X}$, consists of four variables measured on 8 specimens. We have therefore a matrix of order 8×4.

Row	Var 1	Var 2	Var 3	Var 4
1	4.51	2.66	0.42	4.10
2	6.07	0.58	1.77	1.54
3	6.42	1.32	4.65	4.05
4	4.46	2.16	1.41	4.47
5	8.92	2.54	4.66	4.50
6	7.60	2.39	4.14	4.49
7	5.29	1.69	3.66	3.77
8	4.73	2.65	3.29	5.08

Using a program published in that vademecum of FORTRAN-programmers, Press *et al.* (1986), the following singular value decomposition was computed for the above matrix.

[*Table* 2.1 *continued overleaf*

TABLE 2.1.—*Continued*

If we designate the decomposition as $\mathbf{X} = \mathbf{VGU}^T$, matrix $\mathbf{V}$ (8 × 4) is as follows:

Row	Vector 1	Vector 2	Vector 3	Vector 4
1	0.264	−0.558	−0.473	0.122
2	0.261	0.424	−0.602	−0.332
3	0.380	0.262	0.374	−0.518
4	0.281	−0.454	−0.135	−0.505
5	0.482	0.270	−0.115	0.506
6	0.430	0.110	0.007	0.277
7	0.326	0.051	0.288	−0.042
8	0.339	−0.382	0.400	0.130

This matrix is the Q-mode representation of $\mathbf{X}$. The $p \times p$ matrix $\mathbf{U}$ is:

Variable	Vector 1	Vector 2	Vector 3	Vector 4
1	0.744	0.361	−0.560	−0.044
2	0.243	−0.471	−0.047	0.847
3	0.389	0.442	0.788	0.179
4	0.486	−0.672	0.251	−0.500
Singular values g_i	23.333	3.932	2.734	0.684

The matrix of latent vectors $\mathbf{U}$ is the R-mode representation of $\mathbf{X}$.

Let us see how these figures agree with the latent roots and vectors of the minor product matrix $\mathbf{X}^T\mathbf{X}$ and major product moment matrix $\mathbf{XX}^T$ (this terminology was introduced by Horst (1963) and was adhered to by Jöreskog *et al.* (1976)).

The latent vectors of the major product moment matrix are:

Variable	Vector 1	Vector 2	Vector 3	Vector 4
1	0.744	−0.361	−0.560	−0.361
2	0.243	0.461	−0.048	0.471
3	0.389	−0.442	0.788	−0.442
4	0.486	0.673	0.251	0.673
Latent roots	544.43	15.46	7.47	0.47

You will observe that the latent vectors of $\mathbf{U}$ are the same as those of $\mathbf{X}^T\mathbf{X}$, but the latent roots are an order of magnitude greater. Remember, however, that the singular values are the square roots of the corresponding latent roots. Taking the square roots of the above latent roots yields exactly the singular values g_i. The latent vectors of $\mathbf{XX}^T$ are

Row	Vector 1	Vector 2	Vector 3	Vector 4
1	0.264	−0.558	0.473	0.122
2	0.261	0.424	0.602	−0.332
3	0.380	0.262	−0.374	−0.518
4	0.281	−0.454	0.135	−0.505
5	0.482	0.270	0.115	0.506
6	0.430	0.110	−0.007	0.277
7	0.326	0.051	−0.288	−0.042
8	0.339	−0.382	−0.400	0.130

This is the same as matrix $\mathbf{V}$ from the singular value decomposition, the Q-mode representation for vectors 1 to 4.

The present example gives us the useful piece of practical information that if we are interested in a combined Q- and R-mode analysis of a data matrix, proceeding via the algorithm for the singular value decomposition is a realistic way of organizing the computing.

This algebraic relationship has been capitalized on in the methods of correspondence analysis and the Gabriel biplot, which are discussed in Chapter 3.

2.7 Some More Definitions

The J-matrix is one that shows up in the analysis of compositional data. It is defined as the matrix in which all positions are occupied by ones.

Skew-Symmetric Matrices: Matrices of the form $\mathbf{S}^T = -\mathbf{S}$ are said to be skew symmetric. They are square and have all the elements on one side of the diagonal equal to minus their counterparts on the other side; the diagonal elements are zero.

Any square matrix $\mathbf{M}$ can be expressed in terms of a symmetric and a skew-symmetric matrix:

$$\mathbf{S} = \tfrac{1}{2}(\mathbf{M} + \mathbf{M}^T) \tag{2.32}$$

and

$$\mathbf{T} = \tfrac{1}{2}(\mathbf{M} - \mathbf{M}^T).$$

Here, $\mathbf{S}$ is symmetric and $\mathbf{T}$ is skew symmetric. These matrices are used in one of the palaeoecological problems taken up in Chapter 8 involving the analysis of asymmetry.

2.8 Descriptive Statistics

The main statistical entities required in the following chapters may be summarized as follows:

The *Mean Vector* can be expressed in matrix notation as

$$\bar{\mathbf{x}} = \frac{\mathbf{1}^T \mathbf{X}}{\mathbf{1}^T \mathbf{1}} \tag{2.33}$$

where (2.33) expresses the p-component vector of means and $\mathbf{1}$ is a vector of unit elements. This is an example of the convenient fact that many univariate statistics can be transformed to multivariate counterparts by the simple expediency of matrix notation.

Deviate Scores and Standardized Scores: If we denote the raw score as x_{nj}, then the deviate score is defined as:

$$y_{nj} = x_{nj} - \bar{x}_j. \tag{2.34}$$

The sum of deviate scores for a variable is obviously nought. The means of variables expressed as deviate scores are likewise nought. The geometrical effect of converting variable to deviate scores is to shift the origin of each variable to its mean. This is a useful manoeuvre in many situations, including graphical reports.

The Variance

In matrix notation, the variance can be expressed in the following manner:

$$s_j^2 = \mathbf{y}_j^T \mathbf{y} / \mathbf{1}^T \mathbf{1} \tag{2.35}$$

where the $\mathbf{y}_j$ are deviate scores as defined in (2.34). The standard deviation is the square root of (2.35), s_j.

Standardized Scores

It is often useful to express observations in terms of their deviations from the mean, using the standard deviation, s_j, as their unit of divergence. The standardized score, z_j for a particular specimen with respect to variable j, is given by the relationship:

$$z_j = (x_j - \bar{x}_j)/s_j. \tag{2.36}$$

Standardized variables have a mean of zero and a variance and standard deviation of one. The equivalent of (2.36) in matrix notation is

$$\mathbf{Z} = \mathbf{YD}^{-1/2} \tag{2.37}$$

where $\mathbf{D}$ denotes the diagonal matrix formed from the diagonal elements of the covariance matrix, $\mathbf{S}$, defined below (2.39), and $\mathbf{Y}$ is a matrix of *deviate scores*, defined as in (2.34).

You might like to test how well you have grasped matrix notation by writing out equation (2.37) in long-hand.

The Covariance Matrix

The covariance matrix is the basic building stone in multivariate statistics. The covariance is defined as

$$s_{ij} = \frac{\sum\limits_{n=1}^{N} y_{ni} y_{nj}}{N} \tag{2.38}$$

The matrix notation for (2.38) for expressing the covariance matrix is

$$\mathbf{S} = \mathbf{Y}^T \mathbf{Y}/\mathbf{1}^T \mathbf{1} \tag{2.39}$$

Hence, the covariance matrix is the minor product moment of the data matrix, expressed in deviate form and with each element divided by N (or, for an unbiassed estimate, the size of the sample less 1; said as N minus 1 degree of freedom). The elements of the principal diagonal of $\mathbf{S}$ are the variances s_{ii} of the p variables.

The Correlation Matrix

Each element of the correlation matrix is formed by computing

$$r_{ij} = s_{ij}/(s_i \cdot s_j). \tag{2.40}$$

This is the covariance between variable i and variable j, divided by the product of the standard deviations of i and j. For a matrix in standardized form, as in (2.36), the correlation matrix may be represented in matrix form as:

$$\mathbf{R} = \mathbf{Z}^T \mathbf{Z} / N. \tag{2.41}$$

It is instructive to consider the correlation coefficient defined in deviate form:

$$r_{ij} = \frac{\displaystyle\sum_{n=1}^{N} y_{ni} y_{nj}}{\left(\displaystyle\sum_{n=1}^{N} y_{ni}^2 \sum_{n=1}^{N} y_{nj}^2\right)^{1/2}}. \tag{2.42}$$

This indicates that the geometrical relationship (and also that the correlation coefficient) is the minor product moment of two vectors, divided by the product of their respective lengths. This is the same equation as (2.5) for expressing the angle between two vectors. Thus, for standardized variables, the correlation coefficient can be expressed as the cosine between the two vectors:

$$r_{ij} = \cos \phi = \mathbf{z}_i^T \mathbf{z}_j / N. \tag{2.43}$$

2.9 On Transformations

Transformations play an important part in many of the techniques used in this book, particularly those used in Chapter 4. In usual multivariate statistics, for example, the routine of canonical variate analysis is a two-stage transformational procedure. In Chapter 4, the method of principal warps depends on the use of transformations. In the following a few definitions and descriptions of transformations are provided. These are very elementary and for more initiated treatments I must perforce refer you to Bookstein *et al.* (1985) and Bookstein (1986, 1989, 1991) for a discussion, in particular, of the subjects of the decomposition of rigid motions into translation and rotation, non-uniform transformations, representations of **growth gradients**, of common occurrence in biology, inversions, conformal mappings and the diagonalization of shears. The main facts needed for Chapter 4 are listed below.

1. Similarity transformations do not change shape. This is the fundamental property of one category of transformations.

a. *Translations*: In a translation of the plane, each point is moved a fixed distance parallel to the x-axis and another fixed distance parallel to the y-axis. This is said to be a one-to-one **mapping** of the whole plane on to itself. Thus, each figure occurring in the plane retains its size and shape and axial orientation, but it is shifted to a new location in the coordinate system.

b. *Rotations*: Rotations form part and parcel of many multivariate statistical techniques: for example, the methods of principal components and canonical variates. In a simple rotation, each point turns about the origin through a certain angle. This is also a one-to-one mapping of the whole plane on to itself. Each point maintains its distance from the origin and each figure keeps its size and shape intact. It is only the orientation that has been changed.

c. *Rescaling*: In the method of shape coordinates (Chapter 4, section 4.5.5) two landmarks are chosen to form a baseline. What happens if you change the baseline for another? The effect of selecting a new baseline is to translate, rotate and *rescale* the ordinary statistical scatter of the shape coordinates.

2. An *affine or uniform transformation* is of the form

$$x' = a_1 x + b_1 y + c_1$$
$$y' = a_2 x + b_2 y + c_2$$

where the determinant Δ of the coefficients a_1, b_2, a_2, b_2 (the terms c_i denote relative locations)

$$a_1 b_2 - a_2 b_1 \neq 0$$

has two important properties. If $\Delta \neq 0$, it carries parallel lines into parallel lines (i.e. leaves parallel lines parallel) and finite points into finite points: thus, a square maps into a parallelogram. An affine transformation can always be factored into the product of transformations belonging to the above-noted cases.

Strain is defined as the change in relative positions of points in a medium, the change being produced through stressing of the medium. The usual formulations of strain are in three dimensions but in the present applications, we are only concerned with two dimensional strain.

The *principal directions of strain* refer to an undeformed medium in which there are two mutually orthogonal directions which remain mutually orthogonal after the deformation has taken place. They are the directions of the extrema of strain, i.e. the directions of maximization or minimization. Other terms for the same thing are **principal axes** or **biorthogonal directions**. Principal axes are invariant under changes of scale. In terms of the square parallelogram mapping, the principal directions are the directions of the maximum or minimum ratio of length in the parallelogram to corresponding length in the square. Bookstein (1991, p. 153) summarizes these concepts by saying: "The principal axes of greatest and least rate of strain are the same as biorthogonal directions which start and finish at 90°. This is the **one crucial fact** of morphometrics. All else (to quote Rabbi Hillel) is commentary".

The *principal strains* are the elongations in the directions of the principal directions of strain. The actual values of the above-mentioned extrema can be expressed in terms of the representation of uniform strain by a symmetric tensor: a pair of directions and two strain-ratios. This portrayal expresses changes of length without reference to Cartesian coordinates.

The *anisotropy* of the uniform transformation can be defined as the ratio of the axes of the ellipse into which a circle is deformed. We can therefore think of it as expressing the extent to which the transformation acts in different directions, i.e. is a change in shape. As an example of the application of anisotropy in geology, I can cite *Gefügeanalyse* in micropetrology.

3. Other Useful Categories

Conformal Transformation: This is a map that preserves angles. That is, a map such that if two curves intersect at an angle ϕ, then the images of the two curves in the map also intersect at the same angle ϕ. Angles at a point are unchanged.

Inversion: Any curve whose points are the inverses of the points of a given curve is called the inverse of that curve. The inverse of a circle which passes through the centre of inversion is a straight line; the inverse of any other circle is a circle; i.e. circles remain circles. The inversion of a point with respect to a circle concerns finding the point on the radius through the given point such that the product of the distances of the two points from the centre of the circle is equal to the square of the radius. Either of the points is called the **inverse** of the other and the centre of the circle is called the **centre of inversion**.

Projection: Projective geometry is the study of those properties of geometric configurations which are **invariant** under projection (i.e. remain unchanged). Projection encompasses a class of maps that take all straight lines of the plane into straight lines. For morphometry, the *conformal point* of the mapping is important. At such points, the values of the principal strains are equal and the principal axes do not exist. All angles measured at this point are unchanged and those observed at the anti-conformal point are only reversed in sense. A plane map possesses a conformal point wherever its affine derivative is skew symmetric with diagonal terms equal.

Exercises

1. Show that the minor vector product obtained by multiplying the vectors

$$(2 \quad 4 \quad 1)^T$$
$$(3 \quad 1 \quad 5)^T$$

is 15. What are the respective lengths of these two vectors? What is the angle between the vectors? Hint: use equations (2.4) and (2.5) to answer the two questions.

2. How is the correlation coefficient interpreted in terms of vectors?

3. Show that the major vector product obtained by multiplying the two vectors in Exercise 1 is the matrix

$$\mathbf{M} = \begin{bmatrix} 6 & 2 & 10 \\ 12 & 4 & 20 \\ 3 & 1 & 5 \end{bmatrix}$$

4. Show that the minor product moment matrix **A**, multiplied by its transpose, is the matrix **B** below:

$$A = \begin{bmatrix} 3 & 1 \\ 1 & 4 \\ 2 & 2 \end{bmatrix} \quad B = \begin{bmatrix} 14 & 11 \\ 11 & 21 \end{bmatrix}$$

Hint: compute A^TA.

5. Show that the major product moment matrix formed from **A** of problem 4 is the matrix **C** below:

$$C = \begin{bmatrix} 10 & 7 & 8 \\ 7 & 17 & 10 \\ 8 & 10 & 8 \end{bmatrix}$$

Hint: compute A^TA.

6. What can you say about the product of the two matrices **A** and **B**.

$$A = \begin{bmatrix} 1 & 1 & 1 \\ 2 & -3 & 1 \end{bmatrix} \quad B = \begin{bmatrix} 1 & 2 \\ 1 & -3 \\ 1 & 1 \end{bmatrix}$$

Hint: use equation (2.12).

7. Show that if

$$A = \begin{bmatrix} 1 & 2 \\ 2 & 3 \end{bmatrix}$$

and

$$B = \begin{bmatrix} 3 & 7 \\ 7 & 6 \end{bmatrix}$$

then

$$(AB)^T = \begin{bmatrix} 17 & 19 \\ 27 & 32 \end{bmatrix}$$

$$= \begin{bmatrix} 17 & 27 \\ 19 & 32 \end{bmatrix}$$

8. Write a simple program for producing the determinant of a positive definite matrix **E**, using its latent roots.

9. Write a simple program for producing the square root of a positive definite matrix **E** using its latent roots.

10. If the vector **1** in formula (2.33) is p-dimensional, work out the size of the sample.

11. The latent roots of the matrix analysed in Table 2.1 are as listed below:

$$\begin{bmatrix} 544.43 & 0 & 0 & 0 \\ 0 & 15.46 & 0 & 0 \\ 0 & 0 & 7.47 & 0 \\ 0 & 0 & 0 & 0.47 \end{bmatrix}$$

which we shall call **D**. Compute $D^{-1/2}$. If element d_{44} were zero instead of 0.47, would you still be able to find $D^{-1/2}$?

12. If matrix **A** is

$$\begin{bmatrix} 0.337 & 0.337 & 0.337 \\ 0.337 & 0.337 & 0.337 \\ 0.337 & 0.337 & 0.337 \end{bmatrix}$$

show that is is idempotent. Find the latent roots of **A**, its trace and its rank.

13. Visualize a soccer ball and what strains would be involved in forcing it to become a rugby ball. Interpret this in terms of transformations.

14. Goodall and Green (1986) suggested using affine transformations when fitting one set of landmarks to another. This permits dilatations (expansions (or magnifications) and contractions). How would you apply this to studying growth in leaves?

CHAPTER 3

Primer of Multivariate Statistical Analysis

Contents

3.1 Introduction

The preparation of this chapter was not made without qualms. As observed in the Preface, this book is not a textbook on multivariate statistical applications in palaeontology, notwithstanding that it is based on applications of multivariate methods. For complete treatments of the standard analytical procedures I ask you to consult any of the texts listed in the references. The question I have asked myself repeatedly is—how far should I go in reviewing multivariate statistical procedures? It can therefore be to the point to explain how I have arrived at the scope and level of the material included here below.

I give a general overview of multivariate statistical analysis in order to demonstrate the natural relationships occurring between the various methods and to set the stage for the palaeobiological analyses. This is explained in detail later on in the chapter. I have then considered the way in which these methods are treated in the standard textbooks on applied multivariate statistical analysis, bearing in mind the factor of availability. Virtually all textbooks view the subject in much the same manner with respect to the way in which major topics are introduced and developed. Thus, the basic steps required for carrying out the calculations are usually presented in a stereotyped fashion and it will normally suffice with brief references to standard volumes for accounts of the usual methods of multivariate analysis. However, for much practical work many improvements have appeared over the last few years which are not generally available in the standard texts. In such situations, I have elected to provide explicit accounts of technical advances in the fields of robust methods and stability of latent vectors and to illustrate their value in biological connexions.

Most of the methods outlined below are programmed in SAS in the Supplement provide by Leslie Marcus at the end of this book.

We start by considering Fig. 3.1. This schematic diagram illustrates the natural relationships between the most commonly used multivariate statistical methods. You may perhaps be surprised to see that I have placed some procedures, usually claimed to be quite different, in close proximity to each other. In such cases, the form of the input data and the methods of calculation are close. Where differences appear in applications, this is due to interpretive philosophies, but also to lack of insight of casual users into the fundamentals of multivariate theory. The ensuing discussion centres around the display in Fig. 3.1. Practitioners tend to associate themselves with particular methods and to decry the value of "competing procedures". This is not the right way of

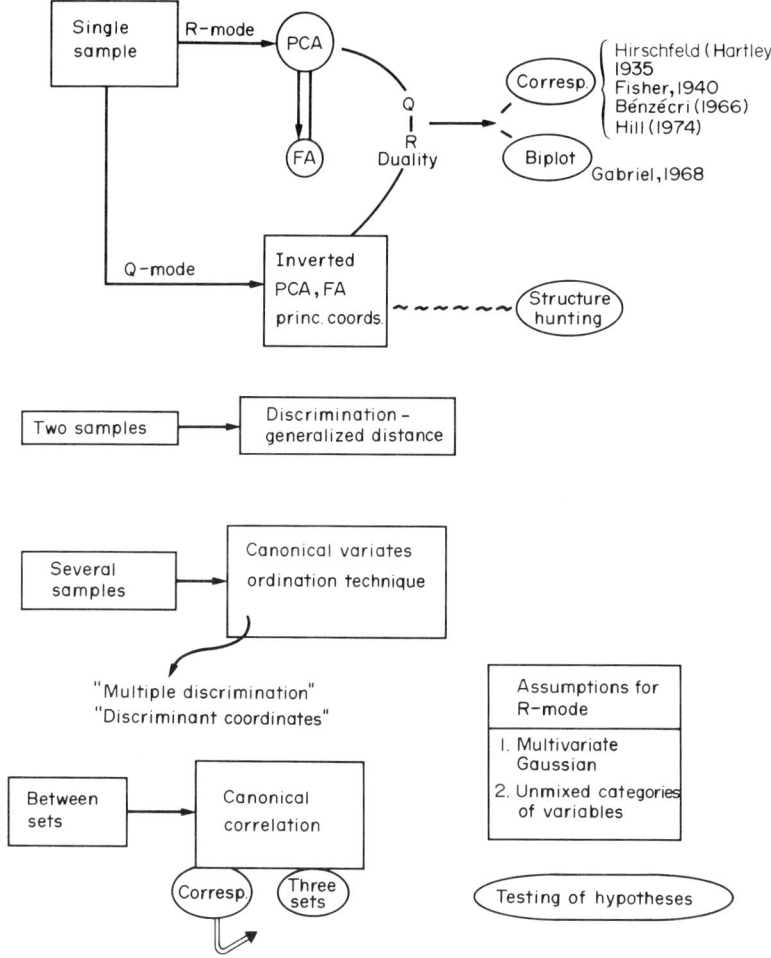

FIG. 3.1. Schematic diagram illustrating the relationships between the most commonly used multivariate statistical methods. Abbreviations. PCA = principal component analysis; FA = factor analysis; CORRESP = correspondence analysis; Princ. Coords. = principal coordinate analysis.

thinking and each method has its own special analytical niche, be it that some may be more versatile than others.

3.2 The Single Sample

I begin this section by standardizing my notation. The data will usually be in the form of a *data matrix*, which I denote as **X**. This was introduced in Chapter 2, but for ease of reference, I repeat the following. Each row of the data matrix consists of a vector of observations containing p elements. The data matrix has N rows, i.e. there are N specimens and p variables or characters. A typical element of **X** can be written as x_{ij}.

The things we measure are called variables, characters, attributes, traits, etc. These all amount to the same thing and the choice of word may depend on the lexical mood of the person. I use all four words in this book. As already noted, Greek letters are used to denote population quantities (i.e. the "theoretical" value to be estimated from a sample). Roman letters are reserved for sample quantities.

We shall be concerned with various matrices and vectors in statistical connotations. It will here suffice with a few definitions. The **mean vector**, or **centroid**, is the name applied to the vector composed of the means of each of p variables. The analogue of the variance of univariate analysis is called the **covariance matrix** (an alternative term is variance-covariance matrix which, however, loses on loquaciousness). In terms of the sample, we designate this matrix **S**. The matrix of correlations corresponding to it is denoted as **R**.

Some other commonly used matrices in multivariate analysis need also to be mentioned. The matrix of sums of squares and cross-products, denoted **A**, and adjusted for the means, occurs in the analysis of several samples. The sum of two or more of these matrices is referred to as the pooled sum of within-groups matrices and is usually given the letter **W**. The matrix of total sums of squares for the entire set of pooled data of k groups is usually written as **T**. The complementary matrix of sums of squares and cross-correlations between groups is conveniently written as **B**. Clearly,

$$\mathbf{T} = \mathbf{W} + \mathbf{B}.$$

3.2.1 *Notes on Some Multivariate Methods*

The most widely used method of multivariate analysis is that known as **principal component analysis**. It is the simple application of the extraction of latent roots and vectors of either the covariance matrix or the correlation matrix. This is an R-mode analysis, a term deriving from the realm of Psychometry and reflecting the fact that the main starting point for the quantitative methods in popular use in that field is the correlation matrix, the standard sample designation of which, as we have just said, is **R**. In many respects, principal component analysis is perhaps the most useful of all the multivariate methods. A recent treatise on principal component analysis is the book by Jolliffe (1986). Another is the text on applications in meteorology and oceanography by Preisendorfer (1988).

Closely related mathematically to principal components is the procedure called **factor analysis**. Factor analysis saw the light of day in psychological connexions, to which attests its very name, to wit, "the factors of the mind". In the form in which factor analysis entered into psychology, a reasonable case can be made for the validity of the interpretational philosophy employed by psychometricians. However, biologists, technometricians, palaeontologists, geologists and mining geologists began to use factor-analytical packaging in a quite different analytical situation. We shall return to this question later. Suffice it here to say that Jöreskog *et al.* (1976) devoted an entire book to the semantic confusion attaching to attempts to use factor analysis in descriptive natural science. As practised in the natural sciences, factor analysis is no more than a variant of principal component analysis, fitted with some of the appurtenances of true factor analysis. This variant was termed *principal component factor analysis* by Jöreskog *et al.* (1976). Bookstein *et al.* (1985) use a variant of factor analysis/path analysis introduced by S. Wright. Crespi and Bookstein (1989) give an account of the application of Wright's factor analysis and path coefficients to a problem in quantitative genetics. I shall say more anon of this.

Still within the framework of the analysis of a single multivariate sample, we have methods that can be usefully thought of as being inverted versions of the foregoing procedures. These are the Q-mode methods, so named because Q comes before R in the alphabet. An R-mode method employs associations between the variables, whereas an inverted, or Q-mode method, uses associations between the specimens, individuals, objects of the sample. Thus, when one speaks of R-mode space, it is the space mapped out by the p variables to which reference is being made. When the reference is to Q-mode space, it is the space mapped out by the N observations that is meant. The term "inverted factor analysis" is due to Bartlett (1965).

An early inverted multivariate procedure on the scene is that of Q-mode factor analysis in which the general methodology of principal components, with various factor-analytical appurtenances borrowed from psychometry, is applied to an $N \times N$ matrix of pseudo-correlation coefficients computed between the individual specimens of the sample. A surrogate of the Pearsonian correlation coefficient is in common use for doing this, the cosine theta measure of association (cf. Jöreskog *et al.*, 1976). The rationale for this measure was touched upon in Chapter 2 in the section on the correlation coefficient.

Another development is the method known as **principal coordinates**, developed by Gower (1966), and which has proved its overall usefulness in many situations. It is, in essence, inverted principal component analysis which preserves distances, and although it can be used with the usual correlation coefficient, it is most applicable where some kind of similarity matrix linking the N specimens of the sample is employed.

Gower (1971) proposed a general similarity matrix which has the advantage that it permits the simultaneous treatment of data comprising continuous, discontinuous, dichotomous and qualitative variables (see Chapter 2). The latter are characters that can not be given any standard categorization but which can be numbered in some arbitrary fashion.

We need to establish an important point of statistical procedure here concerning the manner in which the results of R- and Q-mode analyses can be used. The latent roots and vectors of principal components occur in many connexions in multivariate work, including tests of statistical significance. The latent vectors of Q-mode analyses do not have the same scope and it is not valid procedure to attempt a reification of the elements of the latent vectors in the manner often done in R-mode analyses. Usually, the latent roots of a Q-mode analysis cannot be used in significance testing. The exception is that if a principal coordinate analysis is done on Euclidean distances between objects, the axes of the configuration correspond to the principal component axes (one obtains the same configuration for the principal component analysis of the covariance matrix if the centroid of the points lies at the origin of the coordinates). The aim of Q-mode analysis is graphical, to provide a pictorial appraisal of the interrelationships between individual specimens of a sample, such as the existence of natural clusters. Alternatively, a Q-mode analysis can be useful for ascertaining whether we have a reasonably homogeneous set of observations before us (i.e. the data are **Multivariate Gaussian** in the reduced space of Q-mode analytical representation): this does not relieve one of the necessity of seeking atypicalities by more exact methods. A Q-mode representation of a poorly understood set of observations can form an effective point of departure for a complete multivariate study. Let me give you an example of why this may be so.

A biologically homogeneous sample, i.e. a sample composed of individuals from a single interbreeding population of a species may, on analysis, display groupings into separate entities due to differences in size and shape. Such a situation arises naturally in many species of marine ostracods owing to complex polymorphism as well as a mixture of growth stages in a sample. A Q-mode appraisal of the data can constitute a starting point for the recognition of subsets for a multisample study.

3.2.2 R- and Q-*Mode Duality*

The question that now ensues is whether there is a simple mathematical relationship between R-mode and Q-mode methods. Fortunately, there is. The fundamental paper on the subject is that of Eckart and Young (1936) who enunciated what has come to be known as the "Eckart–Young Theorem". This theorem states that any rectangular matrix can be decomposed into three matrices. This is the *singular value decomposition of a matrix*. In factor analysis, this decomposition is also referred to as the "basic structure of a matrix". If you need to brush up the mathematics of the singular value decomposition, read the relevant part of Chapter 2 (Section 2.6) again.

There are two methods of multivariate analysis that are based on the singular value decomposition. The first of these methods is called *Correspondence Analysis*. The main features of correspondence analysis were recognized by H. O. Hirschfeld (later H. O. Hartley) (1935). Presumably unaware of Hartley's work, Fisher (1940) produced an explicit procedure for the discriminant analysis of a table of contingencies. Since then the method has been rediscovered many times. For example, it occurs in ecological work where it is known as the "method of reciprocal averaging". In ignorance of the work of anglophones, French workers derived an identical procedure for analysing linguistic data arranged in contingency tables (Bénzécri, 1973). Hill (1974) established the statistical background of correspondence analysis. Another important reference is the book edited by Jongman *et al.* (1987). The francophone terminology is different and on first encounter, one can be lured into believing that something entirely new has evolved. The French call their procedure "l'analyse factorielle des correspondances", which, naturally enough received the fuzzy English translation of "correspondence analysis". As Greenacre (1984) pointed out, the two words are morphemically similar, but are not identical in meaning. Although the method was originally designed for discrete data, this distinction was only hazily upheld, right from the outset and analyses of continuously distributed morphological data appear early in francophone publications; for example, the monograph on fossil lemurs by Mahé (1974).

At about the same time as the francophones were developing their "analyse des correspondances", under the leadership of Bénzécri, Gabriel (1968) introduced his method of *biplots*. This technique also makes use of the Eckart–Young theorem. It is basically graphical and differs from correspondence analysis in that it does not employ any particular way of scaling axes, which is one of the hallmarks of correspondence analysis. There is also an asymmetry in the way in which the rows and columns of the data matrix are treated, which is lacking in correspondence analysis. For a long time Gabriel's biplot was relegated to the statistical curiosity closet but of late it has been undergoing something of a revival (Gordon, 1981).

Jolliffe (1986) expressed the opinion that the foregoing methods work best in particular cases and that each does best with the appropriate data. Thus, a contingency table is best suited for treatment by correspondence analysis. Principal coordinate analysis works best with similarity or dissimilarity matrices and standard matrices obtained from N observations on p variables can most suitably be treated by standard principal component analysis. This is sound advice and "to each his own" is the most efficient way of approaching a multivariate analysis.

These are, then, the methods in general use for studying the properties of a single multivariate sample. One of the ideas central to the graphically oriented techniques is that distance between objects should not be seriously distorted by the reduction in dimensionality. An informative (although rarely invoked) means of gauging distance relationships in a two-dimensional representation of a highly multivariate situation is to superimpose a *minimum spanning tree* on the plot. A good reference for this procedure is Gordon (1981, pp. 37–38).

The method comes from the theory of networks, which is of paramount importance in computer technology.

3.3 Two Samples

Notwithstanding that a great deal of multidimensional palaeobiological work is done within the confines mapped out in the foregoing section, many studies entail two or more samples drawn from two or more populations. Moreover, there are situations in which the appropriate methodology rightly requires a two-sample procedure, rather than an incorrectly applied one-sample analysis.

The best known of all the multivariate statistical methods is that called the *Linear Discriminant Function*. This method was originally designed by Fisher (1936) in answer to a taxonomical problem in Botany. It was later applied to anthropometrical data. As is so often the case, an idea may arise simultaneously with different workers, seemingly quite independently. Thus, while Fisher was looking at the problem of discriminating between two populations on the basis of multiple measurements, Mahalanobis, in India, was trying to find some ways of expressing statistically relevant distance between k populations on the basis of multiple measurements. Independently of the British and Indian work, Hotelling in the U.S.A. was concerning himself with finding a multivariate statistical analogue of the univariate t-test. All of these solutions entered the literature within a few years of each other, although without any evidence of cross-referencing. An historical résumé is given in Reyment *et al.* (1984).

Surprising as it may seem today, the close connexions between all three methods were not realized at first and, in fact, it is rather astonishing to read the opinions on the relative merits of the methods published by their respective protagonists. To make matters even murkier, the Pearsonian coefficient of racial likeness became enmeshed in the discussion. This coefficient was produced by Karl Pearson (né Charles: Pearson's doctorate, obtained at Göttingen, was in Germanic Studies: like so many erudite British of the times, he was a well-trained mathematician) to assist anthropologists compare skeletal parts. It turns out to be exactly equivalent to the Mahalanobis Generalized Statistical Distance if all covariances are nought. The various difficulties and misunderstandings were eventually sorted out, but only after the expenditure of much rhetoric and printer's ink.

None of these methods, as originally conceived, took account of the **structure** of the data, apart from the requirement that the samples be drawn from multivariate—normally distributed populations. Little attention was paid to such questions as robustness, geometrical properties of the ellipsoids of dispersion, and the choice of variables. As I shall show further on, these seemingly minor matters can, and do, have a notable effect on the outcome of an analysis. Anderson (1958), in the first edition of his textbook on multivariate analysis, the formative importance of which can hardly be overestimated, addressed himself to the question of statistical distances and linear discriminant functions when the covariance matrices are unequal. His solution was to provide a straight generalization of the Behrens–Fisher problem for the univariate case of unequal variances. This is easy to compute, but does not always give you the answer you really want. A further reference is the paper by Anderson and Bahadur (1962) in which the subject is treated more exhaustively, leading to the Anderson–Bahadur solution of the multivariate Behrens–Fisher problem (N.B. Behrens and Fisher published identical solutions to the univariate problem at about the same time).

Another way of approaching the problem posed by unequal covariance matrices is to use a *Quadratic Discriminant Function*. A quadratic discriminant can accommodate non-linearity, but it does not yield a reifiable vector of discriminatory coefficients, and thus cuts off the user from the possibility of utilizing a number of biologically valuable analytical accessories. In studies concerning the allocation of new specimens, the quadratic discriminant function can prove very useful. It is often employed in medical statistics for relating patients to disease and treatment.

Can the analysis of two samples by discriminant functions be linked in any meaningful manner to the methods reviewed in the preceding section? The answer to this question is yes, for there is an obvious connexion in terms of the latent roots and vectors of the individual sample covariance matrices.

3.4 Several Samples

The step from two to three or more samples is perhaps not as straightforward as one might have wished. The reason for this is that the methods of calculation become more complicated and it is no longer possible to use simple matrix manipulations for producing the analogues to the discriminant function coefficients. The method for k samples, based on p variates, is referred to as **Canonical Variate Analysis**. Another name is "Discriminant Coordinates" (Gnanadesikan, 1977; Seber, 1984), which exposes the fact that the method is considered by some workers from the aspects of ordination and graphics. There is not a little semantic disorientation involved in the terminology surrounding canonical variates which can only appear confusing to the tiro. For this reason, I give below the usage to which I adhere (cf. Reyment *et al.*, 1984).

Canonical Variate Analysis is defined as the set of techniques connected with the study by latent roots and vectors of the within-groups matrix, **W** and the between-groups matrix, **B** for samples drawn from k populations and upon which p variables have been measured. For most purposes in this book, we are concerned with compatible variables, although in some ecological work, there may be a mixture of categories (Digby and Kempton, 1988). A complete canonical variate analysis should be aimed at providing information on the following:

(1) Do the *centroids* differ significantly? This is the question posed by the Multivariate Analysis of Variance, usually referred to by the acronym MANOVA. The usual turn taken by MANOVA is in the nature of a generalized one-way ANOVA.

(2) One will usually be interested in ascertaining how many of the generalized latent roots are significant, noting that there cannot be more latent roots than there are samples. Thus, if $k < p$, there can be no more than k latent roots.

(3) Attention may be directed towards studying discrimination between samples, assignation (usually given the misnomer classification in statistical books) of new individuals to one of the k populations, the identification of specimens and whatever. This is often referred to as Multiple Discriminant Analysis. The elements of the r significant latent vectors are the coefficients of the discriminant functions. Note, that these vectors are not generally reifiable in the manner of principal components, although this is often attempted.

(4) The graphical display of the **scores** for each of the individuals encompassed by the analysis is usually of interest. This is what is being more and more frequently known as "Discriminant Coordinates", but it is no more than a byproduct of the standard analytical procedure. The plot of the transformed means of canonical variates can be conveniently furnished with a minimum spanning tree (cf. Gordon, 1981).

There are some particular problems attached to canonical variate analysis to be taken up. I begin by considering the uncomplicated situation in which we have k species recognized in some objective manner and we want to examine distance relationships between them by means of p diagnostic characters measured on them. In this illustration, the observations occur in groups that have been established *a priori*. If, however, the data are poorly understood and we have no idea how many natural groups occur in it, we run into operational difficulties. The commonly adopted procedure is to begin with the entire sample of observations and to examine it by one of the Q-mode methods already referred to. This is known to the trade as "zapping". Hopefully, this will disclose whether the data are composed of two or more clusters. This could be the case if the data consist of several growth stages, size morphs, sexual dimorphs, etc. These clusterings can then be exploited to produce sub-samples, which in turn, can be "zapped" for the occurrence of hetero-geneities. Eventually, a number of sub-samples may be obtained which can then be fed into a canonical variate analysis.

Although the above procedure can often be useful, it does contain an element of manipulation and the analysis will tend to accentuate the sub-divisions in the data and hence the ordination yielded by the plots of the canonical variate scores. Thus, you are getting no more than you paid for; an ordination achieved by this approach can be referred to as a **secondary ordination**.

3.5 Relationships Between Sets

Some studies seek relationships between disparate sets of variables. For example, one may wish to see whether a set of morphological measures is correlated with a set of ecological factors. Most applications you are likely to see will try to do this via principal component factor analysis or principal components: one set of variables will be made to fill the first s locations in the data matrix and the other set will be placed in the $s + 1$ to p locations.

A technique that was especially derived by Hotelling (1936) for computing between-set correlations is termed **Canonical Correlation**. For detailed discussions of the method, and the problems incurred in attempting to interpret results, I refer you to Love and Stewart (1968), Cooley and Lohnes (1971) and Pimentel (1979). Canonical correlation analysis produces a suite of correlations between various linear combinations of the two sets of variables. A weakness in the procedure is that the correlations are not necessarily linked to the "importance" of the linear combinations and it is therefore required procedure to carry out a good deal of manoeuvring in order to produce a reasonable analytical interpretation. You need a lot of experience to use the method successfully and it is for this reason that many practitioners shy clear of it.

Canonical correlations can be computed between more than two sets of variables and there is, moreover, a variant that appears in special applications of correspondence analysis (Greenacre, 1984; Jongman *et al.*, 1987; ter Braak, 1985, 1986). In usual canonical correlation, the number of observations is required to be greater than the total number of variables, but this is not the case in the canonical correspondence analysis of ter Braak.

3.6 Exemplification of Methods

I shall now exemplify the main properties of the methods outlined in the foregoing sections by simple numerical examples and in so doing will recapitulate some of what was reviewed in Chapter 2. This section concludes with a fully worked morphometrical analysis of traditional stamp. Inasmuch as my treatment of the statistical methods employed must perforce be brief, I refer the reader in need of further information to the professional literature on the subject. I have given preference to the comprehensive text by Seber (1984), to which I can direct you with full confidence. I begin with a résumé of the backbone of multivariate analysis, the algebra of latent roots and vectors.

3.6.1 *Basic Algebra for the Latent Structure of Single Samples*

The ensuing discussion connects to that of Section 2.5.

Consider a real symmetric matrix $\mathbf{R}$, which might be, for example, the minor or major product moment of a data matrix, the covariance matrix, or the correlation matrix. A latent vector of $\mathbf{R}$ is a vector $\mathbf{u}$ given by (3.1)

$$\mathbf{Ru} = \mathbf{u}\lambda \tag{3.1}$$

where λ is an unknown scalar.

Equation (3.1) can also be written

$$(\mathbf{R} - \lambda\mathbf{I})\mathbf{u} = \mathbf{0} \tag{3.2}$$

where $\mathbf{0}$ is the null vector.

Equation (3.2) implies that the unknown vector $\mathbf{u}$ is orthogonal to all row vectors of $(\mathbf{R} - \lambda\mathbf{I})$, which is a system of homogeneous equations. In order to exemplify this statement, we can look at the situation for a 2×2 matrix $\mathbf{R}$. Writing out the equations in full we have that:

$$(r_{11} - \lambda)u_1 + r_{12}u_2 = 0$$

$$r_{21}u_1 + (r_{22} - \lambda)u_2 = 0$$

The first step in solving for $\mathbf{u}$ and λ requires setting the determinant of $(\mathbf{R} - \lambda)$ equal to nought. Thus,

$$|\mathbf{R} - \lambda\mathbf{I}| = 0.$$

There are some useful points to be held in memory. Firstly, if we keep to the bivariate case, we have that

$$\lambda_1 + \lambda_2 = r_{11} + r_{22}$$

that is, the sum of the diagonal elements of the real symmetric matrix $\mathbf{R}$ is equal to the sum of its latent roots. This sum is known as the *Trace* or *Spur* of the square symmetric matrix, as has already been mentioned in Chapter 2.

Secondly, recall that the product of the latent roots of matrix $\mathbf{R}$ is equal to the determinant of the matrix $\mathbf{R}$ (cf. 2.23). Thus,

$$\lambda_1\lambda_2 = r_{11}r_{22} - r_{12}r_{21}.$$

The required latent vectors are easily obtained by means of equation (3.2). In practice, the latent vectors are normalized, which means that they are constrained to be of unit length (cf. 2.19).

If the latent roots λ_i ($i = 1, \ldots, p$) are made to be the diagonal elements of a diagonal matrix $\mathbf{\Lambda}$, and the latent vectors are collected as columns into the matrix **U**, then equation (3.1) can be expressed more generally as

$$RU = U\Lambda. \tag{3.3}$$

The matrix **U** is square orthonormal (cf. 2.19).
Post-multiplication of (3.3) by **U** yields equation (3.4), below:

$$R = U\Lambda U^{T}. \tag{3.4}$$

This equation tells us that the symmetric matrix **R** can be represented in terms of its latent roots and vectors. I shall now illustrate the foregoing presentation by reference to data on Rock Crabs.

Example 3.1. Application of Principal Component Analysis
Variation in the crab-genus *Leptograpsus*

Introduction

The genus *Leptograpsus* (the Rock Crab) is widely distributed throughout the southern, warm-temperature areas of the Indian and Pacific Oceans. In Australia, the genus ranges from North West Cape in Western Australia to the eastern coast of Queensland. Representatives of the genus also occur in the northeastern part of Tasmania and it is also known to occur in the western Pacific Ocean, around New Zealand, and along the eastern coast of South America.

In Australian waters, two taxa occur which are regarded by some as morphs of the same species, *L. variegatus*. These morphs may be referred to as the Orange Morph and the Blue Morph. Recent laboratory work (electrophoretic studies of blood proteins) tends, however, to support the contention that the orange and blue crabs belong to to distinct species. Campbell and Mahon (1974) established morphological criteria for the statistical identification of the two entities with Australia, independently of colour, using a multivariate approach. In the following, I shall refer to the two taxa quite simply as the "Blue Species" and the "Orange Species".

Statistical Analysis

Campbell and Mahon measured the following characters on the carapace of the crabs: (a) the width of the frontal region (lip of the carapace), immediately anterior to the anterior tubercles and denoted FL; (b) the width of the posterior region of the carapace, the rear width and denoted RW; (c) the length of the carapace along the median line, CL; (d) the maximum width of the carapace, CW; (e) the depth of the body, BD. The measurements were made to the nearest millimetre.

TABLE 3.1. *Principal component analysis for the Rock Crabs*

Covariances (upper triangle) and correlations (lower triangle) for 50 specimens of crabs

	FL	RW	CL	CW	BD
FL	8.8447	6.7292	17.1719	19.2543	8.0254
RW	0.9630	5.5202	13.4066	15.0541	6.2345
CL	0.9892	0.9775	34.0725	38.0765	15.9254
CW	0.9896	0.9794	0.9971	42.8007	17.7903
BD	0.9805	0.9641	0.9913	0.9880	7.5751

We shall begin by comparing the results for the principal component analyses of the covariance and correlation matrices for orange females. The correlations and covariances for the five variables and 50 specimens are displayed in Table 3.1. Note that the upper triangle contains the variances and covariances and the lower triangle, the correlations.

The high correlations in Table 3.1 are rather typical of what is found for many groups of crustaceans. Apart from the fact that all correlations are very high, you will no doubt see that there is considerable variability in the sizes of the variances, aligned along the principal diagonal of the array. The first, second and fifth variances are homoscedastic with respect to each other and the third and fourth are homoscedastic in relation to each other. However, the latter two are five to six times greater than the three former. We shall need to bear this in mind when all four samples of crabs are considered in a later example.

The latent roots and vectors for the correlation and covariance matrices are summarized in Table 3.2. You will see in Table 3.2 that the latent roots are

TABLE 3.2. *Latent roots and vectors for the covariance and correlation matrices of Table 3.1*

Latent roots for the covariance matrix; values in brackets denote the percentage of each root in relation to the sum of all roots.

98.1820 (99.36), 0.2659 (0.27), 0.1756 (0.18). 0.1119 (0.1133), 0.0778 (0.0788).

Latent vectors for the covariance matrix

	1	2	3	4	5
FL	0.2976	−0.4075	−0.6236	0.5815	0.1358
RW	0.2325	0.8425	0.0233	0.4611	0.1516
CL	0.5885	−0.1164	0.3270	0.1355	−0.7175
CW	0.6597	0.1202	−0.2892	−0.6269	0.2715
BD	0.2753	−0.3100	0.6481	0.1948	0.6083

Latent roots for the correlation matrix; percentages in brackets.

4.9280 (98.56), 0.0428 (0.86), 0.0197 (0.39), 0.0069 (0.14), 0.0026 (0.05).

Latent vectors for the correlations matrix

	1	2	3	4	5
FL	0.4467	−0.3702	0.6941	0.4261	−0.0112
RW	0.4432	0.8586	0.0148	0.2569	−0.0104
CL	0.4497	−0.1036	−0.0771	−0.4557	−0.7573
CW	0.4496	−0.0404	0.0797	−0.6195	0.6372
BD	0.4469	−0.3367	−0.7112	0.4012	0.1424

aligned in decreasing order of magnitude. The values in brackets are the percentage of each root to the sum of the roots. This is what we have learned to call the trace of the matrix. In the jargon of applied multivariate statistics, the analyst says that the percentages express the proportion of the variability resident in the corresponding latent vector, or principal component. You will doubtless have observed that the first latent root "accounts for 96.3% of the variability in the covariance matrix". This kind of situation turns up when the correlations between variables are very high, as in the present case in which not one of the correlation coefficients is less than 0.96.

Recalling what was said in the chapter on Linear Algebra, we can hardly avoid noticing that the covariance matrix we have before us is dangerously close to being of rank $< p$.

The latent vectors of the covariance matrix are next on the list. These are arrayed in Table 3.2. There are several points I ask you to pay attention to here. Look at the first latent vector. You will observe that elements 1, 2, and 3 are almost equal in size and they are, moreover, appreciably smaller than elements 3 and 4. What we in fact see is a reflection of the magnitudes portrayed in the diagonal of the covariance matrix, to wit, the sizes of the five variances. Many practitioners are wont to invest the elements of the latent vectors with some particular meaning. The first vector of principal components is often interpreted in terms of *size*. One way of avoiding many of the embarrassing effects of linkage to the magnitude of the entries in diag **S** is to perform the analysis on the logarithms of the observations. This "stabilizes" variances, and is, incidentally, the only way to fit the isometric model, as was shown by Jolicoeur (1963).

Some statisticians ask why interest should be so firmly glued to the largest principal component (e.g. Gower, 1967)? Why not be equally as enthused by the smallest principal component, particularly if its corresponding root is almost zero. This must perforce indicate an invariant relationship in the data and should therefore be pregnant with vital biological information.

We turn now to the correlations. I begin, as before, with the latent roots, displayed in Table 3.2 in descending order of magnitude. The relative proportions of the latent roots do not differ much from the results for the covariance matrix. The latent vectors do, however, diverge markedly. Are these really the same data, you might well ask? The first latent vector is now quite different from that yielded by the covariance matrix. There are, however, some agreements to be seen. The second latent vectors are quite close, also the third, and to a certain extent, the fourth. The fifth vectors differ. We return to this enigmatic situation further on.

Before leaving this example, I must point out one more thing, namely, that I have used four decimal places in citing results in Tables 3.1 and 3.2. Obviously, the computations do not possess this level of accuracy, since the crabs were only measured to the nearest mm. Be on the lookout for spurious accuracy in your own work, but also in that of others. I commented on computing accuracy in Chapter 1.

Much the same result as given by the correlation matrix would have been yielded by analysing the covariance matrix of logarithmically transformed observations. This is because of the fact that the logarithmic transformation stabilizes the variances and hence reduces the effects of disparate ranges. The indiscriminate application of the logarithmic transformation to morphometric data can, in some cases, be perilous as it may mask the occurrence of legitimately occurring heterogeneities. If, however, it is desired to fit the isometric model in an analysis, the logarithmic transformation is mandatory. Bookstein *et al.* (1985) discuss this subject.

There is an interesting property of principal component analysis that is useful to know if you are interested in reifying latent vectors, namely, that the principal components derived from a correlation matrix do not depend on the absolute values of the correlations but *only on their ratios*. Hence, multiplication of all the diagonal elements of a correlation matrix by a constant leaves the latent vectors of the matrix unchanged.

3.6.1.1 *How Many Principal Components are Valid?*

The above question is perhaps one of the most frequently asked of the biometrical consultant. The example in which *influential observations* are treated (cf. p. 291) includes one way of answering the question. Jolliffe (1986, p. 95) proposed an *ad hoc* procedure which is worth considering. Jolliffe suggests the cut-off point

$$l^* = 0.7(\Sigma l)/p \qquad (3.5)$$

where the l are the sample latent roots of the covariance or correlation matrix and p the number of dimensions. He reported that the procedure seems to work better with the correlation matrix.

If we apply equation (3.5) to the data for the orange females of the Rock Crab (Example 3.1), we find that $\Sigma l/p = 1$, hence $l^* = 0.7$. This indicates that it is only permissible to keep *one* latent root and vector. The argument in support of this is that any principal component associated with a very small variance, such as any of the subsequent ones, contains less information than the original variables and is, consequently, not worth retaining. Jöreskog *et al.* (1976, p. 75) gave an approximate method for assessing the number of principal components.

3.6.2 *Correlation Between Variables and Principal Components*

It is often useful to examine the correlation between the j-th variable and the k-th principal component. Hotelling (1931) used the normalization

$$\hat{\boldsymbol{\alpha}}_k^T \hat{\boldsymbol{\alpha}}_k = \delta_k$$

where the j-th element of $\boldsymbol{\alpha}_k$ is the desired correlation. He showed that

$$\hat{\boldsymbol{\alpha}}_k = \delta_k^{1/2} \mathbf{a}_k$$

where a_k is the latent vector and δ_k the corresponding latent root.

To appreciate this, note that

$$\text{var}(z_k) = \delta_k \text{ for } k = 1, \ldots, p$$

for the principal components defined by the transformation

$$\mathbf{z} = \mathbf{A}^T \mathbf{x}$$

for covariance matrix $\boldsymbol{\Sigma}$. The vector of p elements $\boldsymbol{\Sigma}\boldsymbol{\alpha}_k$ has the covariance between x_j and z_k as its j-th element.

By (3.1) we have that

$$\boldsymbol{\Sigma}\boldsymbol{\alpha}_k = \delta_k \boldsymbol{\alpha}_k.$$

Hence, the covariance between x_j and z_k is $\delta_k \alpha_{kj}$, from which follows that the correlation between x_j and z_{kj} is

$$\delta_k \alpha_{kj} / [\text{var}(x_j)\text{var}(z_k)]^{1/2}$$

$$= \delta_k^{1/2} \alpha_{kj} / \sigma_{jj}^{1/2}$$

where $\sigma_{jj} = \text{var}(x_j)$.

In the case where the principal components are those of the correlation matrix, $\sigma_{jj} = 1$ so that α_{kj} expresses the correlation between the j-th variable and i-th principal component. Further information on principal component theory can be found in Seber (1984, p. 176, and following pages, and Jolliffe, 1986).

3.6.3 *Scaling*

What we have been discussing up to now is what one might perhaps wish to call "raw principal component analysis". There are however, many applications in which the latent vectors are scaled so as to give, it is hoped, a truer picture of the structural relationships.

The easiest way of introducing the process of scaling in principal component analysis is by listing the steps required for doing this.

1. Compute the covariance matrix **S**.
2. Compute the k largest latent roots L_k and the corresponding vectors $\mathbf{U}_k$ of the covariance matrix.

3. Compute $\mathbf{A} = \mathbf{U}_k \mathbf{L}_k^{1/2}$; this amounts to scaling each latent vector so that its squared length equals the latent root. Note that $\mathbf{A}^T \mathbf{A} = \mathbf{L}_k$.

3.6.4 Q-*mode Analysis and Principal Coordinates*

Gower (1966) introduced the term principal coordinate analysis for a Q-mode principal components type of analysis of a matrix of associations or similarities, including the common correlation coefficient. The procedure seems to have been originally thought up by Young and Householder (1938) and applied to psychometrical work by Torgerson (1958). As noted previously, one of the strengths of principal coordinates is that the method allows one to mix quantitative, qualitative and dichotomous data in the same analysis, using a suitable measure of similarity such as that of Gower (1971). Note that the terms quantitative, qualitative and dichotomous are used here in the spirit of Gower (1966, 1971). A more stringent terminology is given by Gordon (1981); see also the relevant section on variables in Chapter 1.

Principal coordinate analysis is a technique for providing a geometrical representation of "distance" between individuals (that is, how unlike each of N individuals of a sample in relation to each other assessed on p variables). The calculations proceed by means of the extraction of the latent roots and vectors of a matrix of associations between the N individuals of the sample to yield an $N \times N$ symmetric matrix, the rows and columns of which are adjusted so as to sum to zero. The measure of association can be the inverted correlation coefficient (the role of specimen and variable reversed), or some particularly constructed measure, such as Gower's (1971a) coefficient. The latent vectors are scaled so that the sum of squares of the elements of each vector equals the corresponding latent root. This is step 3 in the section headed "Scaling". The scaled latent vectors give a set of coordinates providing an accurate representation of the distances between individuals when most of the latent roots are very small. Usually, the number of variables is very much less than the number of specimens. For fuller discussions, I refer you to Gower's papers, Blackith and Reyment (1971), Gordon (1981), Jöreskog *et al.* (1976) and Reyment *et al.* (1984).

The Q–R-mode duality between principal components and principal coordinates only exists for Pythagorean distances which means that if it is your intention to make use of this relationship for reasons of computer storage, you are restricted to a rather unrealistically small field of possible associations. Moreover, for the Q–R-mode duality to hold, the squared distances between points must be exactly squared interpoint distances in some Euclidean space.

It is often useful to be able to tack on some new observations to an existing results. Gower (1968) provided a procedure for adding extra points to the analysis of non-Pythagorean similarity arrays. You would want to use this if you had already completed an analysis and had subsequently obtained some more data.

Principal coordinates can also be used for samples drawn from k populations. The generalized statistical distance can be computed for each pair of universes to give a $k \times k$ symmetric matrix. A principal coordinate analysis of the array of D^2-values is equivalent to a canonical variate analysis with an unweighted between-groups matrix of sums of squares and cross-products. This method is useful for programming multivariate procedures.

Example 3.2. Application of Principal Coordinate Analysis

For the purposes of illustrating the scope of the method of principal coordinates, I have called on two samples of the Rock Crab, to wit, males and females of the Orange Species. There are ten individuals in each sample; these were pooled for the analysis.

Three latent roots of the 20×20 association matrix of Gower (1971a) were extracted. As only to be expected from the high degree of integration indicated by the principal component analysis of one of the samples, the first few roots represent a large proportion of the trace of the association matrix. After two latent roots had been extracted, the percentage residual was found to be 28.79% and after the extraction of the third root, 20.86%. Clearly, most of the distance information is contained in the first two latent roots. The details of the analysis are summarized in Table 3.3. These results were used to prepare the graph for the first and second axes, shown in Fig. 3.2. If you have had any experience of Q-mode analyses, you will probably recognize the strange shapes formed by the points. The first is a parabola, or "horseshoe", shown here. Other projections give other geometrical figures.

TABLE 3.3. *The first three principal coordinates for Orange Crabs*

	1	2	3	SEX
1	−0.46	−0.01	0.17	female
2	−0.45	0.03	0.19	female
3	−0.11	0.27	0.04	female
4	0.09	0.21	−0.06	female
5	0.18	0.19	−0.14	female
6	0.41	−0.06	−0.06	female
7	0.49	−0.21	0.07	female
8	0.45	−0.14	0.00	female
9	0.53	−0.30	0.16	female
10	0.54	−0.32	0.18	female
11	−0.71	−0.43	−0.41	male
12	−0.60	−0.21	0.00	male
13	−0.57	−0.13	0.09	male
14	−0.39	0.08	0.19	male
15	−0.21	0.23	0.11	male
16	−0.12	0.27	0.05	male
17	0.13	0.22	−0.13	male
18	0.28	0.11	−0.16	male
19	0.19	0.17	−0.14	male
20	0.32	0.06	−0.14	male

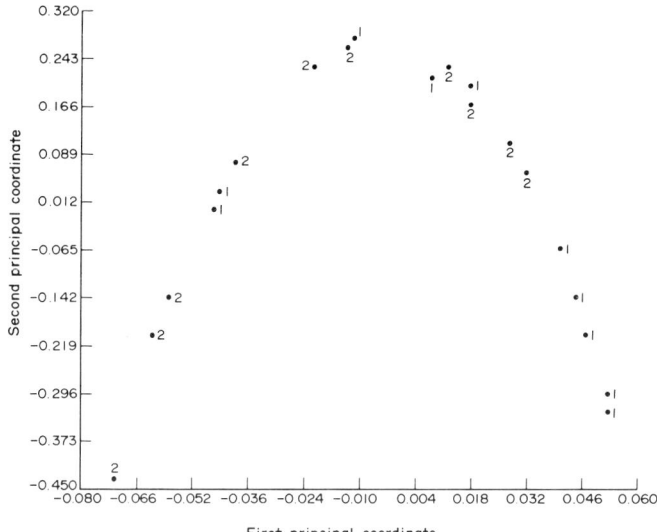

Fig. 3.2. Principal coordinate plot for the first and second axes for ten specimens of females (marked 1) and ten specimens of males (marked 2) of the "Orange Species" of the Rock Crab (*Leptograpsus*). Note the "horseshoe-shape" of the plot.

Figure 3.2 discloses that the separation into males and females is not good at all. One tail of the parabola is occupied by males and the other by females, but the middle zone is mixed. Thus, on the basis of the measurements used here, the principal coordinate analysis does not make a very good job of recognizing specimens of the two sexes.

I mentioned "*horseshoe*"-plots. Seber (1984, p. 250) has given a useful discussion of the phenomenon. Such figures usually appear where there are very high correlations between the variables that are approximately equal. The problem is also discussed in some of the textbooks on linear algebra. Greenacre (1984, pp. 226–233) devoted an entire section to the "horseshoe effect", and other figures, yielded by multivariate ordinating techniques. Another geometrical effect is the sinusoidal "folium", a third, the right lemniscate. You can find these in any table of curves and their pedals. It is sometimes claimed to be useful to "detrend" a plot, for example, in canonical correspondence analysis (see Chapter 7), but the advisability of this has been questioned by several competent specialists. Ter Braak (1985) gave a clear demonstration of the value of detrending from a theoretical viewpoint. Hill (1974) proved that the geometrical figures are produced when there is a quadratic relationship between scores on the first and second axes.

Jöreskog *et al.* (1976) reviewed a method called Q-mode factor analysis and supplied examples. I do not make use of this technique in the present volume and I do no more in the present connexion than refer the interested reader to the relevant chapter in the above-cited text. This Q-mode method embraces rotations of axes and several other procedures that no doubt can prove useful in some studies.

3.6.5 *Correspondence Analysis*

The Simultaneous Recognition of Objects and Variables: The methods described up to now, from the aspect of *ordination*, focus on obtaining a graphical representation of a set of objects as a configuration of points. Similarly, variables can also be represented as points in Euclidean space and this is a quite familiar technique in psychometric factor analysis. A natural development of the graphical sides of multivariate analysis has led to the appearance of methods for making a simultaneous representation of objects and variables on the same plot.

In Correspondence Analysis, the data matrix is standardized so that the entries sum to one. The use of the method is, of course, inapplicable if the variables in the data set are not commensurable (i.e. are of mixed type). It is appropriate for the analysis of presence/absence data and counts in a contingency table. It is, however, also applied to continuously distributed variables, the observations on which are arrayed in a contingency table. The entries in the data matrix are further scaled by being divided through by the square roots of their respective row and column totals. This gives a redefined data matrix. A succinct account of the use of this data matrix is to be found in Gordon (1981, p. 118).

References are Bénzécri (1973), Hill (1974), Gordon (1981, pp. 117–120), Greenacre (1984), Jongman *et al.* (1987) and Jöreskog *et al.* (1976, pp. 107–115). In the latter book, there is a worked example using the same set of data for correspondence analysis and Q-mode methods.

Example 3.3. Application of Correspondence Analysis

We shall continue with the orange species of the Rock Crab and, as before, consider a pooled sample of ten females and ten males. The results are listed in Table 3.4. The plot for the first two axes of this analysis is displayed in Fig. 3.3.

The results listed in Table 3.4 show that 88% of the "inertia" is concentrated to the first two axes. Note that the terminology of correspondence analysis in its French garb is larded with jargon quite unfamiliar to anglophone statisticians. The term inertia is used in francophone literature by analogy with its definition in the applied mathematics of the moment of inertia, which is the integral of mass times squared distances to the centroid. Greenacre (1984, p. 66) provides useful comments on how French terminology can be reconciled with the statistical literature at large.

Figure 3.3 discloses that the correspondence analysis has made a better job of ordinating males and female Rock Crabs than was done by principal coordinates. All males are segregated into their own field and all females into

TABLE 3.4. *Correspondence analysis for male and female crabs of the orange species*

SIMILARITY MATRIX

	FL	RW	CL	CW	BD
FL	0.1449	0.1319	0.2035	0.2152	0.1367
RW		0.1205	0.1853	0.1961	0.1245
CL			0.2860	0.3025	0.1921
CW				0.3200	0.2032
BD					0.1291

	LATENT ROOT	PERCENTAGE INERTIA	CUMULATIVE PERCENTAGE
1	0.0003134	68.81	68.81
2	0.0000878	19.27	88.08
4	0.0000369	8.09	96.17
5	0.0000182	3.83	100.00

CORRESPONDENCE AXIAL LOADINGS

Variable	R-MODE I	R-MODE II	Specimen	Q-MODE I	Q-MODE II
FL	−0.017	0.021	females		
RW	0.045	0.007	1	0.024	0.001
CL	−0.006	−0.005	2	−0.002	0.015
CW	0.001	−0.005	3	−0.008	0.006
BD	−0.011	−0.007	4	0.028	0.004
			5	0.010	−0.005
			6	0.009	0.002
			7	0.025	−0.007
			8	0.010	0.004
			9	0.024	0.001
			10	0.014	−0.003
			males		
			11	−0.015	0.025
			12	−0.009	0.000
			13	0.006	0.010
			14	−0.016	−0.002
			15	−0.022	0.014
			16	−0.028	−0.002
			17	−0.013	−0.017
			18	−0.005	−0.015
			19	−0.029	0.009
			20	−0.011	−0.006

Notes: The computations were made using a program written by J. E. Klovan and R. A. Reyment. They were checked using the GENSTAT program listed by Greenacre (1984). Bénzécri (1973) provides a FORTRAN IV listing of a very complete version of correspondence analysis. The ultimate in advanced computing techniques is the program by ter Braak (1987) and marketed under the name of CANOCO. Although this latter program is mainly directed towards plant-ecological studies, it can be used to great effect in many other connexions.

theirs. The first axis is clearly successful in identifying sexual dimorphic differences. The second axis spreads out the range of variability of the carapace, whereby it shows that males are dispersed more broadly than females.

What do the points for variables in Fig. 3.3 tell us? Specimens with relative high values on certain variables will tend to fall in the region around the points

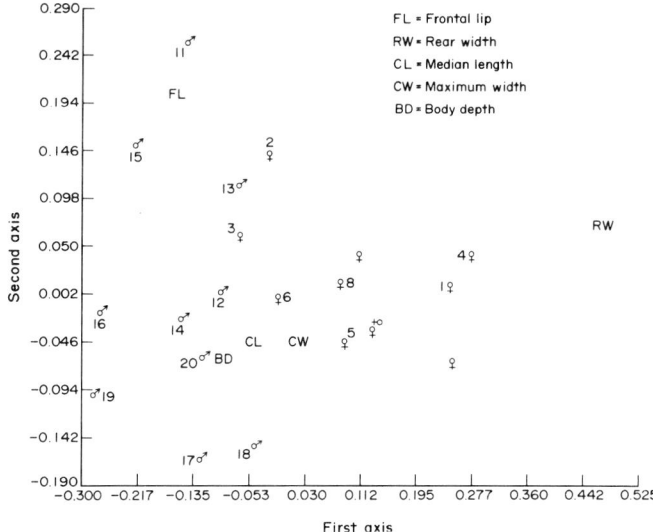

FIG. 3.3. Correspondence analysis of the same set of specimens as analysed in Fig. 3.2 for the first and second axes. The points corresponding to the 20 specimens are displayed together with their respective sexes. The letters denote the five variables. In interpreting the figure, the groupings of the points are of interest as also the locations of the variables. The significance of the biplot can be assessed from variable *BD* and point 20, which lie adjacent to each other. This specimen has a high loading for body depth. Females are numbered 1–10 and males, 11–20.

for these variables. Thus we see that the character BD, body depth, attracts many male carapaces, as does also FL, frontal lip. Females are slightly drawn to the region around RW, rear width, but for the most part, these specimens seem to behave neutrally. Reyment *et al.* (1984, p. 27, fig. 3.7) gave a full analysis of 200 specimens of males and females of the orange and blue species. The excellent ordination into categories is even more clearly displayed in that figure. To be fair, it should be mentioned that the principal coordinate analysis of the same data gave an almost equally as satisfactory ordination (Reyment *et al.*, 1984, p. 26, fig. 3.6).

3.6.6 *Two Samples*

We shall now consider the case of samples drawn from each of two populations. The linear discriminant function was devised by Fisher (1936) for dealing with the following situation:

It is desired to allocate specimens to one of two populations by means of a linear function, assuming that the specimens come from either of these. The linear discriminant function between two samples may be defined as:

$$y = (\bar{\mathbf{x}}_1 - \bar{\mathbf{x}}_2)\mathbf{S}^{-1}\mathbf{x} \tag{3.6}$$

where the $\mathbf{x}_i$ are the means of the respective samples, $\mathbf{S}$ is the pooled sample covariance matrix (or some constant times that matrix), $\mathbf{x}$ is a vector of variables and y is the discriminant score corresponding to the observational vector $\mathbf{x}$.

The coefficients of the linear discriminant function are defined as

$$\mathbf{a} = \mathbf{S}^{-1}(\bar{\mathbf{x}}_1 - \bar{\mathbf{x}}_2) \tag{3.7}$$

where $\mathbf{a}$ is the vector of discriminant coefficients. These coefficients give an approximate idea of the relative importance of each variable to the efficiency of the discrimination if the variances of the variables are roughly equal. If the variances differ greatly, the coefficients can be stabilized by dividing them by their standard deviations.

The linear discriminant function is connected to the Mahalanobis D^2 by the following relationship:

$$D^2 = (\bar{\mathbf{x}}_1 - \bar{\mathbf{x}}_2)^T \mathbf{a} \tag{3.8}$$

where $\mathbf{a}$ is as given in equation (3.7). The generalized statistical distance of Mahalanobis is connected to the Hotelling T^2, a generalization of Student's t, as follows:

$$T^2 = \frac{D^2(N_1 \cdot N_2)}{(N_1 + N_2)} \tag{3.9}$$

where N_1 and N_2 are the respective sample sizes and D^2 is defined as in equation (3.8). Hotelling's T^2 is easily converted to the variance ratio and can therefore be used as a test for significance in the difference between two mean vectors.

It is generally assumed that the covariance matrices of both samples are compatible, that is, homogeneous. This can be a rather bold assumption, as we shall experience further on. Seber (1984, pp. 293, 333, 338) pays close attention to various facets of discriminant analysis, including the systematic selection of variables.

Example 3.4. Application of the Linear Discriminant Function

In order to exemplify the main ideas underlying the linear discriminant function and associated statistics, we shall analyse 50 specimens each of males and females of the orange species of Rock Crabs.

The pooled covariance matrix for the two samples was used for these calculations. The linear discriminant function obtained from this matrix and the difference vector of means, to wit,

$$(0.968, 2.570, 0.930, 1.848, 0.318)^T$$

is

$$y = 1.18\text{FL} + 7.11\text{RW} - 3.79\text{CL} + 1.28\text{CW} - 1.04\text{BD}.$$

In standardized form, this becomes as follows (and we note that the variances for the third and fourth characters are many times greater than those for the other variables)

$$y_{stn} = 3.83FL + 16.16RW - 25.70CL + 9.64CW - 3.28BD.$$

The associated value of the generalized statistical distance is $D^2 = 17.92$, which yields a variance ratio of $F = 85.92$ which for 5 and 94 degrees of freedom is highly significant. We can therefore accept the hypothesis that there is strong sexual dimorphism in the carapace of *Leptograpsus*. We can also compute the probability of misidentification of a specimen. It is $P = 0.017$; that is, in less than two cases out of a hundred will a specimen really belonging to one of the sexes be identified as belonging to the other. The likelihood of misidentification can also be examined by seeing how well the specimens of the two samples used for constructing the linear discriminant function are identified by the function. When I did this, I found that three males were wrongly referred to the sample of females (i.e. 6% wrongly identified), but no female carapaces were wrongly allocated.

A slight improvement was achieved by the use of the corresponding Quadratic Discriminant Function in that only two males were wrongly grouped with females. No females were wrongly allocated by that function.

3.6.7 *Several Samples*

There is no particular difficulty attaching to extending the methodology of discrimination of the foregoing section to more than two samples, and this was done by Fisher, shortly after having developed the two-sample technique. It is interesting to note that distribution-theory was not considered by Fisher (1936), nor was he concerned with the question of homogeneity of variances and covariances.

At the same time as Fisher was developing a linear function for taxonomic purposes, P. C. Mahalanobis in India was wrestling with the same general problem, though by attempting to compute statistical distances between categories. His particular problem in conjunction with the anthropometrical survey of India was to make use of a suite of cranial measures for typifying various castes, ethnic entities and religious groups in a manner favourable to comparing and contrasting eventual morphological differences. His solution to the problem was the generalized statistical distance, equation (3.8); the distances were portrayed graphically in wire-and-ball figures in the earlier publications. A classical reference for the various calculatory techniques of the pre-computer era is the book by Rao (1952).

The generalization to more than two populations can be made by pairwise comparisons of linear discriminant functions about a common covariance matrix: the standard reference is still Anderson (1958). A more expeditious approach is offered by the "flagship" of multivariate statistical analysis, namely, *Canonical Variate Analysis.*

Canonical variate analysis aims at probing the interrelationships between three or more populations simultaneously (for two populations, it is identical with the usual linear discriminant function) with the end in view of representing the connexions graphically in just a few dimensions, ideally two or three. The axes of variation are chosen to maximize the separation between the populations, relative to the variation within each of the populations. Successive axes are chosen such that the resulting plots of individual specimens in the new coordinate systems should be uncorrelated within each population. A good overview of the geometrical interpretation of canonical variate analysis is given by Campbell and Atchley (1981, p. 271).

The geometrical representation of canonical variates is essentially the same as for the linear discriminant function. Consider three or more statistical populations. The individual observations of the samples can be projected on to lines representing all possible directions in the relevant plane. The first canonical vector is given by that direction for which the separation between the means of the projected points for each population is greatest, relative to the scatter of the projected points within each population. The geometrical significance of this statement is shown in Fig. 3.4.

In algebraic terms, the first canonical variate is that linear combination which maximizes the ratio of the between-groups sums of squares to the within-groups sums of squares for a one-way analysis of variance of the canonical variate scores. The ratio of the between-groups to within-groups sums of squares is termed the canonical root for that combination. Subsequent

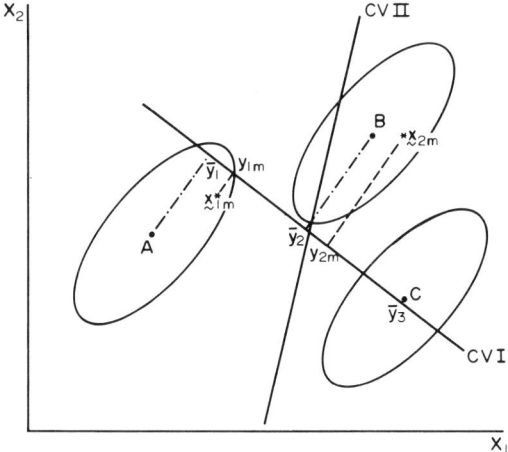

Fig. 3.4. Representation of canonical vectors for three groups and two variables (Campbell and Atchley, 1981). The group-means A, B, and C are shown. CVI and CVII are the two canonical vectors. The points y_{1m} and y_{2m} denote the canonical variate scores corresponding to the first canonical vector for two observational vectors.

canonical variates satisfy the same criterion, subject to being both uncorrelated within groups and between groups with the previous ones.

Following the presentation in Reyment *et al.* (1984, Chapter 7), we have for k groups and p variables the canonical variate scores:

$$y_{ij} = \mathbf{c}^T x_{ij} \qquad (3.10)$$

where x_{ij} denotes the i-th of N_i observations for the i-th group. The first canonical vector is derived so as to maximize the ratio

$$f = \mathbf{c}^T \mathbf{Bc}/\mathbf{c}^T \mathbf{Wc} \qquad (3.11)$$

where $\mathbf{B}$ is the between-groups matrix of sums of squares and cross-products and $\mathbf{W}$ is the within-groups matrix of sums of squares and cross-products. The canonical vectors $\mathbf{c}$ and canonical roots f satisfy

$$(\mathbf{B} - f\mathbf{W}) = \mathbf{0}. \qquad (3.12)$$

We have already met this kind of relationship earlier on in this chapter.

The canonical vectors are usually scaled so that

$$\mathbf{c}^T \mathbf{Wc} = N_w$$

where N_w is the within-groups degrees of freedom. There are $min(k - 1, p)$ non-zero canonical roots.

The steps involved in computing canonical variates can be looked at from another angle, also geometrical. We can think of the calculations as expressing a two-stage principal component analysis, that is, a two-stage rotational procedure. The first stage concerns rotation of the original characters to their principal components with the latent roots and vectors being those of the pooled within-groups covariance or correlation matrix. The correlation matrix has attractive interpretational properties as it has the effect of standardizing variances and hence transforming the ellipsoidal distributions of the original data to spheroidal distributions. If the correlation matrix is used instead of the covariance matrix, the group-means must be standardized by the corresponding within-groups standard deviations.

The principal components are then scaled by their standard deviations, here the square roots of the latent roots (cf. Section 3.6.3), to have unit standard deviations. This rotation and scaling to give orthonormal variables thus characterizes the first stage. The second stage uses these rotated and scaled axes as its reference axes. The variation between the population means is considered in this coordinate system. The second phase consists of a principal component analysis of the group-means for the new orthonormal variables. The resulting latent roots give the usual canonical roots, and the corresponding latent vectors are the canonical vectors for orthonormal variables. The canonical vectors for the original characters are found by first reversing the scaling and then the rotation.

Once the calculations have been done, the next step in the analysis is to make graphical displays of the results and to determine the variables that contribute most to the differentiation between groups. Plots of the group-means for the first canonical variates display the main differences and similarities in the material; these displays can often be enhanced by the superposition of a minimum spanning tree. Useful information can be obtained from plots of the canonical variate scores, which yield an impression of the degree of separation achieved by the various projections.

Notwithstanding the undoubted value of canonical variate analysis, it is not without its weaknesses; these should be understood if maximum value is to be obtained from the method. In many applications, the groups have been obtained *a priori* by some ordinating procedure, usually a Q-mode one. Obviously, differences that may have been slight at the outset are now strengthened by what is done to them in the canonical variate calculations. In cases where groups have been obtained by some quite objective procedure, serological, genetic, and the like, the above *caveat* does not normally apply. An informative counterpart to a canonical variate analysis is to compute the principal components of the pooled within-groups covariance matrix, **W** and plot the principal component scores.

The latent vectors are sometimes given a size-shape interpretation in morphometric work, analogously to what is done in principal component analysis, but this is a moot point and there are good reasons for treating such a reification with a good dose of prudence.

Look now at Fig. 3.5. This is a schematic representation of the plot of the canonical variate scores for 200 specimens of the Rock Crabs projected on to the plane of the first two canonical variates (the complete diagram was published by Reyment *et al.* (1984, p. 18, fig. 3.3). The ellipsoids of scatter, although approximately equally inflated, differ markedly in orientation. This is a question we shall be giving attention in several practical connexions. Suffice it to say that the so-called Bartlett test (Maurice Bartlett confided in me years ago that he bore no responsibility whatsoever for this test) of homogeneity of multivariate dispersion is very sensitive to departures from normality, atypical observations, etc. As far as my experience goes, it does give a useful result where the ellipsoids are identically oriented and differ only in inflation.

In the present illustration, you will no doubt have observed that the within-groups correlations are very high, which can sometimes lead to quite misleading interpretations. This may occur where a subset of highly correlated characters produces a latent vector with a very small latent root and, at the same time, the between-groups variation in the direction of the latent vector is also small. The effect of this situation can be that the components of the canonical vector for the subset of characters are similar in magnitude but different in sign. If one or more of the variables of this subset is deleted, or if the principal component corresponding to the latent vector with a small latent

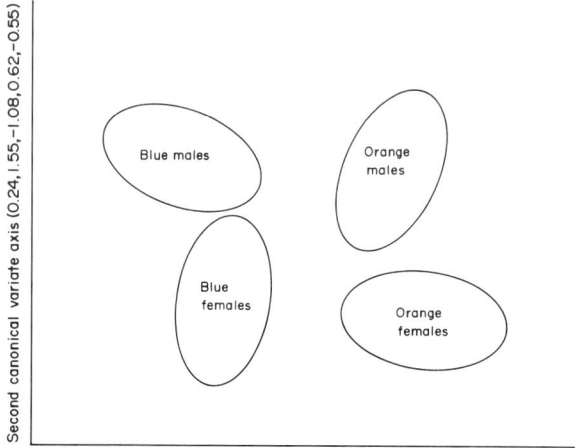

Fig. 3.5. Schematic plot of the canonical variate scores for 200 specimens of Rock Crabs, there being 50 males and females of each of the "Orange Species" and the "Blue Species". The projection on to the plane of the first two axes is shown. Each axis is labelled with the loadings corresponding to the five variables. You will observe that the ellipsoids of scatter are approximately equally inflated but the orientations of the principal axes differ. The ellipsoid for orange females has about the same orientation as blue males, whereas the ellipsoid for orange males have approximately the same orientation as that of blue females.

root is eliminated, there will usually be little change in the canonical roots, but a marked change in the canonical vectors corresponding to the *subset of characters* of interest (Campbell, 1978).

3.6.7.1 *Shrunken Estimators and Stability of Canonical Vectors*

The subject of achieving stability in canonical variate analyses has been taken up by Campbell (1978, 1979) in papers of special importance for applied multivariate statistical analysis and particularly in morphometrical connexions. His methods were applied to variation in a species of Cretaceous foraminifers (*Afrobolivina afra*) by Campbell and Reyment (1978). In this paper, implicit computing instructions are supplied. Seber (1984, p. 166) gives a helpful account of the procedure.

3.6.7.2 *Description of the Method*

One begins with the within-groups matrix of sums of squares and cross-products, **W**, and the corresponding between-groups matrix, **B**. These are the familiar points of departure for usual canonical variate analysis. The matrix of sample-means is also required. Matrix **W** is then standardized to the

TABLE 3.5. *Canonical Variate Analysis of males and females of the two species of Rock Crabs*

1. CORRELATIONS AND COVARIANCES FOR **W**
   (Correlations in the lower triangle)

	FL	RW	CL	CW	BD
FL	1879.69	1330.32	4046.05	4558.45	1842.02
RW	0.9624	1016.62	2905.01	3282.53	1324.83
CL	0.9922	0.9687	8845.91	9957.46	4028.65
CW	0.9914	0.9708	0.9983	11,246.97	4538.60
BD	0.9870	0.9632	0.9929	0.9921	1860.95

2. BETWEEN-GROUPS MATRIX OF SUMS OF SQUARES AND CROSS-PRODUCTS **B**
   (upper triangle only).

	FL	RW	CL	CW	BD
FL	551.5	292.91	802.25	730.42	513.59
RW		300.59	350.35	350.33	239.80
CL			1245.07	1153.43	756.67
CW				1108.86	674.73
BD					487.20

3. THE LATENT ROOTS AND VECTORS FOR THE DETERMINANTAL EQUATION
   These are for $\mathbf{W}^{-1}\mathbf{B}$. Standardized counterparts are given in Table 3.6.

	Latent Vectors		
Variables	I	II	III
FL	4.7298	−0.7533	−5.1102
RD	1.3497	−3.5438	1.0998
CL	1.2806	7.0920	5.7011
CW	−11.5256	−4.5892	−5.7540
BD	4.3545	1.7524	3.4966
Latent roots	7.5668	3.3102	0.1522

4. The canonical variate *means* corresponding to the three latent vectors in section (3) above are (N.B. $\min(k - 1, p) = 3$.)

	Mean corresponding to canonical vector		
Group	I	II	III
1	3.076	−1.768	−0.337
2	2.260	2.424	0.273
3	−2.016	−1.699	0.483
4	−3.321	1.043	−0.419

corresponding correlation matrix, with similar scaling for **B**. These steps can be summarized as follows:

$$\mathbf{W}^* = \mathbf{S}^{-1}\mathbf{W}\mathbf{S}^{-1}$$

and

$$\mathbf{B}^* = \mathbf{S}^{-1}\mathbf{B}\mathbf{S}^{-1} \tag{3.13}$$

where **S** is a diagonal matrix formed from the square root of diag **W**. The latent roots e_i and corresponding latent vectors $\mathbf{u}_i$ of $\mathbf{W}^*$ are then extracted. The corresponding orthogonalized variables are the principal components.

With

$$\mathbf{E} = \text{diag}(e_1, \ldots, e_p)$$

and

$$\mathbf{U} = (\mathbf{u}_1, \ldots, \mathbf{u}_p),$$

we have that

$$\mathbf{W}^* = \mathbf{U}\mathbf{E}\mathbf{U}^T. \tag{3.14}$$

The next step is to scale each latent vector by the square root of its corresponding latent root. This transformation brings about within-groups sphericity. A technique borrowed from multivariate regression is then invoked, to wit, *ridge-regression*, or estimation by *shrinkage* constants. Shrunken estimators are formed by adding shrinkage constants k_i to the latent roots e_i before scaling the latent vectors.

$$\text{If } \mathbf{K} = \text{diag}\,(k_1, \ldots, k_p),$$

we can define a matrix

$$\mathbf{U}^* = \mathbf{U}(\mathbf{E} + \mathbf{K})^{-1/2} = \mathbf{U}^*_{(k_1, \ldots, k_p)}. \tag{3.15}$$

Form now the between-groups matrix in the space of the within-groups principal components:

$$\mathbf{G}_{(k_1, \ldots, k_p)} = \mathbf{U}^{*T}_{(k_1, \ldots, k_p)} \mathbf{B}^* \mathbf{U}^*_{(k_1, \ldots, k_p)}. \tag{3.16}$$

Let g_i denote the i-th diagonal element of $\mathbf{G}$, the between-groups sum of squares for the i-th principal component. Extraction of the latent roots and vectors of $\mathbf{G}_{(0, \ldots, 0)}$ yields the usual canonical roots f and canonical vectors for the principal components, $\mathbf{a}^U$. The usual canonical vectors $\mathbf{c}^U$ are yielded by

$$\mathbf{C}^U = \mathbf{U}^*_{(0, \ldots, 0)} \mathbf{a}^U.$$

Shrunken (or generalized ridge-) estimators are determined directly from the latent vectors $\mathbf{a}^s$ of $\mathbf{G}_{(k_1, \ldots, k_p)}$. A generalized inverse solution results when $k_i = 0$ for $i \leq r$ and $k_i = \infty$ for $i > r$ where r corresponds to the first elements of $\mathbf{a}^U$. Regarding the choice of values for k_i, if ∞ is selected, this leads to the complete elimination of principal component i. If k_i is put equal to 0.5, for example, this signifies that the between-group contribution from component i is diminished. Further notes on the method are provided in Section 8.6.

The results of the analysis of the four samples of Rock Crabs are displayed in Table 3.6. We note that the fifth principal component of $\mathbf{W}^*$ is very small. The associated latent vector has two large elements, the third and the fourth, and these bear opposite signs. Moreover, there are several small values in diag $\mathbf{G}$, including the value of g_5. When this component is eliminated from the analysis, marked changes occur in elements three and four of the first

TABLE 3.6. *Shrunken estimators for canonical analysis of Rock Crabs*

Between-groups matrix of sums of squares and cross-products (upper diagonal) and correlation matrix for within-groups (lower diagonal)

	FL	RW	CL	CW	BD
FL	551.6	292.9	802.2	730.4	513.6
RW	0.9624	300.6	350.3	350.3	239.8
CL	0.9922	0.9687	1245.0	1153.0	756.7
CW	0.9914	0.9708	0.9983	1109.0	674.7
BD	0.9870	0.9632	0.9929	0.9921	487.2

Within-groups standard deviations

3.066	2.255	6.651	7.499	3.050

Latent roots and vectors for within-groups matrix (correlations)

Variables	I	II	III	IV	V
FL	0.45	−0.27	0.74	0.41	−0.04
RW	0.44	0.89	−0.02	0.08	0.02
CL	0.45	−0.20	0.05	−0.46	0.74
CW	0.45	0.15	0.05	−0.57	−0.67
BD	0.45	−0.26	0.67	0.54	−0.05
Roots	4.928	0.0496	0.0130	0.0078	0.0016
diag **G**	0.1810	2.831	0.3531	6.438	1.223

Usual canonical roots and standardized vectors

			Variables			
Vector	FL	RW	CL	CW	BD	Root
I	4.728	1.350	1.275	−11.52	4.355	7.564
II	0.754	3.544	−7.092	4.590	−1.753	3.310

Estimate of canonical variates with g_5 deleted.

			Variables			
Vector	FL	RW	CL	CW	BD	Root
I	5.332	1.568	−5.183	−6.258	4.743	6.795
II	0.740	3.542	−7.097	4.644	−1.770	3.309

canonical vector. Clearly, any attempt at reifying the first canonical vector is bound to fail if the elements are not stable under conditions of repeated sampling.

What happens if we delete the directions corresponding to the smallest diagonal entry of **G**, the first one? The value is only 0.181 and its elimination has little effect on the canonical roots. Nor are the canonical vectors altered much by this operation.

Remarks

The method just presented requires a fair amount of experience for successful results. A few guidelines can be of help. Look closely at the elements of the within-group principal components, particularly if the last one or two latent roots do not greatly differ from nought. If the corresponding vector, or vectors, have one or more very large elements (>0.6), differing in sign, then there is a good chance that the variables concerned will be affected by instability.

3.6.7.3 *Robust Estimation*

The performance of classical multivariate statistical procedures may be adversely influenced by atypical values. A robust method is defined as one that produces an estimate that is little affected by observations that deviate from the main body of the data: obviously, robust estimation is an appealing complementary utility in biometrical work.

What alternatives are there is the data contain observations that are quite valid from the biological point of view but which are statistically awkward. For example, in many groups of crustaceans, size polymorphism occurs in which the variability is so broad that it becomes difficult to account for heterogeneities in the data. It would therefore be open to discussion if some of the specimens were deleted from the analysis merely on the grounds of statistical incompatibility. In favourable situations, it is possible to pick out observations which are markedly atypical in a single character by applying appropriate univariate screening techniques to one variable at a time (Barnett and Lewis, 1978). However, for multidimensional data, observations are often only found to be atypical when the value is scrutinized in relation to other variables and specimens.

Robust multivariate statistical methods can be regarded as simple modifications of the usual techniques. The principle followed in most cases is that the contribution of an observation to the statistic of interest (or statistics) is given full weight if it is a "reasonable observation", otherwise its contribution is downweighted: i.e. adjusted so as to have less influence. A robust estimator can be defined by introducing a Weight Function which depends on the discrepancy between the observation and some robust average value, relative to a robust measure of scatter. The weight function is inversely connected to the *influence* of an observation, which can be done formally by means of the Influence Function of Hampel (1974) and bounding the influence of observations with unduly large discrepancies. The concept of robust estimation is summarized in Seber (1984, pp. 156 and 443). The underlying aims are summarized in Hampel's own words as follows:

1. To build in safeguards against unexpectedly large amounts of gross errors.
2. To put a bound on the **influence** of hidden contamination and questionable outliers.
3. To isolate clear outliers for the possibility of separate treatment.
4. To retain near-optimality at the strict parametric model.

Two useful ideas deriving from the theoretical study of robustness are (a) the *breakdown-point*, which can be defined as the fraction of outliers that can be tolerated without the estimator running the risk of breaking down and (b), the **influence function**, which quantifies the influence of contamination on an estimate. Robust estimation procedures have opened up new areas of development in multivariate statistics and it is to be expected that the subject will be greatly expanded in the near future. An example of robust estimation is given in Chapter 8.

3.7 Canonical Correlation

The method of canonical correlations can be a valuable technique in palaeoecological studies but it remains one of the least used procedures in the repertory of multivariate statistics. The reason for this is, as I have already noted, that the interpretation of results if far from straightforward. The intuitive appeal of the method should be self-evident—it is conceptually an extension of the correlation between two variables to correlation between two sets of variables.

Consider a vector of measurements $\mathbf{v}$, composed of two subvectors, $\mathbf{s}$ and $\mathbf{t}$. Let vector $\mathbf{v}$ be p-dimensional, subvector $\mathbf{s}$ m-dimensional and subvector $\mathbf{t}$, n-dimensional; moreover, note that $p = m + n$.

The variables encompassed by the first subvector might be morphological measurements on some organism and the elements of the second subvector could be determinations made on chemical components of the shell of each specimen. In many respects, there is a clear connexion with multiple regression, notwithstanding that a predictor-criterion relationship is not normally involved. In multivariate regression, a single variable is contrasted with a set of variables. In canonical correlation, a rôle of the single variable has been replaced by a set of related variables. Thus, the mathematics underlying the derivation of canonical correlations resembles that of multiple regression, which can therefore be regarded as a special case of canonical correlation.

The calculation of canonical correlations (Hotelling, 1936) has as its starting point the covariance or correlation matrix of the entire set of p variables. The first step is to partition this matrix into four submatrices, corresponding to the subvectors. This means that there will be a matrix $\mathbf{R}_{11}$ $(m \times m)$, a matrix $\mathbf{R}_{22}$ $(n \times n)$ and a matrix $\mathbf{R}_{12}$ $(m \times n)$, and its transpose, for the correlations between sets. The coefficients of canonical correlation are derived from the square root of the latent values of the determinantal equation:

$$|\mathbf{R}_{21}\mathbf{R}_{11}^{-1}\mathbf{R}_{12}\mathbf{R}_{22}^{-1} - d_i| = 0. \tag{3.17}$$

The second set of canonical correlation vectors $\mathbf{b}_i$ are obtained from the relationship:

$$(\mathbf{R}_{21}\mathbf{R}_{11}^{-1}\mathbf{R}_{12}\mathbf{R}_{22}^{-1} - d_i\mathbf{I})\mathbf{b}_i = \mathbf{0}. \tag{3.18}$$

The coefficients $\mathbf{a}_i$ are obtained from the equation

$$\mathbf{a}_i = \mathbf{R}_{11}^{-1}\mathbf{R}_{12}\mathbf{b}_i/d_i. \tag{3.19}$$

Here $\mathbf{a}$ and $\mathbf{b}$ are appropriately standardized.

For fuller acounts of the various aspects of canonical correlation, I can refer you to the book by Cooley and Lohnes (1971) and the text by Pimentel (1979), who gives an interesting ecologically oriented example. Another important reference is the monograph by Love and Stewart (1968). I have applied the method to eco-geochemical data (Reyment, 1972) and ostracods (Reyment, 1975), for example. One of the difficulties attaching to the reification of the results is the troublesome fact that the canonical correlations do not portion off successively important parts of the total association between sets in a manner directly equatable with what occurs in, say, principal component analysis in relation to the sum of the diagonal elements of the covariance (equivalent) matrix. Hence, it is quite possible to find the most informative relationship between sets expressed in some combination of variables connected to a canonical correlation of lesser dignity than the largest value.

Another difficulty is that the results often do not make much sense. It is for this reason that Cooley and Lohnes (1971), and their coworkers, introduced a procedure they called *redundancy* analysis. This is rather complicated to apply and not always crystal-clear in what it is trying to say with respect to the reification of the elements of the subvectors. A good case can often be made for using redundancy analysis and I have found it serviceable on many occasions. Bookstein and Sampson (1990) provide a demonstration of how Partial Least Squares (the singular value decomposition of $\mathbf{R}_{12}$) can provide a useful interpretation of relationships between sets by maximizing covariance instead of correlation. However, much of the dissatisfaction with the method seems to be coming from the effects of instability in the vectorial elements in somewhat the same fashion as we have encountered for principal components and canonical variates. Another moot point concerns the rigidly linear assumptions, a point which has been discussed in the publications of ter Braak (see Chapter 7). I am therefore going to be so daring as to suggest that before you decide to try your hand at a canonical correlation analysis, you should begin by examining the following aspects of your data:

1. Are your data multivariate normally distributed? This situation should apply at least for each of the subsets of observations.
2. Are your correlations approximately equal and are they all very high? If this condition applies, you could consider trying a robust canonical variate analysis analogous to a robustified principal component analysis.
3. Recalling what has been stated earlier on in this chapter, your trouble could be coming from atypical observations or influential values. You might just be able to achieve stability and hence enhanced interpretational possibilities by deleting a strongly divergent observation. Truncating has not been advocated up to now, but *necessitas non habet legem.*

I have already mentioned that a common way around the problem addressed by the method of canonical correlation is to treat the entire set of observations in the one connexion and to analyse them in a principal component analysis (or surrogate). This does give some kind of a result which perhaps sanctifies the action. It can however not be denied that this way of going about treating the problem is sub-optimal. Seber (1984, p. 256) discusses the theory of canonical correlation and some of its ramifications. Another good theoretical reference is Anderson (1984).

Example 3.5. Canonical Correlations for a Species of Cretaceous Ostracods

Some data on Cretaceous ostracods will be used to provide a simple example of the steps pertaining to the calculation of canonical correlations.

We have not met these data before, so let me make a brief presentation of the material. Abe *et al.* (1988) described secular variation in two species of ostracods from the Santonian of Israel. For the purposes of the present illustration, we consider seven characters measured on the carapace of *Veenia fawwarensis* Honigstein: four of these, here denoted as variables $x_1, \ldots, x_4$, describe variability in the anterior part of the carapace and the location of the adductorial tubercle. Three measures, betokened as $x_5, \ldots, x_7$, report on the variability of the posterior. Our problem is to estimate the strength of integration between the two sets of variables. The sample comprises 260 specimens.

We begin by writing down the 7×7 correlation matrix for all of the characters. This is:

$$
\begin{bmatrix}
1.00 & 0.74 & 0.53 & 0.91 & | & 0.86 & 0.86 & 0.68 \\
0.74 & 1.00 & 0.62 & 0.68 & | & 0.86 & 0.50 & 0.77 \\
 & & \mathbf{R}_{11} & & | & & \mathbf{R}_{12} & \\
0.53 & 0.62 & 1.00 & 0.37 & | & 0.55 & 0.55 & 0.46 \\
0.91 & 0.68 & 0.37 & 1.00 & | & 0.80 & 0.80 & 0.64 \\
\text{—} & \text{—} & \text{—} & \text{—} & | & \text{—} & \text{—} & \text{—} \\
0.86 & 0.86 & 0.55 & 0.80 & | & 1.00 & 0.63 & 0.75 \\
 & & \mathbf{R}_{21} & & | & & \mathbf{R}_{22} & \\
0.86 & 0.50 & 0.55 & 0.80 & | & 0.63 & 1.00 & 0.44 \\
0.69 & 0.77 & 0.46 & 0.64 & | & 0.75 & 0.44 & 1.00
\end{bmatrix}
$$

Inspection of this partitioned matrix discloses that there are some very high correlations but there is a considerable range and some values are fairly low. This is certainly different from what we have learnt to expect from the rock crabs, likewise crustaceans, in which the characters of the carapace are very highly correlated. The quite high level of integration is further indicated by the multiple correlation coefficients which for the variables of the smaller set

TABLE 3.7. *Results of the canonical correlation analysis of* Veenia fawwarensis *from the Cretaceous of Israel*

Canonical correlation	Left-hand coefficients				Right-hand coefficients		
	1	2	3	4	5	6	7
0.972	0.65	0.09	0.16	0.21	0.48	0.51	0.15
0.723	0.56	−1.64	0.57	0.45	0.81	1.14	0.36

REDUNDANCY ANALYSIS

Correlation	Left-hand structure				Right-hand structure		
	1	2	3	4	5	6	7
1	0.99	0.81	0.63	0.91	0.91	0.88	0.74
2	0.06	−0.57	0.03	0.05	0.36	0.47	−0.47

(characters five to seven) are 0.93, 0.92 and 0.79, all of which are very highly significant.

The multiple correlations for the larger set (variables one to four) are equally as high, being 0.78, 0.94, 0.78 and 0.94. All of these are very highly significant.

The extraction of the latent roots and vectors proceeds by a two-phase sequence (in fact, the same computer programs as were used for finding the canonical roots and vectors were employed for the present computations). There are three latent values of which two are statistically highly significant. Thus, we can expect to have to unravel two sets of linear combinations. You will probably have divined by now that the number of canonical correlations is equal to the dimensionality of the "smaller" set. The latent roots and associated subvectors are listed in Table 3.7.

The first juxtaposition of subvectors shows that character 1 dominates in the anterior set and variables 5 and 6 provide most of the action in the posterior set. With respect to the second canonical correlation, all variables of the first set are engaged, with a strong influence from variable 2. The second set of variables is again dominated by characters 5 and 6.

These results are not unreasonable in that given that the level of integration is quite high over the entire carapace, it is to be expected that posterior and anterior co-variation is highly correlated. This is indeed indicated by the plot of the scores for the two sets of scores (Fig. 3.6). What is difficult to accept out of hand is that the integration lies solely with one variable of the anterior set.

Without wishing to enter into the details of the calculations, for which I refer you to Cooley and Lohnes (1971), the "redundancy analysis" gives a more satisfying result. These coefficients are also listed in Table 3.7. We see that all variables of both sets are shown as participating in the covariation, for both groups. As regards the second canonical correlation, only variable 2 participates in the covariation with the set of posterior variables. The plot of the corresponding scores is shown in Fig. 3.7.

Ter Braak (1986 and references therein) includes an "analysis of redundancy" in his canonical correspondence analysis. This selects the linear

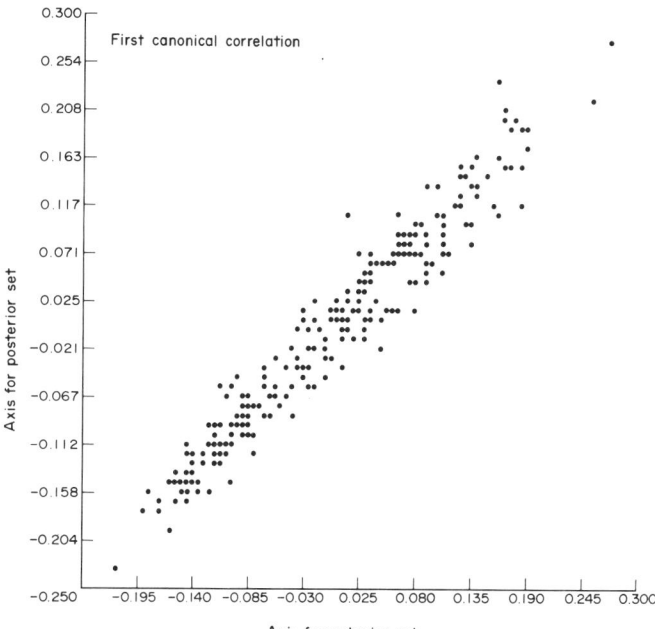

FIG. 3.6. Plot of the scores of transformed observational vectors for the first canonical correlation of the ostracod species *Veenia fawwarensis*. The "anterior set" are the morphological dimensions measured on the anterior part of the carapace, and the "posterior set", denotes the variables determined in the posterior region of the carapace. The plot supports the suggestion that the anterior and posterior development of the shell is well integrated.

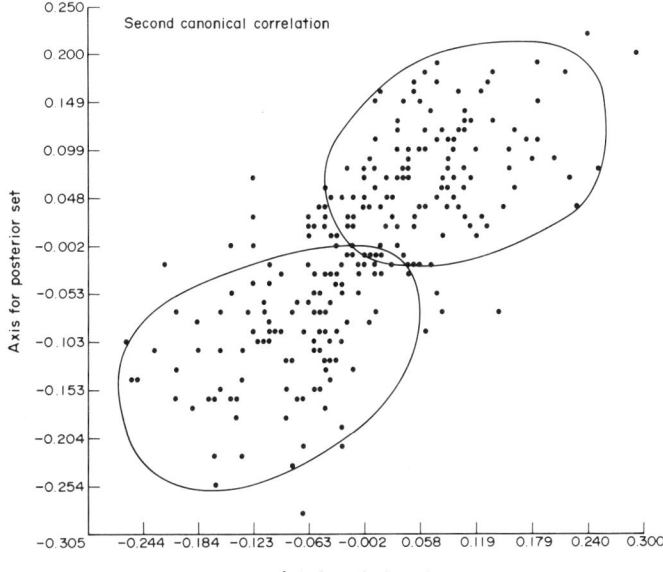

FIG. 3.7. Plot of the scores of transformed observational vectors for the second canonical correlation for the ostracod species *Veenia fawwarensis* from the Santonian of Israel.

combination of external (for example, the "right-hand" set of variables) that gives the smallest total residual sum of squares. This contrasts with usual canonical correlation analysis in which the aim is to maximize the correlation between the left- and right-hand sets of variables. A serviceable account of the method is given in the textbook by Jongman *et al.* (1987).

3.8 Some Special Multivariate Topics

3.8.1 *Multivariate Normal Distribution*

I include here a brief section on some of the properties of the multivariate normal distribution because of the use made of these concepts in Chapter 8 (Example 8.1). Seber (1984, p. 54) summarizes the main points of interest. He takes up the related question of graphical techniques for multidimensional plots on p. 127 in his book and on p. 138 reviews the topic of transforming to multivariate normality. This discourse takes up pitfalls associated with such transformations and the fact that parameters associated with transformed data may not be so meaningful as those associated with the original data.

The density function of a p-element normal distribution is given by the following expression:

$$(\mathbf{X}) = Ke^{-0.5(\mathbf{X} - \mu)\Sigma^{-1}(\mathbf{X} - \mu)} \tag{3.20}$$

where K is chosen so that the integral over the entire p-dimensional Cartesian space is unity. $\mathbf{X}$ denotes a random vector of observations and μ is its mean in the population; Σ is the population covariance matrix. The symbolism in wide use for saying that X is normal with centroid μ and dispersion Σ is $N(\mu, \Sigma)$. For $N(\mu, \Sigma)$,

$$K = (2\pi)^{-0.5p} |\Sigma|^{-0.5}.$$

Empirical distributions are discussed in terms of ellipsoidal scatter. The reason for this is that the surface of the multivariate normal distribution in p-dimensional Cartesian space on which the density function is a constant is an ellipsoid which is specified by the quadratic form:

$$d^2 = (\mathbf{X} - \mu)^T \Sigma^{-1} (\mathbf{X} - \mu). \tag{3.21}$$

There are three important things to note here:

1. The centre of each ellipsoid is at μ;
2. The orientation of each ellipsoid is determined by the covariance matrix;
3. The size of each ellipsoid is determined by the value of d^2.

We have, consequently, an infinite number of similar and similarly oriented ellipsoids around a common centroid. For all points on the surface of a given ellipsoid, d^2 and $P(\mathbf{X})$ are constant. The value of d^2 is also a measure of *distance*—it indicates how far the surface of the ellipsoid is from the centre of the swarm of points relative to distances from the centroid to surfaces of other ellipsoids in the nest.

There are three classes of subsidiary distributions of interest to us:

1. Marginal distributions
2. Conditional distributions
3. Component distributions.

A *marginal distribution* is the univariate distribution for any single element of a vector of variables. If the vector of variables is multivariate Gaussian (i.e. multivariate normally distributed) in distribution, then each of its marginal distributions is normal. The converse is not necessarily true: each of the marginal distributions may be Gaussian without the variable-vector being multivariate normal. This condition is one that must be kept in mind in testing for normality in multivariate work.

One rather obvious reason for showing curiosity for the properties of marginal distributions is that part of a normal multivariate morphometric analysis will include tests on the elements of the vectors of observational variables.

A *conditional distribution* is the predicted distribution for a particular marginal element, z_j, is given the known distribution of the remainder of the vector variable, $\mathbf{z}$. There is an important theorem of statistics that states that if a vector of variables is multivariate Gaussian, then every conditional distribution defined on it is Gaussian.

A *component distribution* is the distribution of any arbitrary linear function of a vector variable. If a vector-variable is multivariate Gaussian, then every component defined on it is multivariate Gaussian. The paradigm here is the concept of principal components (see p. 44, this chapter).

There is a statistical theorem which states that all linear components defined on a vector of variables that is multivariate Gaussian are normal; the condition is reversible because it is also true that any vector of variables for which every possible linear component is normal, is multivariate Gaussian. An example of how to assess multivariate normality in large samples is given on p. 283 in Chapter 8. An important reference is Mardia (1970).

3.8.2 *Estimation of Mixing Proportions*

A topic requiring mention concerns the problem of mixed data. There is a sample from a mixture of several populations and it is desired to estimate the unknown mixing proportion on the basis of this sample. It is not known to which population any observation in the sample belongs. Reference data of known origin are available. Do and McLachlan (1984) considered this problem for seven species of Malaysian rats on which four distance measures had been taken. These data were the reference material. The sample of observations consisted of rat skulls collected from regurgitated owl pellets. The aim of the study was to determine the diet of the owls in terms of the estimated proportions of each species of rat consumed.

The analysts found that they were required to take careful account of atypical observations in order to obtain useful assessments of multivariate normality for the population distributions. The use of the linear discriminant function for assessing proportions is greatly simplified if multivariate normality of the observations can be assumed. The results obtained by maximum likelihood estimation and discriminant analysis were about the same. The diet of the owls consists mainly of one species with a second species making up the difference. Explicit details for doing the calculations are given by Do and McLachlan. The general approach is similar to that of Campbell (1980a, b) in that M-estimators and the use of Q–Q probability plots play a very important part in the analysis.

3.8.3 *The Mantel Non-Parametric Test for Correlations*

The following discussion is based on the presentation in Manly (1985). The Mantel test can be used for testing for similarity in patterns of matrix correlations, that is correlation coefficients computed using the corresponding elements of the two matrices as paired observations (Lofsvold, 1988, p. 56). Mantel's test is a permutation procedure in which the sampling distribution of an index of a matrix similarity under the null hypothesis that any observed similarity is random is obtained empirically by randomizing the elements of one of the two matrices and recomputing the index. The null hypothesis is tested by means of a special statistic.

The same procedure can also be applied to the analysis of polymorphism. Suppose that a population is sampled from, say, four demes and the proportions of the two morphs occurring determined. These proportions can be used to construct a 4×4 distance matrix where the entry in the i-th row and j-th column is the "morph distance" between deme i and deme j. Hence, two demes in which the proportions are similar will be separated by a small distance; if, however, the proportions differ greatly, then the distance between the demes will be great.

This can be shown simply as follows:

The matrix of morph distances may have the following appearance:

$$\mathbf{M} = \begin{bmatrix} m_{11} & m_{12} & \cdots\cdots\cdots & m_{12} \\ \cdots\cdots\cdots\cdots\cdots\cdots\cdots \\ m_{41} & m_{42} & \cdots\cdots\cdots & m_{44} \end{bmatrix}$$

This is a square symmetrical matrix, the diagonal elements of which are noughts.

The question we ask frequently in actuopalaeontological studies on, say, ostracods, in which we suspect ecophenotypy to be the driving force for observed morphological variability, is whether the observed variability can be linked to environmental effects. We can also construct a distance matrix for some ecological factors—say chemical and physical components of the environment.

Such an ecological matrix could be expressed as

$$\mathbf{E} = \begin{bmatrix} e_{11} & e_{12} & \cdots\cdots & e_{14} \\ \cdots\cdots\cdots\cdots\cdots\cdots \\ e_{41} & e_{42} & \cdots\cdots & e_{44} \end{bmatrix}$$

This matrix is symmetrical with noughts down the diagonal.

Mantel's test consists of finding if the elements in $\mathbf{M}$ and $\mathbf{E}$ are correlated. This is done by calculating

$$Z = \Sigma\Sigma m_{ij}e_{ij} \tag{3.22}$$

which is compared with the distribution of Z obtained by taking the demes in random order for one of the matrices.

Lofsvold (1988) used Mantel's test to compare matrix correlations for the additive genetic covariance matrix and the matrix of locality effects for four species of *Peromyscus*.

Let us consider for a moment what the Mantel test tries to do. The basic idea is that the elements of the two matrices we are comparing, $\mathbf{M}$ and $\mathbf{E}$, are quite unlike each other. If this is true, matrix $\mathbf{E}$, say, will resemble one of the randomly ordered matrices, $\mathbf{E}_R$, and the observed value of Z will be typically randomized. If, none the less, the elements of the two matrices are positively correlated, then the observed value of Z will be greater than the values given by randomization. A negative correlation would lead to a low observed Z-value compared with the randomized distribution.

The test uses the mean $E(Z)$ and the variance, Var(Z), of the randomized distribution of Z to compute

$$g = [Z - E(Z)]/\{\mathrm{var}(Z)\}^{1/2} \tag{3.23}$$

which can be treated as a standard normal variate.

Manly (1985, p. 424) gives a FORTRAN program for doing the required calculations. The Pearsonian coefficient of correlation between elements is part of the output supplied by this program. The matrix correlation can be used as an approximate measure of the similarity in the arrangement of the elements of the matrix.

A simple test of significance of the correlation coefficient is not permissible owing to the fact that the values in $\mathbf{M}$ and $\mathbf{E}$ are not independent. The procedure can be quite useful in studies of polymorphism in such situations as represented by colour-banding in the land-snail *Cepaea nemoralis*.

3.9 Factor Analysis or Principal Components?

3.9.1 *Introduction*

No doubt some readers will have hunted, with puzzled mien, for applications of factor analysis, whereas others have triumphantly observed that I must have come to my senses. Robert Blackith and I (Blackith and Reyment, 1971) observed some 20 years ago that the subject of Factor Analysis tended to provoke heated exchanges. The situation has hardly changed since then. Jöreskog *et al.* (1976) produced a detailed appraisal of the entire field of factor analysis *sensu lato* in which the reasons for most contentious rumblings were shown to lie with a spate of misunderstandings and misconceptions, in part of a semantic nature. It is my opinion that it is naïve to dismiss this branch of multivariate analysis out of hand on the grounds of disliking or not understanding some minor part of the theoretical corpus. The review given here is more detailed than those for most other techniques considered in this book. The reason for this is that factor analysis has been largely shunned, until quite recently, by professional statisticians and the accounts provided in standard texts on multivariate analysis are seldom informative. Of recent years, many gifted statisticians have begun to work in psychometry. Moreover, it has been shown by Bookstein *et al.* (1985, pp. 78–102) that factor analysis can be correctly applied to morphometrical data via the theory of **path analysis** of Sewall Wright (1968), the celebrated geneticist and mathematician. A useful reference for conveying the scope of modern theory and imaginative research is the book compiled by Fornell (1982).

The term Factor Analysis is almost always loosely and often wrongly applied in the Natural Sciences and in many cases, the correct designation should be principal component analysis, or some variant thereof. Factor analysis can be easily confused with principal components; it has, however, as its main goal, the extraction of a lower-dimensional linear structure from the data in order to explain the correlations between variables. Principal component analysis cannot be regarded as modelling—it is data-transformation to new measures. Over the years, the subject of Factor Analysis has come to encompass several mathematically similar techniques. It is convenient to work within the confines of clarifying definitions, since this will greatly contribute towards teaching you what the factor model really is.

I start by defining the scope of *R*-mode factor analysis *sensu lato*. There are two techniques involved, to wit, *component* analysis and what must, for a better name, be referred to as **true factor analysis.** In all work I know of in the natural sciences, the first, descriptive variant "component analysis" is what is employed and the fundamental, inferential model of psychometry is just not available. In order to standardize the basis for discussion, our familiar principal component analysis is presented below within a factor-analytical framework.

3.9.2 *Fixed and Random Cases*

The **Fixed Case** refers to the application of factor analysis to analysing one particular collection of data (a sample), without regarding it as a sample drawn from a statistical population. There is then no interest in estimating population quantities and the results obtained can only be interpreted with respect to the specimens in the sample. This kind of application is not uncommon in analytical chemistry. It also dominated the early history of psychometry and is the dominant variant in geology and biology.

The **Random Case** is, as the term applies, the situation that pertains when our sample is conceived of as a random sample drawn from some specified population. Interest is now directed towards estimating population quantities on the basis of what can be deduced from the properties of the sample. The random case requires assumptions about the multivariate distribution of the variables as well as a consideration of sampling theory. In most cases, however, it has been found that the procedure is moderately robust to deviations from theory.

3.9.3 *Discussion of Models*

3.9.3.1 *Fixed Case*

The appropriate mathematical model for operating on an isolated sample is most succinctly developed in terms of the data matrix. The appropriate mathematical model is:

$$\mathbf{X}_{(N \times p)} = \mathbf{F}_{(N \times k)} \mathbf{A}^T_{(k \times p)} + \mathbf{E}_{(N \times p)} \tag{3.24}$$

where
$\mathbf{X}$ is the data matrix
$\mathbf{F}$ is the matrix of factor scores
$\mathbf{A}$ is the matrix of factor loadings
$\mathbf{E}$ is the matrix of residuals (or error terms)
and k is a scalar denoting the number of factors to be used. It is less than or equal to the number of variables. This scalar may be postulated at the outset of the analysis, or determined as a part of the computations.

For any row $\mathbf{x}^T$ of $\mathbf{X}$, we can write, in transformed form,

$$\mathbf{x} = \mathbf{A}\mathbf{f} + \mathbf{e}. \tag{3.25}$$

Equation (3.25) is the *fundamental model* for all forms of R-mode factor-analysis. Here, $\mathbf{x}$ is a vector representing one of the objects of the data matrix. Equation (3.25) states then that **each observed variable** (i.e. an element of $\mathbf{x}$) can be expressed as a weighted sum of factors, plus an error term, or residual. What does this imply? It intimates that the matrix vector product $\mathbf{A}\mathbf{f}$ yields a vector of estimates of $\mathbf{x}$ and that the vector $\mathbf{e}$ is the difference between this estimate and the observed vector. It is usually assumed that the residuals are uncorrelated with the factors.

The above point is quite important for understanding the way in which factor-analytical interpretation is carried out. Let us say that we tried a two-factorial solution for a set of data, but found that the approximations to the observed values were poor. The logical next step is to try a model with three factors, and so on. This procedure can be automatized in a computer program.

3.9.3.2 *Random Case*

The usual derivation of the model for the random case is by way of the variances and covariances of the variables. In terms of the data matrix in deviate form, which we denote $\mathbf{Y}$, the factor model can be written

$$\mathbf{Y} = \mathbf{F}\mathbf{A}^T + \mathbf{E}. \tag{3.26}$$

If we now employ the variances and covariances, we obtain a didactically useful representation of the various terms.

$$\frac{1}{N}\mathbf{Y}\mathbf{Y}^T = \mathbf{A}\left[\frac{1}{N}\mathbf{F}^T\mathbf{F}\right]\mathbf{A}^T + \mathbf{A}\left[\frac{1}{N}\mathbf{F}^T\mathbf{E}\right] + \left[\frac{1}{N}\mathbf{E}^T\mathbf{F}\right]\mathbf{A}^T + \frac{1}{N}\mathbf{E}^T\mathbf{E}. \tag{3.27}$$

As the sample-size N approaches infinity so does each term in equation (3.27) converge towards its population value. Hence, at the level of the population

$$\mathbf{\Sigma} = \mathbf{A}\mathbf{\Phi}\mathbf{A}^T + \mathbf{\Psi}. \tag{3.28}$$

Here, $\mathbf{\Sigma}$ is the $p \times p$ population covariance matrix, $\mathbf{A}$ is the $p \times k$ matrix of factor loadings, and $\mathbf{\Psi}$ is the $p \times p$ residual covariance matrix (which by definition is diagonal: i.e. the residual terms are uncorrelated).

Differential Diagnoses: Let us look at some of the major conceptual differences between principal components and factor analysis.

1. In principal components, the "factors" are determined so as to account for **maximum variance** of all the observed variables. The common use of principal component analysis is to reduce the dimensionality of a data set by replacing the p measured variables by a much smaller number of principal components.
2. In "true" factor analysis, the factors are defined to account maximally for the **intercorrelations** of the variables.
3. The residual terms, e_i, are assumed to be small, or just ignored, in standard principal component analysis; but this is not so in "true" factor analysis.
4. Both methods assume the residuals to be uncorrelated, but in the factor-analytical model, the residuals are taken to be uncorrelated among themselves.

3.9.4 *Principal Component Factor Analysis*

I now introduce principal component analysis in factor analytical attire. In the fixed case, the principal component model can be fitted by the method of least squares to the data matrix $\mathbf{Y}$ (which you will recall is now in deviate form). This means that matrices $\mathbf{F}$ and $\mathbf{A}$ in equation (3.26) must be determined such that the sum of squares

$$\mathbf{E} = \mathbf{Y} - \mathbf{F}\mathbf{A}^T$$

is as small as possible. The solution to this problem is yielded by the singular value decomposition of the matrix product (see Section 2.6).

$$\hat{\mathbf{F}}\hat{\mathbf{A}}^T = \gamma_1 \mathbf{v}_1 \mathbf{u}_1^T + \cdots + \gamma_k \mathbf{v}_k \mathbf{u}_k^T \tag{3.29}$$

where k denotes the number of relevant factors and the hats denote the corresponding estimated matrices. You will perceive that (3.29) tells us what the product of $\mathbf{F}$ and $\mathbf{A}$ is but it does not disclose what each of these is on its own. In fact, it can be easily shown that an infinite number of solutions to this problem exist. Two ways of determining these matrices are available. These turn out to be essentially the same and differ only in the way in which the factors are scaled.

Solution 1:

Write $\mathbf{V}_k$ as the estimate of $\mathbf{F}$ and $\mathbf{U}_k \mathbf{\Gamma}_k$ as the estimate of $\mathbf{A}$. In practice, the computations are as follows:

1. Compute the covariance matrix $\mathbf{S}$ (in standardized form this will, of course, be the correlation matrix $\mathbf{R}$).
2. Extract the k largest latent roots, $\mathbf{\Lambda}_k$ and the corresponding latent vectors of $\mathbf{S}$, to wit, $\mathbf{V}_k$.
3. Compute then

$$\hat{\mathbf{A}} = \mathbf{V}_k \mathbf{\Lambda}_k^{1/2}. \tag{3.30}$$

   This is no more than standard procedure in principal component analysis. We note that the covariance matrix of the factors is

$$\mathbf{V}^T \mathbf{V} = \mathbf{I},$$

   the identity matrix.
4. To find the estimate of $\mathbf{F}$ the required calculation is

$$\hat{\mathbf{F}} = \hat{\mathbf{Y}} \mathbf{A} \mathbf{\Lambda}_k^{-1}.$$

Solution 2:

This solution quite simply lets $\mathbf{U}_k$ be the estimate of $\mathbf{A}$ and the estimate of $\mathbf{F}$ becomes

$$\hat{\mathbf{F}} = \mathbf{V}_k \mathbf{\Gamma}_k.$$

With this choice, the covariance matrix of the factors is Λ_k. That is the covariance matrix of the factors is a diagonal matrix, the elements of which are the latent roots of S. This is also a standard method of principal component analysis. For practical purposes, the computations required may be summarized as follows:

1. Start with the covariance matrix S.
2. Find the k largest latent vectors of S and let these be the estimate of A.
3. The estimate of F is obtained by multiplying Y by the estimate of A.

3.9.5 *True Factor Model*

I have just demonstrated two ways of expressing principal component analysis in a factor-analytical form. For true factor analysis, that is, the correct model for psychometrical studies, we start as before with equation (3.28). Matrix Ψ is now a diagonal matrix; its diagonal elements are not necessarily small. In true factor analysis, the elements of this matrix must be estimated from the data, together with the factor-loading matrix A. The logical procedure is to minimize the sum of squares

$$\mathrm{tr}(S - AA^T - \Psi)^2$$

which is mathematically equivalent to choosing the *communalities* of $S - \Psi$. In some canned computer programs, the traditional, superseded method of doing this is used. This is done by letting the community of each variable be chosen in relation to the squared multiple correlation coefficient of that variable with all other variables. This amounts to selecting

$$\hat{\Psi} = (diag\,S^{-1})^{-1}$$

This is called the *principal factor method* of estimation. This and similar methods are no longer used in professional work. The method of Maximum Likelihood Estimation, as developed by K. G. Jöreskog, requires that we estimate the following:

$$\hat{\theta} = \frac{1}{p - k} \sum_{m=k+1}^{p} \lambda_m$$

to wit, the average of the $p - k$ smallest latent roots of S^*, where

$$S^* = (diag\,S^{-1})^{1/2} S (diag\,S^{-1})^{1/2}$$

i.e. the average of the $p - k$ smallest latent roots of S^*.

The least-squares estimate of A corresponding to the above is

$$\hat{A} = (diag\,S^{-1})^{-1/2} U_k (\Lambda_k - \hat{\theta}I)^{1/2}$$

where U_k is the matrix of order $p \times k$, the columns of which are the orthonormal latent vectors of S^*.

More efficient, scale-free methods are now available, including the method of generalized least squares, and estimation by a more complete model for maximum likelihood (cf. Jöreskog *et al.*, 1976).

3.9.6 *Summary and Discussion*

Is there a genuine need to attempt to fit morphometrical data to such a complicated model as factor analysis you might well ask? The estimation of factors in the manner appropriate to psychometric work does not seem justifiable in very many situations in the Earth Sciences and an unencumbered principal component solution is most likely to yield what is wanted. In actual fact, this is what most people do, although they believe they are using a true factor-analytical model. No harm is done in most cases. The difficulties compound themselves when it is decided to enhance the analysis with some form of rotation to *simple structure*. This can no doubt be motivated mathematically and logically in true factor analysis, but it is often difficult to find a convincing reason for rotating the latent vectors of a principal component in order to seek out a neater clarification of results. In a genuine factor-analytical model, there are two fundamental requirements that can be said to lie at the root of the rotational procedure.

1. The number of references necessary to display the data will be determined by the approximate rank of the data matrix.
2. The axes are located according to some mathematical criterion. The latent vectors of principal components constitute a uniquely positioned set of factors. But, they are only one of an infinite number of sets of factors that will describe the configuration of the data equally as well. This indeterminacy of a "correct" solution underlies much controversy.

Recently, the rationale for rotating principal components as an aid to reification has been taken up by statisticians. Jolliffe (1986, 1989) gives an enlightening account of the reasons and justifications for doing this. In particular, rotation of principal components has proved to be beneficial for interpreting small components with roughly equal variance, i.e. they are ill-defined (Jolliffe, 1986, Chapter 7). Rotation of all components has been found useful in regression analysis and also in detecting outliers. But, do not be beguiled into believing that rotation of principal components causes a metamorphosis into a variety of true factor analysis; this is not so.

3.9.7 *Illustrations*

3.9.7.1 *True Factor Analysis*

Consider six variables, $x_1, x_2, \ldots, x_6$, measured on a very large population of organisms, so large indeed as to represent the true values of the statistical population to which the variables belong. The correlations between the variables are as follows:

$$
\begin{bmatrix}
1.000 & 0.720 & 0.378 & 0.324 & 0.270 & 0.270 \\
0.720 & 1.000 & 0.336 & 0.288 & 0.240 & 0.240 \\
0.378 & 0.336 & 1.000 & 0.420 & 0.350 & 0.126 \\
0.324 & 0.288 & 0.420 & 1.000 & 0.300 & 0.108 \\
0.270 & 0.240 & 0.350 & 0.300 & 1.000 & 0.090 \\
0.270 & 0.240 & 0.126 & 0.108 & 0.090 & 1.000
\end{bmatrix}
$$

Inasmuch as we are dealing with a very large sample, an element of this array can be denoted as ρ_{ij}; it will be taken to represent a population value.

Using the relationship

$$\Sigma = \mathbf{A}\Phi\mathbf{A}^T + \Psi$$

factors $f_1, \ldots, f_k$ which account for the correlations in the above array were determined. The way in which the reasoning works here is that you ask the following:

1. Is there a factor, f_1, which when extracted from the covariance matrix removes all correlations between variables?
2. If this be so, the partial correlations between any pair of variables, x_i and x_j, must vanish after f_1 has been partialled out. This can be easily done by using the relationships

$$x_i = a_i f_1 + e_i$$

and

$$x_j = a_j f_1 + e_j$$

with e_i and e_j uncorrelated for all $i \neq j$. Under these circumstances, $\rho_{ij} = a_i a_j$, so that the correlations in rows i and j would then be proportional. A glance at the matrix above shows that this is certainly not so and it is therefore clear that more than one factor will be required to account for the intercorrelations.

Assume now that the factors are uncorrelated, a safe assumption. This permits a simplification in our formula to

$$\Sigma = \mathbf{A}\mathbf{A}^T + \Psi. \tag{3.31}$$

Thus, the major product of $\mathbf{A}$ should reproduce *exactly* the off-diagonal elements of Σ. The formula for reproducing any element is

$$\rho_{ij} = a_{i1} a_{j1} + a_{i2} a_{j2} + \cdots + a_{ik} a_{jk}$$

and the *communality for variable* x_i is yielded by the relationship

$$\sigma_{ci}^2 = a_{i1}^2 + a_{i2}^2 + \cdots + a_{ik}^2.$$

Using the computational techniques already reviewed in this section it was ascertained that the matrix **A** below satisfies equation (3.31):

$$\mathbf{A} = \begin{matrix} & \text{Factor 1} & \text{Factor 2} \\ x_1 \\ x_2 \\ x_3 \\ x_4 \\ x_5 \\ x_6 \end{matrix} \begin{bmatrix} 0.889 & -0.138 \\ 0.791 & -0.122 \\ 0.501 & 0.489 \\ 0.429 & 0.419 \\ 0.358 & 0.349 \\ 0.296 & -0.046 \end{bmatrix}$$

The communality of the first variable, for example is

$$\sigma_{ci}^2 = (0.889)^2 + (-0.138)^2 = 0.81.$$

This illustration was extracted from Jöreskog $et\ al.$ (1976) who developed it further in order to show the effects of imposing simple structure, etc.

3.9.7.2 *Wrightian Factor Analysis*

The method of factor-modelling introduced into genetics by Wright (1968) has been recently amplified into a powerful biometrical analytical technique by Bookstein and his associates (Bookstein $et\ al.$, 1985; Crespi and Bookstein, 1989). I shall give a brief introduction to the technique by re-analysing the matrix studied in the foregoing illustration.

Basically, Wright's method is to fit a single-factor model to the data and then to examine and interpret the residuals after the extraction of this factor. The following table lists the loadings on the single size-factor and the residual correlations.

	var 1	var 2	var 3	var 4	var 5	var 6
loadings	0.807	0.740	0.555	0.490	0.417	0.289
var 1	0.349	*0.123*	−0.070	−0.071	−0.066	0.036
var 2		0.452	−0.075	−0.074	0.068	0.026
var 3			0.692	*0.148*	*0.119*	−0.035
var 4				0.760	0.096	−0.034
var 5					0.826	−0.031
var 6						0.916

The relatively large residual correlations shown in italics in the morphometrical pattern of an organism were shown by Wright to bear a genetical message. He interpreted them as expressing growth regulation, organ by organ, rather than all-embracing contrasts over the entire organism. This is an important point in the analysis of size and shape that most of us tend to overlook when we try to analyse large quantities of material by stereotyped procedures.

Wright then went on to introduce a procedure for adjusting the first factor for the information intimated by the larger residuals. This is done by expanding the correlation matrix to encompass not only the diagonals but also the hypothesized residual factor-spaces which contain terms exactly fitting the model. The results of the adjusted analysis then become as indicated below:

	var 1	var 2	var 3	var 4	var 5	var 6
loadings	0.624	0.555	0.606	0.519	0.502	0.325
var 1	0.611	*0.374*	0.000	0.000	−0.043	0.067
var 2		0.692	0.000	0.000	−0.038	0.059
var 3			0.633	*0.106*	0.046	−0.071
var 4				0.730	0.039	−0.061
var 5					0.748	−0.073
var 6						0.894

The coefficients of the size factor have been changed by the new specification and some of the residual correlations have been reduced to zero. One, that between variables 1 and 2 has increased. The results can be succinctly expressed as *path-coefficients* in the following manner:

	organ x_1/x_2	organ x_3/x_4
var 1	$0.617\ (=r_{res\,12}^{1/2})$	
var 2	0.617	
var 3		0.326
var 4		0.326
var 5		
var 6		

The path coefficients give a neat representation of the joint causes that form the observed correlations. In the present illustration, there is a primary size factor and two significant secondary factor, which can be conveniently displayed in a path diagram:

$$
S \begin{cases}
\quad \text{var 1} \\
0.624 \qquad 0.612 > x_1/x_2 \\
\quad \text{var 2} \\
0.555 \\
\quad \text{var 3} \\
0.606 \qquad 0.326 > x_3/x_4 \\
\quad \text{var 4} \\
0.519 \\
\quad \text{var 5} \\
0.502 \qquad > 0.01 > x_5/x_6 \\
\quad \text{var 6} \\
0.325
\end{cases}
$$

3.9.8 *Concluding Comments*

I have taken up several rather complicated and, in the eyes of some, controversial topics in this section. Providing you have a clear idea of what you want to do and how to construct a cause-effect model, I think that there is a lot to be said for modern factor-analytical procedures. Whatever you do, do not dismiss them out of hand as useless manipulations. If you do this you have just not understood the difference between static data analysis and the exploration of a scientifically valid model. If, as I sincerely hope, you have been fired with the desire to come closer to multidimensional model-building in morphometry, I suggest you work carefully through the book by Fornell (1982). Many of the points I have perforce glossed over in this section are presented in great detail in that text.

3.10 Example of Integrated Multivariate Analysis

Statement of Problem

In order to illustrate how an integrated multivariate morphometric analysis of traditional kind, and with rather modest aims, can be constructed, I have chosen the data on the Miocene foraminiferal species *Brizalina mandoroveen-sis*, used several times in this book. These data are not remarkable in any way apart from the fact that they introduce the kind of practical problems often associated with reworked or contaminated material that is common to most palaeontological investigations. Here I use distance measures. The original study on this species (Bookstein and Reyment, 1989) was, however, entirely concerned with shape variability. The problems we meet pertain to the following points:

1. Does the material vary significantly in average size over the levels sampled?
2. Does the material show differences in covariance patterns between samples?
3. How reliable are the data for multivariate work? Are there significant deviations from the multivariate Gaussian distribution?

MATERIAL

Each sample consists of ten haphazardly selected specimens from each of five stratigraphically distinct levels in a borehole drilled at Ikang, Cameroun, West Africa. These are the same specimens as were used in Bookstein and Reyment (1989) for a study of geometric morphometry in the species. There are six variables, these being the natural logarithms of computed Euclidean distances between landmarks (cf. Fig. 4.2). The data are listed in Table 3.8. The

TABLE 3.8. *The six distance measures on* Brizalina mandoroveensis

1	2	3	4	5	6
LEVEL 1					
11.10	10.38	9.78	10.80	11.10	10.26
11.02	10.42	9.84	10.75	11.13	10.23
11.02	10.41	9.82	10.74	11.12	10.19
10.98	10.42	9.94	10.77	11.14	10.25
11.09	10.49	10.08	10.83	11.20	10.20
11.10	10.32	9.83	10.84	11.23	10.23
10.94	10.35	9.92	10.72	11.07	10.10
11.09	10.46	10.09	10.76	11.18	10.20
10.93	10.32	9.77	10.62	11.04	10.16
11.03	10.45	9.83	10.75	11.14	10.27
LEVEL 2					
11.28	10.18	9.88	11.22	11.39	9.78
11.20	10.38	9.73	10.99	11.31	10.26
10.98	10.32	9.81	10.69	11.10	10.16
11.29	10.05	9.59	11.34	11.40	9.53
11.29	10.49	9.98	11.06	11.37	10.24
11.04	10.39	9.81	10.66	11.09	10.15
11.41	10.23	9.86	11.22	11.47	10.01
11.06	10.25	9.87	10.91	11.14	9.78
11.07	10.51	9.80	10.75	11.26	10.62
11.49	10.25	9.97	11.44	11.62	10.01
LEVEL 3					
11.24	10.41	9.78	10.99	11.29	10.14
11.13	10.38	9.88	10.84	11.22	10.19
11.12	10.35	9.68	10.89	11.17	10.04
11.33	10.34	9.69	11.13	11.39	10.11
11.57	10.53	9.90	11.37	11.64	10.35
11.12	10.36	9.93	10.85	11.17	9.99
11.37	10.51	10.11	11.22	11.51	10.33
11.15	10.38	9.77	10.94	11.23	10.12
10.92	10.40	9.85	10.59	11.00	10.14
11.63	10.97	10.34	11.31	11.62	10.54
LEVEL 4					
10.89	10.30	9.86	10.47	10.93	10.03
10.68	10.17	9.58	10.46	10.85	10.06
10.47	10.12	9.62	10.08	10.51	9.71
10.92	10.52	10.06	10.65	11.00	10.11
11.06	10.35	9.93	10.76	11.18	10.23
11.07	10.39	9.71	10.93	11.16	10.06
11.41	10.70	9.95	11.16	11.49	10.50
11.20	10.68	10.20	10.97	11.35	10.47
11.21	10.47	9.87	11.05	11.31	10.20
11.65	11.21	10.42	11.53	11.79	11.01
LEVEL 5					
11.14	10.49	10.00	10.86	11.22	10.21
11.45	10.41	10.06	11.24	11.56	10.33
11.50	10.65	10.09	11.27	11.56	10.37
11.26	10.23	9.69	11.05	11.32	9.99
11.05	10.57	9.96	10.59	11.08	10.29
10.98	10.09	9.51	10.96	11.07	9.62
11.06	10.44	9.95	10.78	11.09	9.98
11.06	10.81	10.34	10.68	11.16	10.50
11.41	10.96	10.23	11.25	11.49	10.61
11.57	10.49	9.82	11.46	11.65	10.26

set of observations was produced by Leslie Marcus, Note that Leslie did not use all the possible distances available.

THE ANALYSIS

Despite the stabilizing effect of the logarithmic transformation on recalcitrant data, many of the variances for the variables of the present material are unequal over levels. This can be seen from the set of bar diagrams displayed in Fig. 3.10 for the range of observations for some of the variables. Much the same tendency can be seen for all of the variables. The material from Level One tends to produce the shortest bars. The immediate interpretation for this situation is that the material from that level is stratigraphically more compact, whereas the samples from the other levels may be more mixed, reworked, contaminated, and/or distorted post mortem. This suggestion is supported for all variables with, perhaps, the exception of variable 3, for which all levels produce about the same bar-lengths. From the multivariate analytical point of view, it can be expected that the covariance matrices of the data are not homogeneous. This can influence an analysis adversely, particularly if it is intended to extract the last drop of information from the data: vectorial comparisons, significance tests, reification of vectors, etc. If our ambitions do not go beyond graphical studies, then my experience is that you can go ahead without running graver risks for doing something silly. A good book to have close by at this stage is that of Gnanadesikan (1977). The subject of transformations of data was taken up by Box and Cox (1964) in one of the earlier articles on the subject.

The news becomes a little better if we look at the results of specific tests for skewness and kurtosis. Many of the variables, over all levels, conform with the properties of the univariate normal distribution with respect to skewness, or almost so. Examples of the exceptions are variable 4 in Level One, which is significantly kurtosic on the 5% level and variable 2 in Level 3, which displays highly significant skewness ($t = 2.55$) and highly significant kurtosis ($t = 6.98$) and the same variable in the sample from Level 4 for which there is significant kurtosis at the 5% level. Variable 3 for Level 4 also shows highly significant kurtosis ($t = 3.16$) and highly significant kurtosis ($t = 10.0$). Generally, then, deviations are more common with respect to the flatness or pointedness of the univariate distributions than to skewness.

The acid test, as it were, is to see what specific tests for multivariate normality say (Mardia, 1970; Section 8.2). The results for all 50 specimens considered simultaneously in tests for multivariate skewness and kurtosis are:

$$B_{1,p} = 10.01$$

and,

$$B_{2,p} = 42.97.$$

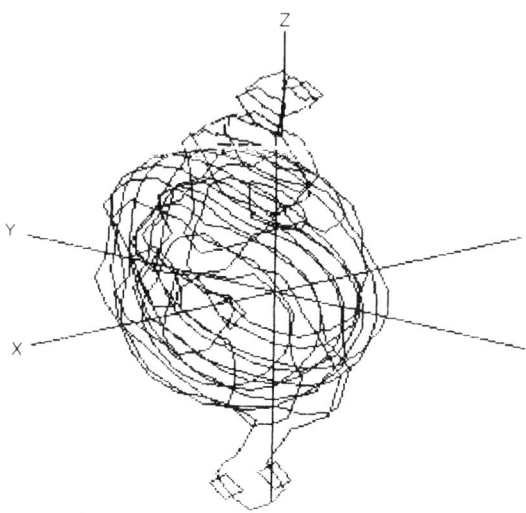

F\ɪɢ. 3.8. Three-dimensional wireline diagram for variables 3, 4 and 6 for all samples of
the *Brizalina* data. This offers a way of expressing nearness to multivariate normality.

The multivariate skewness statistic translates to a value of chi-square of 83
which for 56 degrees of freedom is significant on the 1% level. the multivariate
kurtosis statistic is transformed to a normal standard deviate of −1.815. A
table of the normal standard deviate shows this value to be significant on the
3% level. Hence, when *all variables are considered in the one connexion*, we find
that the material is both multivariate skewed and multivariate kurtosic.

Graphical methods of appraisal can be very informative for probing
multivariate normality. A three-dimensional diagram for variables 3, 4 and 6
for all samples, shown in Fig. 3.8, indicates that the points lie within a roughly
ellipsoidal hull. This way of displaying multivariate observations is very
useful. The figures are called wireline diagrams and I have made my
illustrations using the NCSS statistical package marketed by Dr J. Hintze,
Utah, U.S.A. (see Bibliography). This is a method for fitting wireframe shells
to three-dimensional point clouds. The point clouds are represented as bands
of equal density which are often referred to as density contours. The NCSS
program first scales the three-dimensional data so that they lie inside a square.
This square is then sliced into equal segments along each of the axes to form
a collection of equi-sized square-shaped cells. The number of points falling
into each of these small cells is then counted and these frequencies used in the
subsequent calculations. The cell-counts are smoothed by the method of
averaged shifted histograms which produces smoothed shell counts. These
smoothed counts are used to connect contours of equal value to construct
the wireframe shell. It is usual to show at least two such frames; one frame
marks an outer shell and the other, the inner shell of the shape of the
point-cloud.

TABLE 3.9. *Variances for the six variables, level by level for* Brizalina mandoroveensis

Variables	X1	X2	X3	X4	X5	X6
LEVELS						
1	0.0043	0.0034	0.0137	0.0040	0.0034	0.0025
2	0.0291	0.0200	0.0131	0.0759	0.0295	0.0944
3	0.0485	0.0357	0.0406	0.0582	0.0441	0.0273
4	0.1173	0.1005	0.675	0.0170	0.1296	0.1244
5	0.0482	0.0654	0.0599	0.0816	0.0513	0.0822

A good diagram may be worth a welter of entries in a table. I could have shown more combinations of variables, but the figure used here served to give you a reasonable idea of how the data are distributed and the value of visual aids in a multivariate analysis

We should also look at the variances in more detail. I have listed these in Table 3.9, variable-by-variable and level-by-level. The variance for Level 1 are generally much lower than for the other four levels. This lends support to the notion that the material from that level is stratigraphically more homogeneous than the other samples.

Bivariate Correlations: The correlation coefficients tend to be high for certain pairs and consistently low for others. However, considerable spread can be expected in such small samples and the same observation applies to the results of the preceding section. Correlations in excess of 0.85 are listed in Table 3.10. We note that r_{45} is always very high as is also r_{15}. This seems to be a genuine trait of the material. Moreover, r_{14} is high in all samples except that from Level 1, where, however, it is 0.83.

PRINCIPAL COMPONENTS

The next stage in the analysis is to examine the multivariate properties of each sample in turn by means of the patterns exposed by the method of principal components. This is, as it were, a kind of statistical dissection of the multivariate sample. The most striking feature of the results summarized in Table 3.11 is the reversal of vectors in the sample from Level 2. As I have

TABLE 3.10. *Significant bivariate correlations for the five Levels of* Brizalina mandoroveensis

Level	Correlations			
1	(1–5):0.948		(4–5):0.933	
2	(1–4):0.927	(1–5):0.971	(2–6):0.933	(4–5):0.913
3	(1–4):0.977	(1–5):0.987	(2–3):0.884	(2–6):0.903
	(4–5):0.988			
4	(1–2):0.911	(1–4):0.983	(1–5):0.993	(1–6):0.923
	(2–3):0.915	(2–4):0.895	(2–5):0.910	(2–6):0.955
	(3–6):0.865	(4–5):0.988	(4–6):0.910	(5–6):0.945
5	(1–4):0.911	(1–5):0.992	(2–3):0.912	(2–6):0.919
	(3–6):0.903	(4–5):0.923		

TABLE 3.11. *Principal component analyses for covariances of logarithms*

Level	1		2		3	
	I	II	I	II	I	II
Latent	0.0181	0.0094	0.1786	0.0693	0.2080	0.0390
roots	(58.2%)	(30.3%)	(68.3%)	(26.5%)	(81.6%)	(15.3%)
Vectors						
var. 1	0.324	0.449	0.476	0.305	0.468	0.231
var. 2	0.325	−0.063	−0.395	0.470	0.351	−0.456
var. 3	0.763	−0.546	0.020	0.594	0.329	−0.635
var. 4	0.325	0.389	0.524	0.088	0.492	0.451
var. 5	0.308	0.429	0.459	0.331	0.446	0.268
var. 6	0.082	0.402	−0.362	0.465	0.328	−0.245
	4		5			
	I	II	I	II		
	0.6622	0.0339	0.2303	0.1156		
	(93.3%)	(4.8%)	(59.3%)	(37.5%)		
	0.414	0.243	0.353	0.348		
	0.374	−0.393	0.428	−0.360		
	0.273	−0.664	0.383	−0.397		
	0.496	0.460	0.332	0.615		
	0.437	0.252	0.369	0.363		
	0.421	−0.262	0.547	−0.284		

pointed out in Chapters 3 and 4, the first latent vector of a morphometrical covariance matrix for logarithmically transformed variables can often be interpreted as a size vector just because of the fact that all elements of the vector are positive. The "size vector" is usually the first linear combination of positive vector elements. In the case of the sample from Level 2, however, the second latent vector, bound to only 26.7% of the trace of **S**, has usurped this role. This incites curiosity about the composition of the data from Level 2, which we shall now examine more closely.

The set of techniques used by Krzanowski (1987a, 1987b), known as cross-validation, were applied to the entire set of 50 specimens in order to attempt to pick out atypical and/or influential observations (cf. Section 8.4). The suite of methods is of an *ad hoc* nature and interpretations rely on the judgement of the analyst. For reasons expounded by Krzanowski in the above article, I used standardized data (i.e. I worked in the space of the correlations).

Using the W_m criterion described in Krzanowski's paper, I ascertained that there are three significant principal components. This procedure is based on the amount of predictive information associated cumulatively with the principal components. It produces a cut-off at the point where the addition of a further component adds little to the efficiency of the prediction.

The next step is to see whether any variables are redundant. I found two of the variables to be equally important, which is ascertained by deleting each variable, one at a time, and seeing what the effect is on the sum of squares, established by carrying out a Procrustes analysis on the same configuration of N points, with and without the variable. (See Seber 1984, p. 253 and Gower, 1971b for information on Procrustean (matching configurations) methods.)

A resulting large sum of squares indicates that the variable deleted is important for the analysis. Conversely, a variable associated with a small sum of squares can usually be deleted. Variables 3 and 6 are associated with quite large sums of squares and are therefore important for the analysis. The remaining variables yield about the same sum of squares on deletion and are of lesser importance for the analysis. Their sums of squares are a little less than half of those for variables 3 and 6, but they are not so small as to warrant removal.

Perhaps the most important question to be answered concerns the influence of each of the N observational vectors. The effect of deleting observations successively can be judged by computing the critical angles between the new and old principal component spaces. The larger the angle, the greater is the disturbance resulting from the omission of the corresponding specimen. Krzanowski (1987a,b) gives two angles. One of these is for the principal components associated with the largest latent values and the second is for the principal components matching the smallest latent values. Large angles deriving from "large" latent vectors tend to be connected with disturbances in variances, whereas large angles deriving from "small" latent vectors indicate disturbances in covariances. In the present case, I have already established that there are three significant principal components. Consequently, the angles are to be computed for three vectors. None of the observational vectors produces a large angle on omission.

In conclusion, the cross-validation analysis of the data shows that there is not much wrong with them. The reason for the reversal of order in the latent vectors of the second sample is therefore not likely to be due to atypical specimens.

Cross-validation is widely used in chemometrics. Further references are: Eastment and Krzanowski (1982), Krzanowski (1983, 1987b), Wold (1978), Golub *et al.* (1979) and Stone (1974).

I also checked through all samples using $Q - Q$ probability plots (see Gnanadesikan (1977), and Seber (1984), Appendix C, where probability plotting and order statistics are reviewed; Section 8.3) of the generalized statistical distances computed between each observational vector and the centroid of the sample. This is a good graphical way of singling out atypical observations (see also Campbell, 1980a,b and Seber, 1984) which profiles every specimen in a sample in relation to its fellows. None of the samples contains markedly aberrant specimens, but we must keep in mind that in such small data sets, it is difficult to track down an atypical individual. A further reference is the paper on the use of residuals for the detection of outliers by Brown and Kildea (1979).

Principal component scores

Under the supposition of homogeneity of samples, the shape of the empirical distribution can be examined by means of the scores of principal

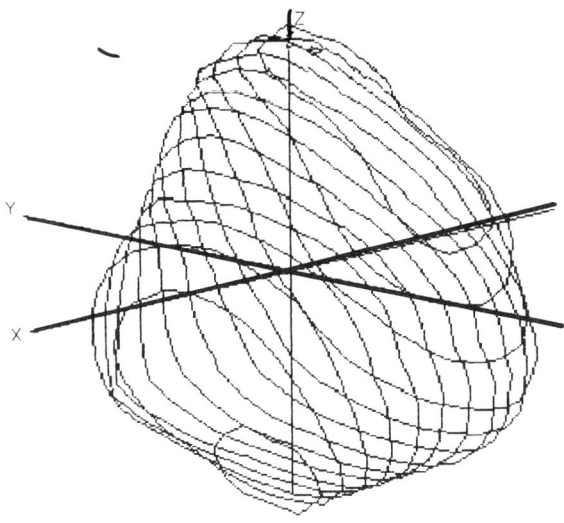

FIG. 3.9. Principal coordinate scores for the *Brizalina* data. This figure shows the inner contouring of the point cloud.

components. This was done for all samples. I actually employed the principal coordinate values rather than the scores of principal components for making the figure displayed in Fig. 3.9. This is the inner contouring of the point cloud. The shape of the shell is not very ellipsoidal and is more like that of a pear. This reflects the difference in variability between the first sample and the four others, which seems to me to be the cause of a good deal of the divergence from the multivariate Gaussian state disclosed by specific tests.

Two-sample comparisons

Comparison of Levels 1 *and* 2: The two samples that contrast most strongly are those from Levels 1 and 2. I checked these against each other by a set of techniques based on the generalized statistical distance and the properties of ellipsoids of dispersion (Reyment, 1969; Blackith and Reyment, 1971). Briefly, the methods examine the ellipsoids with respect to

(a) relative inflation
(b) orientation of the major axes.

The program used is an updated and expanded version of the one published in Blackith and Reyment (1971). See also Seber (1984, pp. 102, 110 and 114) for remarks concerning equality of dispersion matrices and testing the equality of means.

Examining firstly *relative inflation*, the appropriate procedure is to compute the generalized version of the test for equality of variances. I used the information-theoretic variant espoused by Kullback (1959) which we can write in the form usual in the statistics of information theory as

$$2I(H_1 : H_2(\cdot)) = \sum_{i=1}^{r} N_i \ln |\mathbf{S}|/|\mathbf{S}_i| \qquad (3.32)$$

where r denotes the number of groups, the $\mathbf{S}_i$ are the sample covariance matrices and $\mathbf{S}$ is the pooled covariance matrix, each founded on p variables. The best way of ascertaining significance is by means of Fisher's B-distribution, the non-central chi-squared distribution expressed by (3.33), which, unfortunately, has only been tabulated for 7 degrees of freedom. A reasonable approximation in terms of the usual chi-squared distribution is

$$B^2(1 - 2\beta^2/(r-1)p(p+1)) \qquad (3.33)$$

for $(r-1)p(p+1)/2$ degrees of freedom.

For the *Brizalina*, I found that $B^2 = 56.7$ which for 21 degrees of freedom as chi-squared is highly significant. This result indicates that the two ellipsoids of scatter are differently *inflated*, like two balloons with unequal amounts of air in them, as it were. This is by no means unexpected as we have already taken account of the significant differences in variances—consult, for example, the bar diagrams illustrated in Fig. 3.10. We also know that the order of the latent vectors is reversed in the second sample. The application of the test expressed

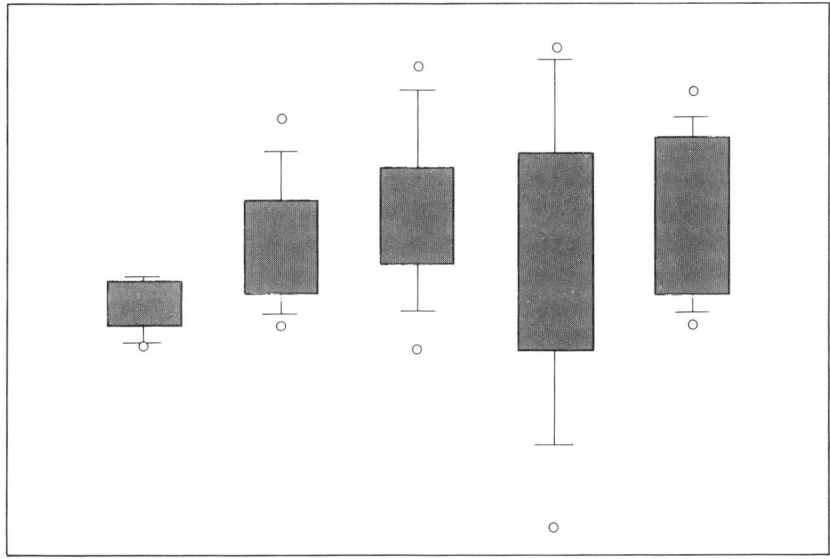

Brizalina
bar – diagram 5 levels variable 2

Fɪɢ. 3.10. Variable 2.

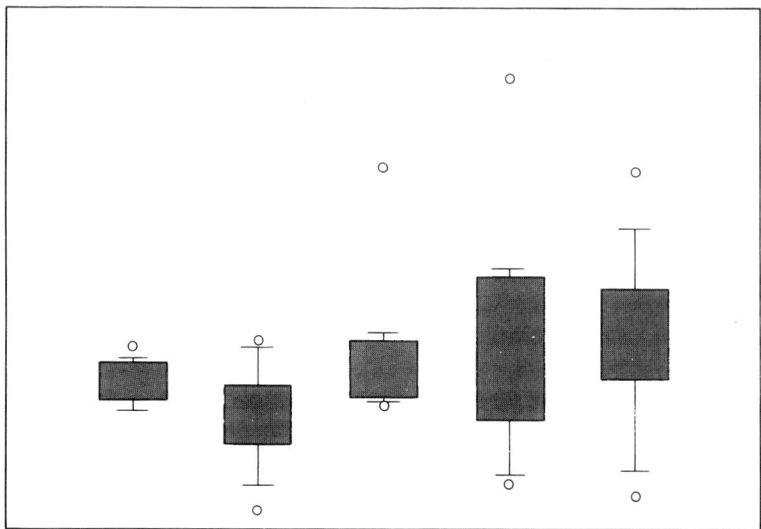

Brizalina
variable 3

FIG. 3.10. Variable 3.

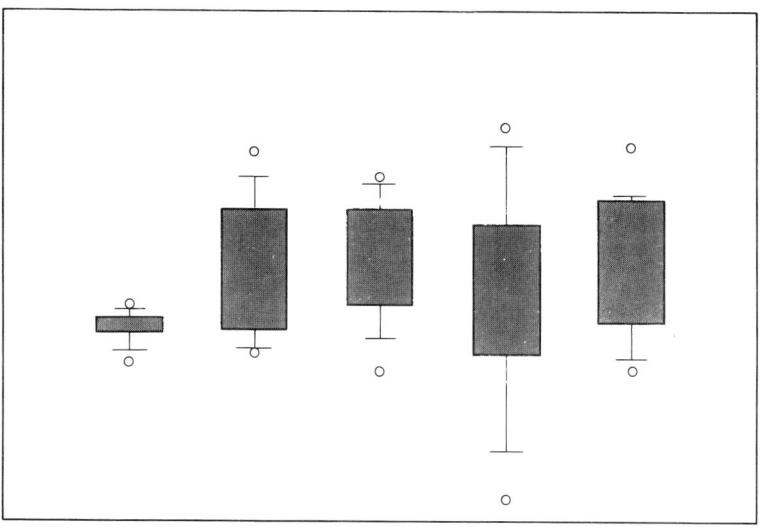

Brizalina
box-plot for variable 4

FIG. 3.10. Variable 4.

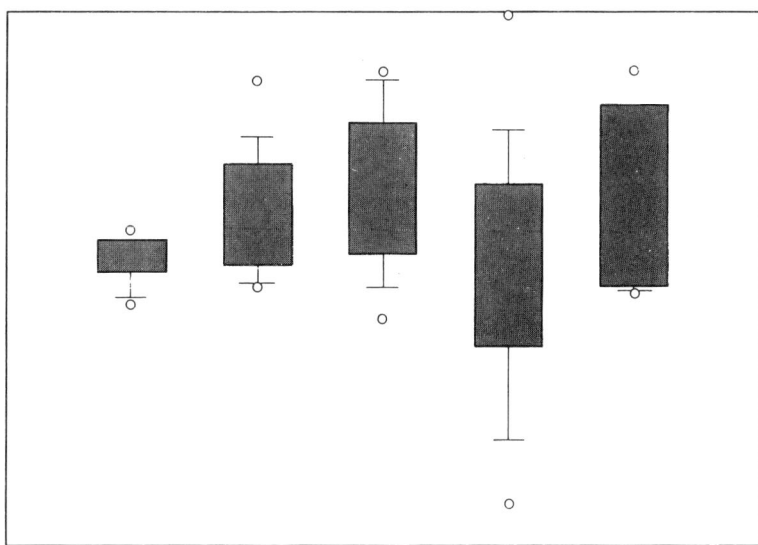

Brizalina
box-plot of variable 5

FIG. 3.10. Variable 5.

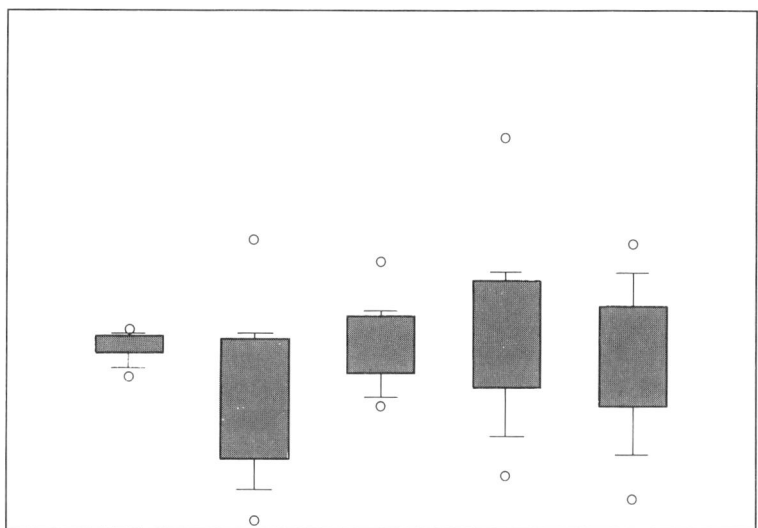

Brizalina
variable 6

FIG. 3.10. Variable 6.
FIG. 3.10. Bar diagrams for showing the variance for the traits measured on *Brizalina*.

in formula (3.33) can be fraught with danger unless the data accord reasonably well with the multivariate normal distribution. Hence, in the present analysis, take care.

The next step is to check the orientations of the major axes of the two ellipsoids. Using an approximate test for orientations (Reyment, 1969) based on Anderson (1963), I found that the three significant axes yield significant values of chi-squared and hence indicate that the ellipsoids are rotated in relation to each other. Three axes are near to zero. This result, obtained independently from the strategy utilized in the set of cross-validation procedures, confirms that there are three significant principal components in our data. The subject of testing agreement in covariance matrices has been taken up by Dempster (1969). His figures for relative positions of concentration ellipsoids (Dempster, 1969, p. 208), and the discussion of their shadow properties, can be recommended as an informative account of the geometrical relationships involved.

Comparison of Levels 3 and 4

These two levels serve to illustrate the other side of the picture. Tests for univariate skewness showed all variables to be well behaved for Level 4 and for significant skewness to occur for Level 2. All but variable 4 are kurtosic for Level 3 and variables 3, 5 and 6 are kurtosic for Level 4. All this kurtosis, what can it mean? I think it is being induced by the way in which the variables were produced—by the final step of taking square roots.

The results of the test for homogeneity of inflation gives $B^2 = 20.53$ for $\beta^2 = 7$ which, as chi-squared $= 18.7$ and 21 degrees of freedom is not significant. The test for orientations of the principal axes shows the first vectors to be differently oriented at the 1% level, vectors 2 to coincide and vectors 3 to differ in orientation at the $2\frac{1}{2}$% level. Hence, the ellipsoids are equally inflated but rotated in relation to each other about the second major axis (the second principal component). *Caveat*: Be aware, that the sample sizes in this tutorial are very small and almost all of the procedures used require sample sizes of between 50 and 100 specimens, being what is usually known as large-sample methods.

Comparing other samples

We can also compare orientations of latent vectors by the direct method of computing angles between them. The latent vectors for the samples from Levels 3, 4 and 5 appear to be most similar. Starting with the first latent vector (principal component), we have that the angle between samples 3 and 4 is 7 degrees. Other results are listed in Table 3.12; all of these values indicate reasonably close agreement in the first latent vectors of the covariance matrices.

TABLE 3.12. *Angles between latent vectors of samples from Levels* 3 *to* 5 *for* Brizalina mandoroveensis

Levels being	Angles	
compared	Vector I	Vector 2
3 with 4	7°	10°
3 with 5	18°	20°
4 with 5	15°	20°

The results for the comparisons of the second latent vectors are likewise listed in Table 3.12. Here, again, the vectors do not differ markedly in the respective comparisons.

ANALYSIS OF ALL SAMPLES BY CANONICAL VARIATES

For the present purposes, I use the latent vectors of $W^{-1}B$ as the basis for the calculations. I list these vectors in Table 3.13. They tell us that we need a two-dimensional representation for the multivariate reduction and that the first and second vectors contain equal information, as indicated by the almost equal characteristic roots which, together, account for more than 95% of the sum of the characteristic roots. This is an unusual situation in morphometric work.

Morphometrical Relationships Between all Levels: Establishing the relative locations in space of the specimens of the samples is one of the main interests of the analysis. This is readily done by computing the minimum spanning tree between each canonical variate mean score and displaying the network in the plane of the first two canonical vectors. I have done this in Fig. 3.11.

The distances between points are the generalized statistical distances which are proportional to the lines joining the respective points. Level 4 acts as a fulcrum for the six distance measures, with Levels 2 and 3 being located furthest from it. There is obviously no morphostratigraphical progression.

Canonical Variate Scores: The 50 scores for the first two canonical variates were plotted in the plane of the first two canonical vectors (Fig. 3.12) and four

TABLE 3.13. *Generalized latent roots and vectors (canonical variate vectors) for* Brizalina mandoroveensis

Variable	Vector I	Vector II
1	−0.234	−5.555
2	1.226	1.204
3	−0.726	−0.461
4	−4.017	−3.112
5	5.057	9.756
6	−1.974	−1.944
Latent root	0.60985	0.52938
Percentage	51.50	43.65

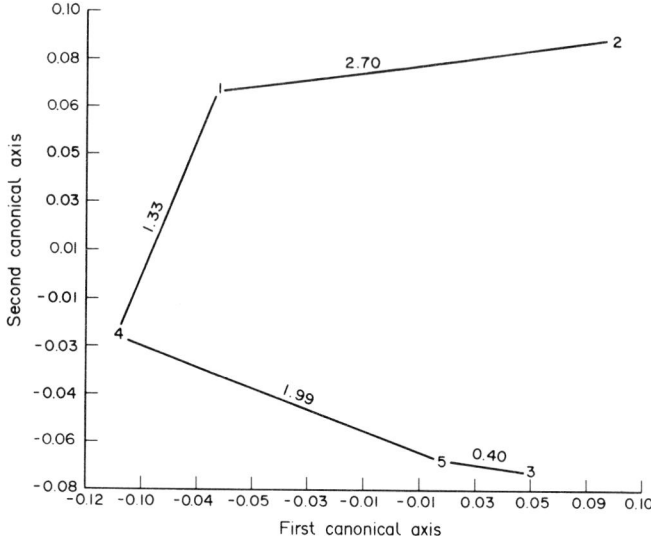

FIG. 3.11. Canonical variate scores and minimum spanning tree for the *Brizalina* data.

of the convex hulls drawn. It will be seen that the sample from Level 2 is almost entirely separated from the other four samples, which largely overlap. The sample from Level 1 has the most compact distribution, which reflects the low variability it displays.

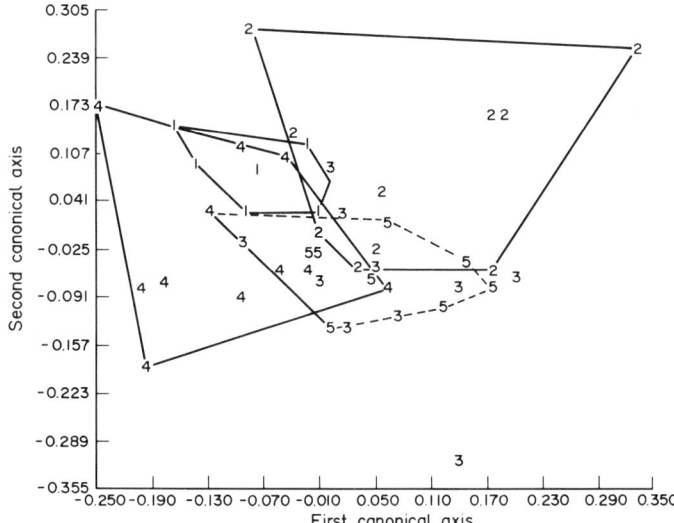

FIG. 3.12. Fifty scores for the first two canonical variates computed for the *Brizalina* data; four convex hulls are shown as an aid to interpreting the diagram.

FINDINGS

It is always a good idea to summarize and comment on the results of a multivariate morphometrical analysis, particularly those that encompass many parts. It is, alas, very easy to fall into slothful habits in this respect in the face of the myriad of relevant and irrelevant masses of information that canned programs generate.

Let us now take stock of what we have found out. I have already pointed out that this is not the greatest data set in the world and that I am using it as a tutorial, despite the small sample sizes. It is, nevertheless, representative of the kind of material we encounter in palaeontological work: somewhat messy and stratigraphically blended. How well has the analysis answered the questions put in the introduction?

1. The material does vary in average multivariate size over the levels sampled. There does not seem to be a systematic component attached to this variation.

2. The material displays differences in covariance patterns. This is again not systematic and seems to be the outcome of haphazard variation and small samples and in part, at least (e.g. Level 2), to be ascribable to reworking. The type of variables used can also have influenced the results (square-root transformation).

In addition, the analysis shows the observations to deviate somewhat from the multivariate normal distribution. The main achievement of our analysis has been to unveil statistical deficiencies in the data and thus to provide a warning to be careful with interpretations.

Exercises

1. The first three latent vectors of the correlation matrix placed in the upper diagonal of the following matrix for the classical data on *Iris* used by Fisher to illustrate discriminant functions

1.0000	0.5259	0.7541	0.5465
0.1052	1.0000	0.5605	0.6640
0.1508	0.1121	1.0000	0.7867
0.1093	0.1328	0.1573	1.0000

are

Vec 1	Vec 2	Vec 3
0.4823	0.6197	−0.4908
0.4648	−0.6729	−0.5398
0.5345	0.3069	0.3401
0.5153	−0.2830	0.5934

Can you write down the latent vectors of the correlation matrix given in the lower diagonal of the first matrix? (Hint: the upper diagonal matrix is a constant times the lower diagonal matrix.)

2. You are faced with performing a multivariate study on a set of observations comprising mixed variables, continuous, discontinuous and qualitative. Why would you be advised to avoid mixing these variables in computing a matrix of variances and covariances?

3. What do you understand by the concept of "morphological distance"?

4. When organisms grow, they change not only in size but also in shape. Growth in our own species is a good example of this. Can you conjure up variables that would be best for measuring size in our growth, and other variables that would best measure changes in shape?

5. The following latent roots and vectors of the correlation matrix for *Iris* (cf. Exercise 1) are:

$$
\begin{array}{ll}
(1) & 2.92639 \\
(2) & 0.54629 \\
(3) & 0.39497 \\
(4) & 0.13235
\end{array}
$$

Use these values to compute the correlations between variables and principal components. (Hint: keep in mind that the variances for the correlation matrix are all equal to one.) Compute the determinant of the correlation matrix using the latent roots.

6. The smallest principal component is often disregarded for interpretational purposes because it is thought to represent no more than "noise". However, many statisticians think otherwise (Gower, 1967; Jolliffe, 1986; Krzanowski, 1987a,b). The possible significance of the smallest component lies with the fact that it is often connected to a very small part of the variance and should therefore point to an invariant relationship between variables. Use this concept to interpret the fifth component of the covariance matrix of *Leptograpsus* (see Table 3.1).

7. The broken stick model is one way of determining the number of useful principal components in an analysis. The idea is that if you have a stick of unit length, and break it at random into p pieces, then the expected length of the k-th longest piece is

$$
l_k^* = 1/p \sum_{j=k}^{p} l/j
$$

Can you suggest how this concept might be applied to determining the minimum number of principal components? (Cf. Legendre and Legendre (1979, p. 406) who provide a table of reference values for l_k^*).

8. Using what you have learned about multivariate analytical procedures, can you form an opinion about the relative merit of a principal component analysis of the pooled within-groups covariance matrix $\mathbf{W}$ in relation to a formal canonical variate analysis of the same set of samples?

9. Can you work out why it is not recommended procedure to mix continuous, discontinuous and categorical variables in the same principal component analysis, but quite acceptable in principal coordinate analysis?

10. What does the operation

$$
\hat{\mathbf{A}} = \mathbf{U}_k \mathbf{\Lambda}_k^{1/2}
$$

amount to doing to the latent vectors $\mathbf{U}_k$ of $\mathbf{S}$?

11. Check that the A-matrix in the illustration on page 81 satisfies the equation (3.31) exactly. Hint: start by showing that

$$
\rho_{12} = a_{11} a_{12} + a_{12} a_{22} = 0.720.
$$

CHAPTER 4

The Study of Size and Shape

Contents

4.1 Introduction

4.1.1 *Overview*

One of the earliest topics inviting interest in quantitative biology was that of the analysis of size and, more particularly, shape. It should by now be known to most readers of this book that one of the earliest and most serious attempts at devising a quantifiable means of examining variability in shape was put forward by D'Arcy Wentworth Thompson (1917). Some of his original drawings appear in books and papers right to the present day. Although interesting as an intellectual feat, Thompson did not provide computational instructions in support of his diagrams, nor did he disclose how he had constructed them. With the knowledge of hindsight, we now know that the diagrams must have been eye-balled, as the saying goes. Of recent years, serious attempts have been made at providing Thompson's grids with a rigorous mathematical framework and these results will be probed further on in this chapter. This new work, mathematically soundly developed, has made the classical multivariate assessment of growth and shape obsolete. Studies on growth and form tend to be concerned with the following questions, which are, in practice, not mutually exclusive.

1. The analysis of differential growth relationships—the field of biometry usually referred to as *allometric growth*.
2. Variation in overall size, usually as a function of time, but also of space. Many palaeontological studies belong in this category.
3. Variation in shape, often considered as a function of time, but also of space.

The quantification of size and shape has traditionally proceeded by algebraic methods and it is these that we shall be taking up first. This might be called the *fait accompli* situation in which we are concerned with describing proportional relationships. This could be thought of as working in the **static mode**, as it were. The more recent quantification by geometrical methods is concerned with presenting a **dynamic** situation by means of a geometrically oriented procedure of some kind—a moment's thought will no doubt convince you that the two approaches are not mutually exclusive and are, in fact, complementary. Even the most evangelical "geometricians" have been seen to work in the static mode, until very recently, if this seemed to suit some particular analytical purpose.

With respect to the disposition of the topics treated in this chapter, the first half deals with what has become "traditional morphometrics", perhaps more vividly referrable to as "pre-Booksteinian morphometrics" for reasons that will become apparent as the theme of this chapter unfolds. These models depend on interpretations of latent roots and vectors based on distance measures (i.e. the familiar traits we are wont to measure on organisms, lengths, heights, breadths, etc.: see Section 1.2 in Chapter 1). The central section is concerned with the analysis of shape in terms of the restricted approach allowed by consideration of the outline alone of the object. Procrustean superposition of landmarks on organisms is a useful proem to the final part of the chapter in which several of the concepts of Geometric Morphometrics are applied to palaeontological problems.

4.1.2 *Huxley's Heterogeny*

The concept of differential growth, as elaborated by Huxley (1932), was restricted to two variables and this is still the form occurring in most studies and, moreover, the most readily interpretable version. Attempts have been made at generalizing the concept of allometry to more than two dimensions. You will be able to judge, further on, how successful these suggestions have been. There is an additional point to be kept in mind, namely, that some animals are asymmetric—we ourselves are a good example of minor morphological asymmetry and this condition has been put to use by Van Valen (1962) who studied correlations between *fluctuating asymmetries* of several widely different groups of animals. This is a factor that can sometimes prove to be important in growth studies. Notwithstanding the valuable work that has been done on the analysis of size and shape and the study of differential growth of late, it is rewarding to go back to the definitive work on the subject, namely the monograph of Huxley (1932). This is a wordy document, typical of the times, and not very well organized. This makes it necessary to sift sections of the text carefully to sort the grain from the chaff. Once you have taken the trouble to do this, you will be rewarded by a wealth of relevant detail, stimulating ideas, some of which are being rediscovered today, again and again. One of the gems, I think, is the perhaps obvious, though seldom invoked, supposition that growth often takes place at different rates in different directions and at different points; hence, we can speak of *growth gradients*, a concept that Huxley placed at the door of D'Arcy Thompson. No single model for the form of this variability is valid in general.

Huxley (1932) seems to be the first to have made a systematic study of constant differential growth ratios. He expressed his ideas in the following form. Let x denote the size of the animal, measured by some linear measure, or by weight (less the weight of the differentially growing organ) and y is the size (or weight) of the differentially growing organ, then the relationship between x and y is given by

$$y = bx^k \qquad (4.1)$$

where b and k are constants.

The role of k is that it implies that for the range over which the formula holds, the ratio of the relative growth rate of the organ to the relative growth rate of the body remains constant. The relative growth rate is defined as the rate of growth per unit of measure (unit weight), the actual absolute growth rate at any instant, divided by the actual size at that instant. There is today a pronounced tendency to restrict the discussion of allometry to measures on characters (distance or landmark), but be aware that the original development of the subject encompassed relationships between internal organs.

Gould (1971) has given informative accounts of the history leading up to the presentation of Huxley's allometric equation (Huxley actually said "heterogeny"), citing earlier works that do not seem to have been known to Huxley. Huxley was quite clear about the interpretation of the constant k. With respect to the constant b he wrote ". . . constant b is here of no particular biological interest, since it merely denotes the value of y when $x = 1$—i.e. the fraction of x which y occupies when x equals unity". Gould (1971, p. 115) noted the existence of a vast body of inconclusive literature concerning the meaning of b.

Gould concluded that b is actually a scale factor that expresses differences in size between comparable animals *of the same shape* on two or more regressions of constant k. **Growth-allometry** refers to the relationships of two morphometrical variables in a growing organism, while **size-allometry** refers to the relationship of such variables in samples of organisms. In palaeontological work, it is the latter category with which we are concerned.

Huxley provided a great number of examples, many of them of special value for the study of relative growth. Such an example is afforded by his account of the crab *Carcinus maenas* in which negative allometric relationships were explored. The outcome of this, and other work, led Huxley to the conclusion that only animals in which all organs are growing at the same rate will preserve their shape unchanged with increase of size—this is conveniently referred to as *isometric growth*. He entertained little doubt that this was an unreal situation in that one must ask oneself what kind of an animal can be expected to possess organs, all of which grow at precisely the same rate.

Huxley (1932, p. 81) was concerned about the desirability of linking a measure of the organ as a whole to its constituent parts by means of a set of perforce bivariate equations, well understanding this solution to have its weak points. We shall see shortly how attempts at doing this as a multivariate exercise have fared.

4.2 Multivariate Representation of Allometry

Various suggestions for generalizing the allometric equation of Huxley occur in the literature. Blackith (1960) proposed a solution which uses multivariate regression. A more widely embraced method based on principal components is that put forward by Jolicoeur (1963), which is nearer to the spirit of the Huxleyan formulation. Kuhry and Marcus (1977) gave a thoughtful account of bivariate linear models in biometry with emphasis on the estimation of parameters in the bivariate linear representation of differential growth. Röhrs (1959) produced a monographic account of problems in allometric studies; this is to be one of the earlier substantial treatments of the subject. There are many other papers I could have cited, but the above references should serve to give you the flavour of the topic. It is important to keep in mind that means of measurements on a sample of organisms are strongly dependent on size fluctuations due to the age distribution of the sample, environmentally controlled differences and, in the case of fossils, post-depositional sorting. These are all potential sources of error.

Let us look at what Jolicoeur's (1963) multivariate generalization implies. This model embraces the assumption that the relative growth rates of all dimensions considered are constant (Flury and Riedwyl, 1988, p. 228). This is a common assumption in bivariate (i.e. classical) allometric work and, as a starting point, it is undeniably useful. None the less, relative growth rates can, and do, change during ontogeny, of which Huxley (1932) was fully aware.

The coefficients of the first principal component, based on the covariance matrix of log variables, were used by Jolicoeur to estimate the relative growth rates, or allometric constants, in a multivariate situation. As in the bivariate case, if the growth rates are identical for all p variables, we speak of isometric growth, with the consequence that all the coefficients of the first principal component will be approximately $1/p^{1/2}$. Naturally, all our reservations concerning the reality of isometry apply here, but amplified. Jolicoeur outlined a large-sample test for ascertaining isometry in a vector, by adapting a procedure for testing latent vectors introduced by Anderson (1963).

In mathematical terms, we are studying multiplicative functions of the original variables. If we denote these original variables as $X_1, \ldots, X_p$, we can introduce the following linear combination of logarithmically transformed variables, to wit, $\log X_1, \ldots, \log X_p$:

$$Y = a_1 \log X_1 + \cdots + a_p \log X_p. \tag{4.2}$$

An equivalent expression to this is reminiscent of the standard allometric equation, namely,

$$Y^* = X_1^{a_p} \times \cdots \times X_p^{a_p}. \tag{4.3}$$

This is quite obvious if you compare (4.3) with (4.1).

We shall illustrate how the ideas involved in the multivariate representation of allometry work by means of some data on voles used in an illustration by Flury and Riedwyl (1988).

Example 4.1. Principal Component Shape Analysis of Voles

I have used the 110 measurements on two living species of voles employed by Flury and Riedwyl (1988) to illustrate multivariate growth. The species are

TABLE 4.1. *The latent roots and vectors for the voles*

Microtus californicus Latent roots		Coefficients of principal components		
		$\log(L_2)$	$\log(B_3)$	$\log(H_1)$
Y_1	0.03087	0.661	0.661	0.356
Y_2	0.00078	−0.617	0.208	0.759
Y_3	0.00022	−0.428	0.721	−0.545
Microtus ochrogaster				
Y_1	0.01472	0.683	0.663	0.308
Y_2	0.00066	−0.322	−0.105	0.941
Y_3	0.00032	−0.656	0.742	−0.142

Microtus californicus and *Microtus ochrogaster*. There are only three size dimensions, to wit: L_2 = condylo-incisive length, B_3 = zygomatic width, and H_1 = skull height. The data used are listed in table 10.16 in the book by Flury and Riedwyl (1988, p. 229).

The Covariance Matrix of the logarithmically transformed observations is (covariances in the upper triangle, variances along the diagonal and correlations in the lower triangle of the following array):

$$\begin{bmatrix} 0.01382 & 0.01332 & 0.00694 \\ 0.97028 & 0.01363 & 0.00729 \\ 0.88826 & 0.93987 & 0.00442 \end{bmatrix}$$

Despite the fact that the observations have been logarithmically transformed, the variance of the third variable is appreciably smaller than those of the first two, which is not a very comforting situation, as it points to marked heteroscedasticity in the variances. The latent roots and vectors of the covariance matrix are displayed in Table 4.1.

Inspection of the elements of the first latent vector for *M. californicus* discloses that they are almost exactly 2/3, 2/3, 1/3. We can therefore write the first principal component as

$$Y_1 = 1/3[2 \log(L_2) + 2 \log(B_3) + \log(H_1)]$$

This equation can be expressed in conventional "allometric form" as

$$\exp(3 Y_1) = L_2^2 \cdot B_3^2 \cdot H_1.$$

These exponents are the desired constants of allometry according to the model used by Flury and Riedwyl. Hence, the relative growth in the directions of L_2 and B_3 is twice the relative growth in the direction of H_1.

In *M. ochrogaster*, the coefficients of the first component are almost, but not quite 2/3, 2/3, 1/3, as before and hence the same interpretation as used for *M. californicus* can be suggested. We note that the angle between the two first latent vectors is no more than 2°. The first component does not fit in with an hypothesis of isometric growth which would require all vectorial elements to be 0.577 (i.e. $1/3^{1/2}$). It is rather common to find the correlation matrix being used for growth interpretational studies (which is in opposition to Jolicoeur's model). This would obscure relationships in the present example as the first latent vector for the correlation matrix of logarithmically transformed variables for *M. californicus* is (0.576, 0.586, 0.570), which is very near to the vector of isometry.

As I have mentioned elsewhere, vectors 2 and 3 would be called **shape vectors** in many practical applications in that a case can be made for the second component representing a contrast between $\log H_1$ and $\log L_2$—this implies that voles with a largish value of H_1/L_2 have a high score on the second principal component. The insignificance of the second latent root suggests, however, that there is no second component at all.

We can formulate a similar interpretation for the third component, but it is wise to pause here and ask whether there would be any real meaning to an explanation of the third vector in a trivariate analysis. None, whatsoever, if the principles of parsimony are supposed to apply. We observe, moreover, that the third component is joined to only 3.8% of the trace of the covariance matrix.

The second species *M. ochrogaster* differs somewhat from *M. californicus* in respect of its second component. (Be aware, again, that this value is very small and there may well be no second component.) The angle between these vectors for the two species is 30°, thus indicating a strong difference in the orientation of the sample scatter ellipsoids. The angle between the third latent vectors is likewise 30°. There is, therefore, a clearly enunciated difference in the orientations of the second and third ellipsoidal axes, notwithstanding that the first principal axes are almost parallel. This could make any procedure based on pooling covariance matrices in some manner or other problematical.

4.2.1 *Hopkins' Factor-Analytical Model*

Somewhat distrustful of the principal component model for the generalization of the allometric equation, Hopkins (1966), elaborating on Wright's (1932) paper, put forward what can be regarded as a refinement of that method. He made explicit assumptions about what the covariance matrix would have to have in the nature of structure for a principal-component type of solution to be logically acceptable. A salient feature of his method is the necessity for reduced rank of the covariance matrix. Sprent (1972) has pointed out that there are inherent difficulties in Hopkins' factor-analytical model. The first of these concerns the formalization involving the use of a straight line in p dimensions for the logarithms of the observations as the concept of simple allometry. For an exact linear relationship to be valid, the covariance matrix of the logarithms must be of rank 1. With random biological variation about the line, this cannot apply. All we can discover from a factor analysis, or principal component analysis, when the covariance matrix is not of rank 1 (which would imply that a factor model was inappropriate), is whether the variation is mostly in a space of dimension less than p. This is not very useful for studies of size and shape if the pattern in that space is unknown.

Let the vector of measurements be x_i for p characters. The population covariance matrix is assumed to be of the form

$$\mathbf{S} = \mathbf{P} + \mathbf{D} \qquad (4.4)$$

where $\mathbf{S}$ contains the systematic covariance and $\mathbf{D}$ is random variance of individual x_i. For isometry to exist, $\mathbf{P}$ must be of unit rank.

The vectors of $\mathbf{P}$ are the vectors of $\mathbf{S}$ for random variance of $\log x_i$. The first latent vector of $\mathbf{S}$, the sample covariance matrix, provides an estimate of the underlying population growth or size pattern. In an alternative situation, if the random variance of each $\log x_i$ is proportional to systematic variance, the main diagonal elements of the estimated covariances, $\mathbf{S}$, are proportional to those of $\mathbf{P}$.

Hopkins' procedure proceeds by the principal component factor analysis of $\mathbf{S}$ for estimating the allometric coefficients, after screening any variates with size-dependent relative growth. If $\mathbf{P}$ is of rank 1, there is a unique pattern of relative growth or size. Although this model is neat, it has not turned out to have had much biological relevance as it seems to be applicable in few practical situations. One of the reasons for this concerns the ambiguity in concepts of what constitutes factor analysis in the Natural Sciences, a subject that was reviewed in detail by Jöreskog *et al.* (1976).

Another article worth mentioning here is that by Shea (1985) who provided a thoughtful account of multivariate allometry in growth, with empirical evidence in support of Jolicoeur's (1963) generalization; he used data on chimpanzees and gorillas to illustrate his ideas.

4.2.2 *The Size-Elimination Method*

Somers (1986) proposed an *ad hoc* method for extraction of isometric growth from a covariance matrix in order to produce a partition into pure shape variation. The idea depends on whether the principal-component definition of size and shape variability is more or less correct. Somers (1986) noted that size and shape are incorporated into the first principal component of logarithmically transformed variables, as was indeed obvious to Hopkins (1966). This fact is often less than fully appreciated in many analyses I have seen.

It is a moot point whether the method of size-constrained principal components is a fully justifiable procedure. Rohlf and Bookstein (1987) and Bookstein (1991) show that Somers' method is an inferior way of achieving one of Burnaby's (1966) aims in his method of growth-invariant discrimination (see Section 4.4.3).

4.2.3 *Shape Vectors and Size Variables*

Up to the present point, we have been considering rather *ad hoc* ways of quantifying size and shape. We have seen how various attempts at producing a multivariate analogue of the bivariate equation of allometry go part of the way towards finding a solution but that there is always something missing, and there is a fair degree of arbitrariness in what is being done. Mosimann (1970) has taken the question further towards a general solution with his definitions of *shape vectors* and *size variables*. He provided reasonable clarifications of what we mean when we speak of size allometry and formalized ideas implicit in earlier work. We can express the basic ideas in terms of measurements on, say, the carapace of an ostracod. Let these be length (L), height (H) and breadth (B). A measure of shape could be given as the simple ratio H/L, but also H/B is a measure of shape and so is L/B. Ostracod-workers recognize these ratios and know that they often mean something quite definite with respect to sexual dimorphic differences, for example. Biometricians are rightly cautious of the use of ratios for studying dimensions, but the fundamental work on size and shape of Penrose (1954) was expressed in the form of quotients.

A measure of total size would be L, but also H might do quite well (noting that length and height may be highly correlated in ostracods). An areal measure such as $LH^{1/2}$ would also be a measure of total size. All these measures possess a physical dimension—mm, mm². Mosimann generalized these concepts to any joint distribution of positive random variables. In the present connexion, we are interested in considering p dimensions to define shape vectors and size variables associated with a set of p distances between specified points. (This is an important feature and one we shall have to examine further in a later section.) If these distances are measured between corresponding points on two individuals we may denote the measurement vectors by x_1 and x_2. Mosimann defined *sameness* of shape of the individuals with respect to these vectors to imply that

$$x_1 = cx_2 \qquad (4.5)$$

for some scalar $c > 0$.

Consequently, equality of shape depends both on the p dimensions and the components of x_1 and x_2. An example will make this statement clear. If we have an ostracod carapace, the dimensions of which are length = 1 mm and height = 0.5 mm and another species, the length of which is 0.5 mm and the height of which is 0.25 mm, and this is all we know, then we conclude that the two shells have the same shape. If, however, a third dimension becomes available, say, breadth of the carapace and the first species has a breadth of 0.3 mm, whereas the breadth of the other is 0.25 mm, then it is no longer possible to say that the two species have the same shape.

Sprent (1972) pointed out that the above definition of shape is in terms of a p-variate vector with p finite, but when we think intuitively about the shape of organisms, we do so from the point of view of continuous surfaces in three dimensions. At the time of writing, Sprent bewailed the lack of serious attempts at formalizing the study of size and shape on a continuous basis. Fortunately, the situation has changed dramatically over the last few years and there are now several directions of research being pursued.

Mosimann (1970) defined **shape vectors** corresponding to the scalars H/B, H/L etc. of the bivariate case. Any dimensionless vector derived from $\mathbf{x}$ is called a shape vector by Mosimann if it has the form

$$\mathbf{z} = \mathbf{x}/g(\mathbf{x}) \tag{4.6}$$

where $g(\mathbf{x})$ is a *standard size variable*. A standard size variable is any real-valued function $g(\mathbf{x})$ that is positive for all positive $x_i, i = 1, 2, \ldots, p$, and such that

$$g(a\mathbf{x}) = ag(\mathbf{x}) \text{ for all positive } a, x_i.$$

Typical standard size variables are x_i, Σx_i, Πx_i, etc. Shape vectors include the vector of proportions $\mathbf{p} = (p_1, p_2, \ldots, p_p)$, where $p_i = x_i/(\Sigma x_i)$, and $\mathbf{x}/g$, and where

$$g = \Pi(x_i)^{1/p}.$$

If (4.6) holds for two individuals, it follows that their vectors of proportions are equal. If two individuals have the same shape, then every shape vector of the first is equal to the corresponding shape vector of the second. In relation to the results of the next section dealing with Geometric Morphometrics, the size variable in Mosimann's theorem is *centroid size*.

Mosimann also proved that if any shape vector is independent of a particular size variable, then, in particular, the vector of proportions is independent of that variable. Hence, isometry can be defined as independence between a shape vector and a given size variable. Mosimann's theorem tells us that given a random measurement function $\mathbf{x}$ and any non-degenerate random shape vector $\mathbf{z}_1$ and any standard size variable $g_a(\mathbf{x})$ such that $\mathbf{z}_1$ and $g_a(x)$ are independent, then any other shape vector $\mathbf{z}_2$ is also independent of $g_a(\mathbf{x})$ and no shape vector can be independent of any other size variable, say, $g_b(\mathbf{x})$ unless the ratio of standard size variables is a degenerate random variable. For many applications, this is a very useful result.

In summary, we can say that Mosimann derived an important axiomatic approach to the analysis of size and shape, and one that plays a fundamental rôle as well in geometric morphometrics. We have seen that organisms have the same shape in Mosimann's model if all of a set of measurements made on one individual when multiplied by a constant yield the measurements of a second organism. Once a convenient measure of size has been selected, then a shape vector is a unit-free vector of measurements in which each of the measurements is divided by the chosen arbitrary measure of size. An essential feature here is that it is impossible to define a single shape vector that is independent of more than one measure of size. Therefore, in order to "scan" for informative relationships, Mosimann and James (1979), studying geographical variation in blackbirds, defined several size measures for probing the geographical variation of shape in relation to these measures of size. This paper is an important reference for anybody wishing to make practical application of the procedure reviewed here. Another important reference for the application of Mosimann's theorem is that of Bookstein *et al.* (1985, pp. 30–32). The concepts involved are well illustrated with diagrams in that monograph.

Armed with the foregoing solution, we can now look back at the principal component representation of multivariate allometry. In terms of the present argument, it implies independence between any shape vector (the second latent vector of logarithmic variables, for example) and the standard size variable (the first principal component of logarithmic variables). It also becomes apparent that the designation of shape- and size-variables in a principal component solution is arbitrary, for if one component is said to be a size measure, all others are shape measures by default, and hence can tell us nothing about how shape covaries with size. This is clearly an undesirable property.

Isometry is rare in the bivariate case, as Huxley (1932) underscored. It is perforce even rarer in multivariate data. A simple illustration of the application of some of the ideas just expounded will now be given using work on Cretaceous ammonites (Reyment, 1983).

Example 4.2. Size and Shape Variation in Some Japanese Ammonites

I have chosen this illustration because it exposes some of the problems attaching to studying variation in ammonites, which, on account of the rigidity imposed by the logarithmic spiral, usually exhibit very high correlations between distance-variables. Mosimann (1970) provided a test of the independence of size and shape based on the square of the multiple correlation coefficient. The hypothesis tested is in effect a test of isometry—i.e. that shape is independent of size.

Reyment (1983) analysed morphometrical variation in four shell-dimensions of Japanese Late Turonian *Subprionocyclus* from the Pombets River, southern Hokkaido. The measures used were the usual ones: maximum shell-diameter (D), maximum shell height (H), shell breadth (B), and maximum umbilical diameter (U). The shape association for the four variables is given in Table 4.2.

In this table, for each vector of measurements y_i, the values

$$[y_{i1} - y_{i4}, \ldots, y_{i3} - y_{i4}, \sum_{i=1}^{4} a_j y_{ij}], \, i = 1, \ldots, N$$

have been calculated for shape and size and are listed (see Mosimann and James, 1979). Here, a is some scalar > 0. The multiple correlation coefficient

TABLE 4.2. *Shape association for four variables measured in three species of ammonites. Entries are the multiple* R^2 *of the logarithm of three-dimensional shape* $(y_H - y_B, y_D - y_U, y_B - y_U)$ *with the log size variable*

Species	N	Log of geometric mean of four variables
S. (*Subprionocyclus*) *normalis*	28	0.11
S. (*Subprionocyclus*) *neptuni*	23	0.21
	9	0.37
	39	0.25*
S. (*Reesidites*) *minimus*	35	0.65**

*Denotes significance at the 5%-level; **denotes significance at the 1%-level.

R of the shape vector, given above, and with log G, the size variable, is computed in the usual manner.

Both species of *Subprionocyclus* show poor association between three-dimensional shape and the size measure. *Reesidites* differs markedly in displaying high associations for the log geometric mean as well as with the umbilical diameter, which was also tried as a size variable. The value of $R^2 = 0.70**$ indicates an allometric relationship to occur.

4.2.4 *The Multivariate Linear Model of Veitch*

Veitch (1978) devised a multivariate linear model for describing allometry in size and shape in fiddler crabs, *Uca*. The essential point he makes in his introductory comments is that in the multivariate linear model, the shape variables can be construed as dependent variables to be described in terms of a size variable, or function of the size variable, the size-component being regarded as the independent variable. The general structural relationship to Mosimann's (1970) formulation is fairly obvious. Veitch develops his idea in terms of a multivariate statement of Huxley's allometric equation. The method does not seem to have caught on but seems to be worthy of mature consideration.

4.3 Size Divorced from Shape

In some rather specialized cases, the worker may wish to consider size free from shape. This is referred to as shape-invariant modelling. Stützle *et al.* (1980) provide an iterative method of producing ontogenetic series free from shape-confounding. At first sight, this might not seem to have much palaeontological interest but not so; the use of measurements on fossils for producing a biolog in quantitative stratigraphy would be greatly enhanced by the method developed by Stützle *et al.* (1980). Reyment *et al.* (1984) reviewed the main features of this method.

Sampson and Siegel (1984) have also considered size independent of shape. Their size measure is a weighted geometric mean and can be seen as a generalization of the results of Mosimann (1970). It is also akin to Gould's (1971) measure of size differences for constant shape.

4.4 Shape in Terms of Outline

Over the last few years, several methods have been applied to the study of shape which are based on quantifying variation in the outline of an organism. This is doubtless one way of looking at shape and, in many respects, it is a useful one. Outline-analysis has a clear region of application, namely, for ordination and it is not applicable to the extraction of shape features. In respect of ordination, methods based on landmarks offer no necessary gain, but they have the great advantage that they are amenable to interpretation.

There are several rival schools, unfortunately, because all of these have something to say and it is not useful to the student to be confronted with polemics instead of scientifically meaningful accounts. It is clear to me that there is no universal elixir as yet for the analysis of size and shape. Some methods may be better than others depending on what you have in mind for your particular study. I shall start with the method I did not use myself until recently, mainly owing to confusion in my own mind as to its real usefulness and relevance to my own research. This is the class of methods generally called **Fourier Analysis** in the geosciences and some biological work.

4.4.1 *"Fourier Analysis"*

Fourier Analysis has its roots in a theorem established by Baron Joseph B. J. Fourier (1768–1830), the celebrated French mathematician. This theorem tells us that every curve, no matter what its nature may be, or in what way it was first obtained, can be exactly reproduced by superposing a sufficient number of simple harmonic curves—hence, every curve can be built by piling up waves. This area of mathematics is known as *Harmonic Analysis* and all who have studied the theory of musical instruments will fathom the importance of Fourier Analysis in that connexion. Fourier's theorem also says that one only need use waves of certain specified lengths. If the original curve repeats itself regularly at intervals denoted by some length-measure (say at intervals of 1 dm), we have only to employ curves which repeat themselves regularly 1, 2, 3, etc. times every measure. If the original curve does not repeat regularly, we treat its whole length as the first half-period of a curve which does repeat regularly. This is the more usual form of Fourier's theorem which states that the original curve can be constructed from simple harmonic consituents such that the first has one complete half-wave within the range of the original curve, the second has two complete half-waves, the third has three, etc.

In its simplest representation, a Fourier series is of the form

$$\tfrac{1}{2}a_0 + (a_1 \cos x + b_1 \sin x) + \cdots \tag{4.7}$$

where a_0 denotes the sum from $-\pi$ to $+\pi$ of the sum of $f(x)$, divided by π, a_n is the sum of the same range for the function times $\cos nx$, and b_n is the same sum of the range for the function times $\sin nx$, both divided by π. The period of the Fourier Series is 2π. The morphometrical and geological studies called Fourier Analysis, are applications of an aspect of Fourier series. They are not really the mathematical field of Fourier Analysis, which has more ambitious goals.

We can start by referring to an overview by Rohlf and Archie (1984). These authors were concerned with studying similarities and differences in the outline of the wing-shape of mosquitoes. Now, in this special case, Rohlf and Archie did not think that the usual methods of multivariate morphometrics would be very useful as the differences between specimens are not great and it would require some technique capable of sifting subtle dissimilarities if a meaningful ordination could be hoped for. Hence, methods that capture the entire outline of the wing were desired. Fourier methods were deemed relevant to this aim.

The technique used for digitizing the outline is similar to that employed in the method of *eigenshapes* (see below). Rohlf and Archie (1984) gave a detailed account of how they obtained their values and the form taken by their Fourier analyses (Rohlf and Archie, 1984, p. 306). All methods are mathematically equivalent in that given enough harmonics, they can encode the shape of the wing exactly and so the choice of procedure is likely to be bound to such considerations as computing facilities.

Rohlf and Archie (1984) used the Zahn and Roskies (1972) algorithm for quantifying the outline of the mosquito wing at 100 arbitrary points. This algorithm uses cumulative changes in angle of a vector tangent to the outline of the object as a function of distance around the periphery of the wing. The wing was first scaled so that its perimeter is 2π. One computes then

$$\phi^*(t) = \phi(t) - t \tag{4.8}$$

for the 100 equally spaced values of t. The geometry for the parameters of equation (4.8) is illustrated in Fig. 4.1 for a mosquito wing. Thus, t is the distance along the periphery of the wing, from the starting point (ranging from 0 to 2π radians), $t(0)$ is the angle of a tangent vector at the starting point, and $\phi(t)$ is the angle of a tangent vector at a distance t from the starting point. Rohlf and Archie noted that a limitation of the method of analysis they apply to their mosquito wings is that it deals with overall shape variability only and not with changes in distances between "homologous" points.

Ferson *et al.* (1985) applied the same techniques to the study of variation in shape in *Mytilus edulis*. The main difficulties for automatic recognition by

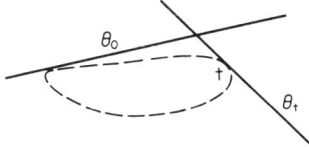

FIG. 4.1. Geometrical relationships of a mosquito wing to illustrate variables used in basic equations of Fourier analysis in shape studies (after Rohlf and Archie, 1984, p. 305).

computer are those of image-segmentation, that is, the isolation of the desired object from its background, and feature extraction, the definition and measurement of a set of characters from the image. The automated technique entails using a television camera and a digitizer to convey the desired quantitative information to a microcomputer. The results obtained should be of interest to quantitative palaeontologists. The Fourier Analysis, supported by a linear discriminant function, distinguished between electrophoretically distinct populations of *Mytilus*. An association between morphology and genotype could be identified, but it could not give a reliable assignation to the population from which each specimen was collected. The above-mentioned method of data-registration seems to be suitable for some work, but detailed studies require a high-resolution digitization tablet.

From the historical aspect, it seems as though a study by Kaesler and Waters (1972) might be the earliest application of Fourier descriptors to the study of morphological shapes in systematics. Rohlf and Ferson (1983) gave a short review of early applications of the method.

4.4.2 *Eigenshape Analysis*

The ordination technique known as Eigenshape Analysis, developed by Lohmann (1983), has begun to catch on with palaeontologists, particularly among people working in deep-sea marine micropalaeontology. It is, in its application and results, quite like a Fourier Analysis linked to some multivariate ordinating procedure, such as principal components. There is currently some uncertainty about when to use these methods and, besides, not a little barbed polemic discussion in the recent literature on the subject. Most applications concern foraminifers; Granlund (1986) applied eigenshape analysis to radiolarians.

An informative introduction to what is involved has been published by Rohlf (1986). The eigenshape technique is really a quite simple ordinating procedure which has as its starting point the x,y-coordinates of a set of p points along outlines of N objects of interest—microfossils of some kind. The essential steps are as follows:

1. Transform the x,y-coordinate designation of each outline into the representation of Zahn and Roskies (1972), to wit, $\phi^*(t)$, already described in

the foregoing section. This yields a normalized net angular change in direction of a tangent $\phi(t)$ to the outline as it is moved around in p steps of equal length, scaled to range from $t = 0$ to 2π. The equation required is (4.8) above. This representation is invariant to size, translation, and rotation of the fossil in the x,y-plane, but it is sensitive to location of the starting point. It also matters whether the outline is traced in a clockwise or counter-clockwise direction.

2. Rotate each outline so as to maximize its agreement to an outline in a standard orientation, unless a unique landmark exists that can be used to define a unique starting point. Lohmann (1983) rotated each outline so as to maximize its correlation, or cross-covariance, with a standard reference-outline. There was in that paper some concern with concepts of "homology", but this seems to have been reduced in emphasis in Lohmann's latest note (Lohmann and Schweitzer, 1990) in the light of justified criticism (Rohlf, 1986; Full and Ehrlich, 1986). These latter authors point out that assuming an homologous starting position on the periphery for all specimens, and measuring a set proportion of the total perimeter does not mean that the same point on each curve is located at the same feature. A standardization used by Lohmann (1983) and Lohmann and Schweitzer (1990) to bring the $\phi^*(t)$ vector to a mean of zero and variance of one is for the most part undesirable as it makes some different shapes indistinguishable from each other.

3. A singular value decomposition is made on matrix $\mathbf{Z}$ formed from the re-oriented and possibly standardized $\phi^*(t)$-vectors; this is a matrix with p rows and N columns.

$$\mathbf{Z} = \mathbf{US}^{1/2}\mathbf{V}^T \tag{4.9}$$

where columns of $\mathbf{V}$ are normalized latent vectors of the $N \times N$ matrix $\mathbf{Z}^T\mathbf{Z}$, $\mathbf{S}$ is a diagonal matrix of latent roots of that matrix and the columns of $\mathbf{U}$ are normalized latent vectors of the $p \times p$ matrix $\mathbf{ZZ}^T$. Columns of $\mathbf{U}$ are considered by Lohmann to be empirical *shape-functions* and the outline of each fossil can be described as a weighted linear combination of these functions. The i-th column of $\mathbf{VS}^{1/2}$ can be interpreted as projections, or weights, of N fossils on to the i-th empirical shape function.

There is an obvious connexion between eigenshapes and some versions of Fourier series that use the same matrix $\mathbf{Z}$ as above. A Fourier analysis can be expressed in the form

$$\mathbf{C} = \mathbf{F}^T\mathbf{Z}$$

where $\mathbf{F}$ is a $p \times p$ Fourier transform matrix (Rohlf, 1986, p. 847).

Often, $\mathbf{C}$ is used as input to principal component analysis, discriminant functions, etc. Given the singular value decomposition of $\mathbf{Z}$, as in eigenshape analysis, the singular value decomposition of $\mathbf{C}$ is:

$$\mathbf{C} - \mathbf{F}^T\mathbf{Z} = \mathbf{F}^T\mathbf{US}^{1/2}\mathbf{V}^T. \tag{4.10}$$

Rohlf (1986) points out that $\mathbf{S}^{1/2}\mathbf{VT}$ gives the transpose of principal components of N outlines with respect to empirical shape functions, identical to those obtained before, and $\mathbf{FTU}$ supplies the empirical shape function in terms of Fourier harmonics rather than an expression of the p individual steps around outlines. Rohlf then concluded that the results usually given in published accounts of eigenshape analysis are identical to those that would be obtained from a Fourier analysis of ϕ^* followed by a principal component extraction, provided all harmonics are retained.

Example 4.3. Illustration of Eigenshape Analysis

Shape-relationships in two species of the ostracod genus *Soudanella*.

The data used here were obtained by Reyment and Aranki (in prep.) for the Paleocene species *Soudanella laciniosa* APOSTOLESCU and *Soudanella ioruba* (REYMENT) from the Paleocene of Nigeria, West Africa. The investigation had as one of its aims to assess the degree of similarity in shape of the two species as part of a taxonomic revision of the genus.

The observation consist of the equally spaced pairs of coordinates located around the outline of 15 specimens of the two species. The starting point for the measuring in each case was at the eye tubercle, which is well developed in species of *Soudanella*. The latent roots of the matrix $\mathbf{Z}^T\mathbf{F}$ fall off very quickly: the first four of these are 13.49528, 0.93554, 0.11701, 0.09858.

Many combinations of plots were tried, but the most informative for taxonomic purposes was that showing the projection of the points on to the second and third axes. All individuals belonging to *Soudanella laciniosa* lie in a kernel-position in the diagram and that there is no differentiation to speak of with respect to origin of specimens. The more heavily costate specimens from Araromi occur interspersed with the less costate specimens from the Ilaro borehole. The material of *Soudanella ioruba* is distributed as a fringe around the kernel containing the second species. Also here, a tendency to differentiation with respect to origin is not manifested.

In summary, the analysis of shape variation in these species of *Soudanella* discloses subtle differences. Examination of the specimens supports the indications yielded by the ordination—the majority of the specimens of both species are similarly shaped.

4.4.3 *On Shape Invariance*

There is a fundamental problem in the analysis of shape, namely, that it is often desired to quantify differences in shape so that these differences will be quite free from all effects of size. Let us consider as a particular example bivalves from different localities and wish to compare then quantitatively for some taxonomical purpose. As soon as you try standard multivariate methods of analysis you find that differences in means can be very deceptive. Bivalves do not have a terminal size and so you cannot preselect adults, as can be done with ostracods, for example, for the multivariate mean of any sample will depend on the fortuitous composition of the sample—one in which half-grown specimens dominate will give a different mean vector from one in which mature specimens dominate, even if the two samples are biologically identical in all other respects. T. P. Burnaby met this problem in his work on Carboniferous lamellibranchs and Cretaceous foraminifers. In the study of ostracods we can, however, run into a similar difficulty as some marine species produce several generations over a season, all of which differ slightly from each other in some minor morphological manner (Abe, 1983; Ikeya and Ueda, 1988).

Now the concepts of size and shape are very differently defined in statistical applications, as you no doubt will have discerned by now. We have yet to meet the geometrical version in its full bloom, but you will have seen enough so far to know that there is not only a good deal of diversity in the way competent workers view the problem but also a great deal of intransigence in the attitudes of many to the thoughts of others. Do you remember the tale of the blind Indian sages and the elephant? All were correct in what they described on touch, but all were wrong in their conclusions about the true appearance of the animal.

The growth-invariant approach is centred around the assumption that size and shape correspond in general factors, linear combinations of variables (cf. Bookstein *et al.*, 1985) rather than single, directly measurable variables. We have already encountered one way of defining a size factor, to wit, the first principal component of a covariance matrix of logarithmically transformed variables. The growth-size-invariant method of analysis is not concerned with estimating size but only with correcting for the effects of size. We have already referred to one method, notably, that of Somers (1986), an *ad hoc* procedure for achieving this result. The method taken up in this section is a multivariate adjustment: there are other ways of extracting size, for example, by regression.

The problem of biologically homogeneous but statistically heterogeneous samples is a bothersome one. The palaeontologist has an even more troublesome situation to master in almost all of the samples he has to study. We have not only mixtures of growth stages, but also mixtures of samples, mixtures that not even the most delicately sampled material can avoid. It should be fairly obvious that standard methods of multivariate analysis cannot really give a valid solution to such data. This is an important and difficult point to which we shall have to return in a later chapter. Thus, as pointed out by Humphries *et al.* (1981), the ordinary discriminant function, canonical variates, etc. do not adequately address this problem nor were they designed to do so. Proper correction for size effects is important since patterns of biologically significant covariation can be masked if this point is not taken into account. Conventional multivariate statistical analysis is not the ideal way of probing shape changes, as I have already pointed out on several occasions, since it does not take sufficient account of *geometrical relationships* in the organism. Any of the multivariate methods reviewed in Chapter 3 takes correlations into account but there is no way for the analysis to express geometrical patterns of development of the organism. Moreover, even if we are successful in choosing "good" measures of shape (by a truss or trellis of variables—Bookstein *et al.*, 1985) a standard multivariate analysis of these variables only makes a new set of variables, but it does not specifically extract information on shape and it cannot depict point-to-point geometrical relationships over the organism. One might say that this is an inadequate use of multivariate analysis—like sending a boy on a man's errand.

Humphries *et al.* (1981) proposed a method they termed "shearing" for achieving size invariance. This has subsequently been put in the closet (Bookstein, 1991), but it is worth examining, since it contains useful thoughts. This employs the first latent vector of a pooled covariance matrix as an index of general size, and therefore is an adaptation of the principal component method of Jolicoeur (1963). We have already observed, I mention *en passant*, that this vector need not be a pure size-indicator at all: all it does is point out the direction of greatest variation within the populations. Rohlf and Bookstein (1987) revised the shearing idea, noting it to be a less adequate approach if *ordination* is the aim of the study. The underlying aim of the method of shearing is factor analytical and its coefficients can be interpreted. The method is best described by Crespi and Bookstein (1989) who clearly expose the connexion between shearing and path analysis (cf. Bookstein *et al.*, 1985, p. 81).

The method of Burnaby can be described as follows (Gower, 1976; Rohlf and Bookstein, 1987): It is desired to find canonical variates among g populations confounded by growth and size effects. With k size differences and p variables, the effects to be eliminated may be represented by a $p \times k$ matrix $\mathbf{K}$, the r-th column of which consists of elements proportionate to the direction cosines of the r-th component. The idempotent symmetric matrix

$$\mathbf{Q} = \mathbf{I} - \mathbf{K}(\mathbf{K}^T\mathbf{K})^{-1}\mathbf{K}^T$$

projects every sample value on to a space orthogonal to $\mathbf{K}$. Here, the sample values are free from growth and size effects reflected in $\mathbf{K}$. For g populations, all with the same $\mathbf{K}$, canonical variates $\mathbf{l}$ can be obtained by solving the following equation:

$$\mathbf{Q}(\mathbf{G}^T\mathbf{G} - \lambda\mathbf{W})\mathbf{Q}\mathbf{l} = 0 \qquad (4.11)$$

where $\mathbf{G}$ is the $v \times p$ matrix of sample means and $\mathbf{W}$ is the pooled within-populations matrix of sums of squares and cross-products. If you refer back to Chapter 3, you will observe that $\mathbf{G}^T\mathbf{G}$ replaces the between-populations matrix of sums of squares and cross-products of classical canonical variate analysis. The solution of equation (4.11) requires the computation of a generalized inverse

$$C = Q(QWQ)^- Q.$$

The required solution is then

$$(CG^T G - \lambda I)l = 0. \qquad (4.12)$$

The squares of the generalized distances between the means of groups i and j when projected on to the Q-space are given by:

$$D^2 = (g_i - g_j)C(g_i - g_j)^T. \qquad (4.13)$$

If we write the idempotent matrix Q as $Q = I - M$, where

$$M = K(K^T K)^{-1} K^T$$

it is easier to appreciate that a p-variate vector y is projected by M on to the space spanned by K and Q projects it on to a space orthogonal to K. Thus vector y can be resolved into components My, confounded with growth and size effects and components Qy which are free from these factors. The Mahalanobis generalized distance can then be partitioned into two parts, D_M^2 and D_Q^2. A casual reading of Burnaby's (1966) paper may give the impression that the generalized distance has been resolved into orthogonal components, but this is not so. The additivity merely reflects the definition of D_M^2, which is interpreted as the distance lost through computing in the Q space (Gower, 1976).

Burnaby was quite insistent about the way in which the size vector was to be estimated. This is to be seen in his published papers and in his correspondence, kept in the Department of Historical Geology and Palaeontology, Uppsala University. Burnaby rightly considered that the size vector should be estimated externally. This places the first latent vector of a covariance matrix in a weak position—particularly if obtained directly from the input. We shall consider a simple exemplification of Burnaby's method, using the computational procedure of Gower (1976).

Example 4.4. Growth-Invariant Study of a Cretaceous Foraminifer

We shall consider an example drawn from the field of micropalaeontology. Five samples of the Late Maastrichtian species of foraminifers *Afrobolivina afra* (REYMENT) are available from a borehole in Western Nigeria (Gbekebo I). There is a total of 104 specimens on which the following characters have been measured: (1) total length of the test, (2) breadth across the test; (3) height of the last chamber, (4) width of the last chamber. The respective sample sizes are for the levels in order from bottom to top: A (21), B (35), C (8), D (31), E (9).

The Problem: There is much size variation in the samples. It is desired to extract this variability in order to produce size-invariant renditions of shape. The method chosen here is that of Burnaby (1966), modified by Gower (1976).

The sums of squares and cross products matrix (lower triangle) and corresponding correlation matrix (upper triangle) are:

$$
\begin{array}{c} \\ 1 \\ 2 \\ 3 \\ 4 \end{array}
\begin{array}{cccc}
1 & 2 & 3 & 4 \\
\left[\begin{array}{cccc}
1.0613 & 0.1650 & 0.5440 & 0.5212 \\
0.1910 & 0.1694 & 0.1303 & 0.1265 \\
0.6487 & 0.1694 & 0.0182 & -0.0680 \\
0.7247 & 0.1766 & -0.1063 & 1.8216
\end{array}\right]
\end{array}
$$

The measure of size to be extracted by Burnaby's method is the vector:

$$\mathbf{F}_1 = (0.57, 0.25, 0.31, 0.72)^T.$$

The squared generalized statistical distances for the untreated data (lower triangle) and the size-adjusted data (upper triangle) are as indicated below:

Mahalanobis Generalized Distances

samples	1	2	3	4	5
1	0	2.57	2.25	7.47	5.21
2	13.43	0	0.20	1.43	1.65
3	4.57	0.24	0	1.55	0.92
4	42.46	8.28	7.34	0	1.16
5	33.23	5.64	4.10	1.55	0

The extraction of the size vector has led to very substantial reductions in the generalized distances, thus disclosing that the amount of variability in size is considerable. The situation encountered by Burnaby (1966) among bivalves can be seen to occur in the present species of foraminifers to a very high degree. You will also see that the degree of reduction in the value of D^2 is unevenly distributed. For example, the contraction for the generalized distance between samples 1 and 2 is almost sixfold, whereas that between samples 2 and 3 is insignificant.

We proceed now to the shape-oriented canonical variate coefficients, or, if you like, multivariate discriminant functions to keep to the spirit of Burnaby's original formulation of the problem. The growth-invariant loadings and untreated canonical variate elements are as follows:

		Vector 1		Vector 2
variable	Total	Size-reduced	Total	Size-reduced
1	4.61	1.43	4.22	3.10
2	−6.93	4.20	5.53	6.55
3	−5.42	4.13	−9.02	−8.04
4	1.12	3.76	−1.35	−1.82

It is quite obvious that the extraction of the "size vector" leads to marked changes in the canonical variate coefficients. This is most outspokenly manifested for the first coefficient in the first canonical variate. The example provided here gives a fair idea of what growth-invariant estimation can do. There are, unfortunately, very few examples published and it is to be hoped that you will be inspired to experiment with this useful procedure.

Up to this point we have been concerned with the algebraical way of approaching the study of shape, without denying that there is an obvious analytic-geometrical component in these methods. We now pass to geometrical solutions. These differ not only in the attempt they make at actually trying to capture shape variability on a point-to-point basis, but also in that they can not be readily connected to many statistical testing procedures, although this can be expected to come as the methods attract more attention. One of the things that has bothered me, without really being able to put my finger on the weak point, concerns the statistical properties of the quantities used in geometrics. However, much recent research has served to allay these fears. Kendall (1989) and Mardia (1989) have been instrumental in probing the statistical properties of the theory of variation in shape.

4.5 Geometric Morphometrics

The topics reviewed in the first part of this chapter can be conveniently grouped under the heading of *Multivariate Morphometrics*, a term introduced by Blackith and Reyment (1971) to embrace the application of multivariate statistical analysis in biological variability in morphological characters, and not only shape variation. This field of research has caught on, as it were, and there are now many excellent practitioners in the field and several textbooks.

Of recent years, a somewhat different way of analysing variability in size and shape has emerged. This orientation directs interest towards elucidating how geometrically expressed patterns vary, with the goals of D'Arcy Wentworth Thompson as their shining light. The geometrical school exhibits a certain tendency to downgrade the value of what they call "traditional multivariate morphometrics" (Bookstein *et al.*, 1985), which is often said, and not without justification, to give misleading, at best, incomplete results. Standard methods of multivariate statistics are accepted solely as a means of compressing the information extracted "from a delicate analysis of actual curved form" (Bookstein, 1978, p. 63).

Thompson (1917) introduced the idea of using Cartesian coordinates in describing shape relationships between organisms. We have already encountered the concept of growth gradients (p. 101) earlier in this chapter in our review of Huxley's ideas. A multidimensional analysis would be of interest if it were to chart the potential for growth throughout the whole organism. Despite its almost 60 years, the monograph by Huxley (1932) is still one of the best general references to the topic and to the thinking that went into the application by Thompson of the principle of Cartesian coordinates to the elucidation and description of growth and form in animals.

Huxley's (1932) treatment of the oft-cited transformation between the fish *Diodon* and its close relative *Orthagoriscus* (to use Huxley's generic assignation) has become an obligatory example in most publications on growth and shape. Huxley (1932) and Bookstein (1978) pointed out the effective use made by Thompson (1917) of diagrammatic sequencing of transforms for displaying the regular succession of deformation in the evolution of the horse. Huxley gave considerable thought to the scientific potential of the Thompsonian grid. He concluded, albeit reluctantly, that the method held little promise for detailed analysis because it cannot yield information on the change of relative proportions with absolute size. He thought it was *static* instead of *dynamic*, and "substitutes the short cut of a geometrical solution for the more complex realities actually underlying biological transformation" (Huxley, 1932, p. 106). This is, of course, not what the geometric morphometrics school would say, nor most other people active in morphometrics today; Bookstein *et al.* (1985, p. 194) claim to have met this objection by means of subtle and suitable models. Let us see how well this endeavour has succeeded.

4.5.1 *Basics of Geometric Concepts*

I lean heavily on Bookstein's publications for this section, and particularly his treatise of 1991. No matter how much experience you have in multivariate statistical applications, you will most likely find geometric morphometrics heavy going the first time you encounter it, and even the second time, and, perhaps, the third time. A good deal of the reason for this lies with the unfamiliarity of these new ideas and the mathematical methods used, and this includes professional statisticians. Do not feel too inadequate if you have trouble with mastering geometric morphometrics. Read what an experienced mathematician, with a solid background in Physics, has experienced with trying to digest the philosophy of shape-variability (Watson, 1989).

Central concepts are *outline* and *landmarks*. We have already met outlines in conjunction with eigenshapes and Fourier methods. The two are however quite different in the way they must be studied and the goals to which they lead.

4.5.2 *Outlines and Landmarks*

Shape can be conveniently, though inadequately, defined by reference to the outline of the object, which in two-dimensional space is a closed curve and in three-dimensional space a surface. Now many shelled organisms are not featurelessly smooth all over—not everything looks like the planktic foraminiferal genus *Orbulina*, I find joy in saying. Consider a cytheracean ostracod. It has ribs, reticulations, bulges and contortions that disturb the smoothness of its outline. These will form corners, vertices and intersections around the outline which correspond to what we could say are special points associated with the biological form in a meaningful manner—we make use of these features in our day-to-day taxonomical work. Bookstein (1978) borrowed a term from craniometrics to label such features; he called them **landmarks**. They are presumed to be "homologous", that is fully comparable in all their histological and topological characteristics from specimen to specimen. Landmarks do not define the form of any edge or surface; they merely provide fixed points of reference on it. The keyword here is **point**.

Bookstein (1978, p. 8) proposed a formal definition of **shape** in E^n (space) as an equivalence class, under the uniform group, of outlines in E^n. In informal terms, a shape is an "outline-with-landmarks" from which all information about position, scale, and orientation has been drained. A change in shape is a map of one shape on to another which sends arcs (or patches of surface) smoothly on to arcs and corners (or edges) on to corners; landmarks on to landmarks. This definition differs starkly from the prevailing concept in biometry with its preoccupation with a notion of shape variables conceived as ratios of sums of distance measurements among landmarks (cf. Mosimann, (1970) and Sprent (1972) for an account of the concepts). The cutting edge of Geometric Morphometrics is that it permits the description of variability as deformation, and variability of deformation in a common geometric context.

The method of *finite element scaling* has been employed in an attempt to achieve some of the aims outlined in Bookstein (1986). An example of this work is the study of macaque skulls by Cheverud and Richtsmeier (1986). Cheverud and Richtsmeier use 2–3-dimensional point-coordinates to describe individual forms, relating their goals to those of Thompson (1917) and to Bookstein (1978). Landmarks are employed in the manner which we have recently discussed and the method of analysis relies on a strain-model employing the "form-tensor" of continuum mechanics.

Over recent years, several attempts have been made at producing a method of constructing Thompsonian grids. An interesting example is that proposed by Tobler (1978). Mardia (1989) has made a useful addition to the scope of the analysis of shape and formalized the concept of what he terms "Bookstein's shape variables".

4.5.3 *Procrustean Superposition*

Siegel and Benson (1982) proposed a method for comparing configurations of landmarks on specimens of phylogenetically interesting genera of ostracods. This is one of the group of procedures known as "Procrustes Superposition" (see Bookstein, 1991, Sect. 7.1; Rohlf and Slice, 1990). The underlying method consists of a simple fit that expresses the differences between two organisms; it only takes account of global parameters such as rotation, translation and scale (see Section 2.9, affine transformations). Hence the procedure superimposes the landmarks of one organism on top of the corresponding landmarks of the other in a manner so as to minimize some measure of net discrepancy between homologues. This measure can, for example, be the summed squared distances, or some "resistant" measure. Differences in shape are expressed by disagreements in position of corresponding landmarks. The residuals thus arising can be used for studying contrasts in shape.

Bookstein (1991) points out a weakness in the Procrustean superposition, namely, that it becomes misleading in the presence of any features of shape of higher order. Rohlf and Slice provide a program Generalized Resistant Fit (GRF) as part of the package of diskettes supplementing the proceedings of the Ann Arbor morphometrics workshop (Rohlf and Bookstein Eds, 1990). I shall say no more of Procrustean superposition, since the procedures of the next section have made the subject largely obsolete.

4.5.4 *Principal Axes of Deformation*

The two directions of maximum and minimum change in a deformation are termed the principal axes of the deformation, as was noted in Chapter 2. It was also stated that the rates of change of length along these axes are termed the *principal strains* at the point. We can denote the larger of these axes as d_1; it is referred to as the major principal strain. The smaller axes, called the minor principal strain, we denote d_2.

The *size component* of the deformation is quantified as the sum of the logarithms of the strains, to wit: $\log d_1 + \log d_2$. The *shape component* of the deformation is measured by the difference of the logarithms of the principal strains, $\log d_1 - \log d_2$.

There are a number of concepts that occur in shape-studies that you will not run into in standard texts on multivariate statistical analysis. Some of them also appear in the publications by Bookstein (1978, 1986). The basic theory underlying the development of the tensor calculus resulted in answer to the need for expressing physical laws independent of any particular coordinate systems used in describing them. Such subjects as relativity theory, differential geometry, mechanics, elasticity, hydrodynamics, electromagnetic theory and, now, morphometry make great use of tensor analysis. An instructive reference for applications is Maxwell (1958).

4.5.5 *Examples of Geometric Morphometrics Applied to Palaeobiological Problems*

4.5.5.1 *Introduction*

In the foregoing section I have reviewed the analysis of shape in general terms in the light of geometric morphometrics (Bookstein *et al.*, 1985; Bookstein, 1991). I shall now consider some examples drawn from recent studies with which I have been associated, and all of which have been carried out with Fred Bookstein. These are all drawn from the realm of micropalaeontology, not only because of my personal interests, but more so because microfossils can often be collected in a manner such as to inspire a certain level of confidence in the statistical validity of the analyses carried out upon them. Bookstein's work was induced originally by clinical studies on deformations of the human skull, but the procedures are quite general. There are several techniques that can be applied to the analysis of landmarks and their surrogates, all of which are entirely or largely due to Bookstein.

1. The analysis of shape triangles in relation to the coordinates of a baseline and the coordinates of landmarks.
2. The use of biorthogonal grids to depict a single geometrical scheme of principal strains that are graded as they pass through the form.
3. A refinement of (2), the method of thin-plate splines to represent differences in shape in terms of principal warps.

I shall begin by reviewing variation through time in the Miocene bolivinid foraminifer *Brizalina mandoroveensis* from a sequence in Cameroun, West Africa. The second example treats species of Santonian (Cretaceous) ostracods from Israel. The third essay, likewise devoted to ostracods, is an account of geographical variation in the ornament of living species of Australian ostracods. Further examples could have been included, but these three each contribute some special feature to the presentation.

Example 4.5. Microevolution in Miocene *Brizalina*

The summary account given here was extracted from Bookstein and Reyment (1989) to which article I refer you for a more detailed account of the analysis.

Presentation of the Problem: Brun *et al.* (1982) studied several species of bolivinids from boreholes in West Africa in the light of hypotheses concerning evolutionary changes in them. Careful examination of the collections forming the basis of that work at the laboratories of ELF-SEREPCA, Boussens, France in 1983 led me to select *Brizalina mandoroveensis* as being the one most likely to be of interest for quantitative work. A preliminary scrutiny of the published material and the specimens themselves suggested that this bolivinid could yield significant information on time-correlated variation in shape.

Samples from five levels in a deep-test drilled at Ikang were chosen for analysis by the method of shape coordinates. The salient features of this method (Bookstein, 1986) may be presented succinctly in the following words.

The method of shape coordinates

For variations in shape that are not too great, both the analysis and interpretation of changes in the landmarks proceed effectively when the landmarks are considered three at a time in a set of triangles distributed over the form to be studied. For the complete coverage of N landmarks, there must be at least N-2 triangles in a **rigid configuration** (I emphasize this point, as failure to appreciate it lies at the root of most misunderstanding of the method). When this requirement is met, it does not matter which triangles are studied, as the multivariate statistical analysis of any such set is the same to terms of the first order in the variation of shape (Bookstein, 1986). Any mean differences, trends, or statistical components of shape to be found in the landmarks may be reviewed as deformations with the help of the biorthogonal diagrams of Bookstein (1978). This relates point (1) of the introduction of this section to point (2).

The multivariate methods required for analysing the shape of a single triangle for a sample are adaptations of the usual procedures, for example the Hotelling T^2, multiple regression, canonical correlation analysis on exogenous variables, etc. The T^2-statistic may be invoked for testing shape differences between two populations of triangles. The morphometric analysis of the shape of a triangle ABC formed from landmarks A, B, and C, is equivalent to the ordinary normal model analysis of the single complex variable

$$(\varepsilon, v) = (C - A)/(B - A) \tag{4.14}$$

which is the same as the ordinary normal model multivariate analysis of the pair of Cartesian coordinates assigned to landmark C in a Cartesian system for which landmark A is always located at $(0,0)$ and landmark B is always located at $(1,0)$. This construction results in the shape coordinates of the triangle (Bookstein *et al.*, 1985, pp. 230–232 provide practical details for doing the geometrical construction). AB is called the **baseline** of the construction. Usually, for convenience, the two furthest points are made to form the baseline; for example, the line running between the endpoints of the longest diameter across the anterior and posterior margins of the carapace of the ostracod. However, other baselines can be selected and it is often a good idea to expend a little time on testing which baseline gives the neatest result.

Permutation of landmarks A, B and C results to first order of mere rotation and scaling of empirical scatters of the coordinates of shape so that the ordinary multivariate analysis of them is not altered. Increase is measured as the specific fractional change in percent of the starting form. The units of such change are mm/mm. Bookstein *et al.* (1985, Section 2.1) show that when rates of change are measured in this way for distances taken across a triangle of landmarks, there is a single well defined direction of maximum rate of change and another minimum of change and these directions lie at right-angles to each other. The multivariate analysis of the set of all the shape coordinates is almost independent of the choice of baseline in the construction and is equivalent to the multivariate analysis of all possible ratios of size variables. In this respect, it incorporates the multivariate study of allometry; this can be easily achieved by the multiple regression of shape on size. Size for the triangles can be expressed in terms of baseline length alone, but for studying allometry, it is usually better to use the root-mean-square of the edge-lengths of the triangle. The figures accompanying this example use the new coordinates (ε, v) as defined in equation (4.14). More specifically, for digitized points A at (x_A, y_B), B at (x_B, y_B) and C at (x_C, y_C):

$$\varepsilon = \frac{(x_B - x_A)(x_C - x_A) + (y_B - y_A)(y_C - y_A)}{(x_B - x_A)^2 + (y_B - y_A)^2}$$

and

$$v = \frac{(x_B - x_A)(y_C - y_A) - (y_B - y_A)(x_C - x_A)}{(x_B - x_A)^2 + (y_B - y_A)^2}$$

When size information is needed, this can be introduced by means of the centroid-size, the root-mean-square of the landmarks from their centroid.

Measurements: Many species of foraminifers tend to be irregular in shape and it is consequently difficult to define suitable landmarks. Bookstein and Reyment (1989) selected the following six characters as landmarks and pseudo-landmarks, noting that they are not of equal status with respect to accuracy of measurement. (A) aperture; (B) mid-rounding of the proloculus; (C) proximal boundary of the last chamber; (D) bulge of the last chamber; (E) bulge of the second last chamber; (F) proximal boundary of the second last chamber. The locations of these landmarks are illustrated schematically in Fig. 4.2.

The Analysis: For the purpose of shape analysis one may proceed by considering the landmarks, three at a time, if the variations are not too great (see foregoing section). In the present study, there are six triangles, to wit: ABC, ABD, ABE, ABF, ACD, and AEF. Note for complete coverage we require N-2 triangles in a rigid configuration.

An important point to be noted here is that if "pure shape" is our primary interest, size effects must be removed from the data. This is done by standardizing all the triangles to the same baseline length. Should you wish to return size to the analysis, this is done by restoring the length of the baseline as an additional variable in the multivariate procedures. When the shape coordinates are multiplied by the length of the baseline, they become the two Cartesian coordinates of a vector of displacement: the separation of landmark

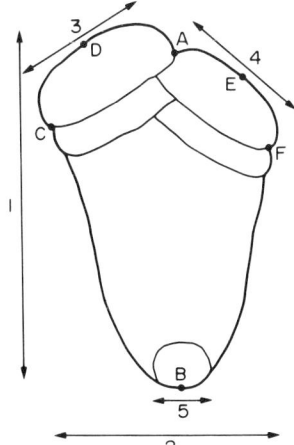

Fig. 4.2. Schematic representation of the landmarks, pseudo-landmarks and distance measures made on *Brizalina*. Variables 1–5 are standard distance measures, *A*, *C*, and *F* are true landmarks and *B*, *D* and *E* are pseudo-landmarks, or helping points.

C from landmark *A* in a coordinate system for which the baseline vector *AB* is horizontal. You would want to do this if you were interested in probing *allometric relationships*. If the root-mean-squares of the edge-lengths of the triangle are used, this measure of size is uncorrelated with shape on a convenient null hypothesis and therefore permits a single *F*-test for the existence of allometry whenever the assumptions of that null model can be shown to be true. The null hypothesis postulates that shapes could have been derived by independent identically distributed circular normal variation about fixed mean locations. Thus, the analysis of allometry becomes the correlation of shape measured by the position of the moving point, the vertex of the triangle, with the size variable.

Findings: The shape analysis is summarized in the pattern of mean shapes depicted level-by-level in Fig. 4.3. This figure shows the apparent motions of landmarks *C*, *D*, *E* and *F* in a coordinate system in which the aperture, *A*, is permanently fixed at locations (0,0) and the proloculus, landmark *B*, is permanently fixed at location (1,0). For all triangles, the scatter of shape for level 3 overlaps scatters for all the other strata so its relationships to them could not be clarified using these six points. The mean position of each landmark in this registered coordinate system is drawn for the four remaining levels as a line of three vectors in the order 5 4 2 1.

Inspection of Fig. 4.3 shows that of the four landmarks considered to be "moving", the mean position for level 4 is close to that for level 1 for all but landmark *F*. Notice too that the change from level 5 to level 4 is not reliably associated with the path from level 4 to level 2 and back. Two landmarks,

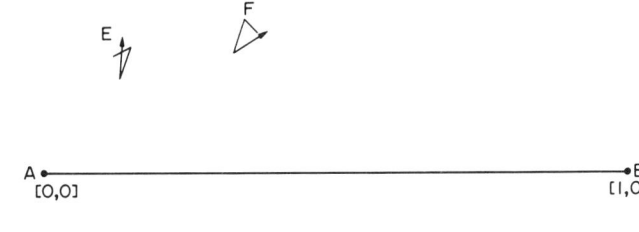

FIG. 4.3. Pattern of mean shapes presented level by level for *Brizalina*. These are trajectories moving, stratigraphically in the direction indicated by the arrow. The unit baseline is *AB*. Redrawn from Bookstein and Reyment (1989).

D and *E*, go off at a new angle to the old trajectory. Point *C* recapitulates part of the previous change. Only point *F*, traversing a triangle, returns to its relative position from level 5.

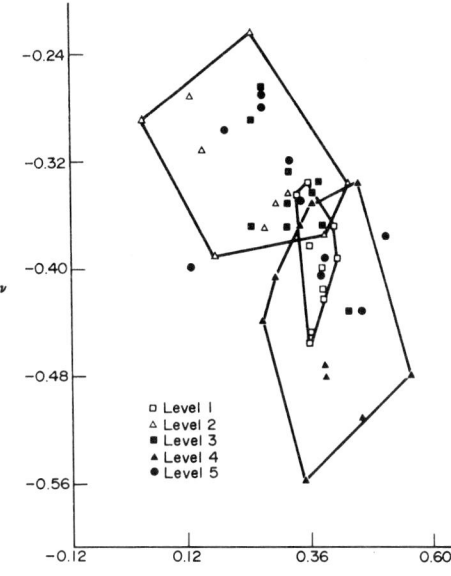

FIG. 4.4. Locations of all 50 specimens of *Brizalina* in the space of shapes of triangle *ABC*. The lines joining outermost points are convex-hulls. Redrawn from Bookstein and Reyment (1989). For an introduction to the concept of convex hulls, see Seber (1984, p. 164).

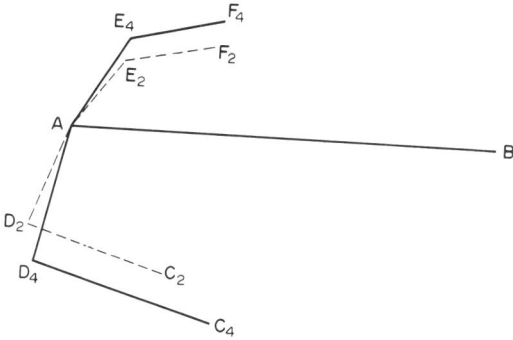

FIG. 4.5. The apparent displacement of bulge points (pseudo-landmarks) *D* and *E* to the usual baseline. Redrawn from Bookstein and Reyment (1989).

These patterns of change are shown in more detail in Fig. 4.4 in which the locations of all 50 specimens in the space of shapes of triangle *ABC* are indicated as well as the outlines of the scatters of subsamples 1, 2 and 4 within that scatter. The scatter, to be precise, represents the positions of the proximal limit of the final chamber in a coordinate system referred to the standard proloculus—aperture baseline. The specimens in the samples from levels 2 and 4 vary in opposite directions about a "core" representing the much lesser variation in the sample from level 1.

It is instructive to consider relations between the shapes of chambers at various sampling levels. Figure 4.5 gives the apparent displacements of bulge points *D* and *E* to the usual baseline. It also indicates the apparent displacements of points *C* and *F*. Figure 4.6, computed separately, shows the connexion between the mean shapes of the triangle *ACD* for levels 2 and 4. The construction of this figure was done by enlarging and rotating the triangle $AD2C2$ so that the edge $AC2$ aligns with the edge $AC4$. The shape change from level 4 to level 2 is then shown by the straight arrow as the displacement from $D4$ to $D2'$, almost directly away from the edge *AC*. You will perceive that the contrast is primarily one of relative height of the chamber, the ratio of this

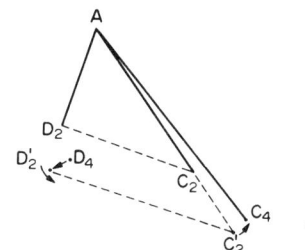

FIG. 4.6. Mean shapes of triangle *ACD* for levels 2 and 4 for *Brizalina*. The arrow indicate the directions of displacement. Redrawn from Bookstein and Reyment (1989).

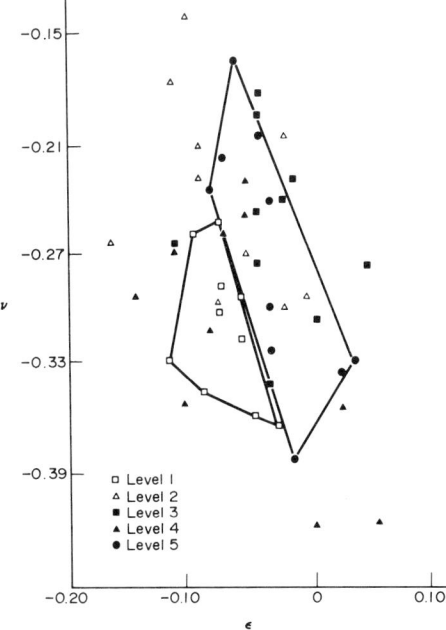

FIG. 4.7. Separation achieved by triangle *ACD* for levels 1 and 5 for *Brizalina*. The plot contains the points for all levels, but only convex hulls for levels 1 and 5 have been drawn. There is complete separation between levels; it is also apparent that the data of level 1 are more tightly distributed than those of level 5. Redrawn from Bookstein and Reyment (1989). Note, that a *convex hull* is drawn so as to unite outermost points; it is not permitted to join two points that lie inside the hull, nor to have concavities.

height to measure 3 of the standard set. Bookstein and Reyment (1989) called this the **Aspect Ratio**. It is greater in the sample from level 2 than in that from level 4. For the general concept of the aspect ratio, see Eisley (1987, p. 282).

We can also consider the separation between samples 1 and 5. The change of relative position of the "bulge" of the final chamber brings about an almost perfect separation between levels in the shape of triangle *ABD*, as shown by Fig. 4.7. In addition to the mean differences in shape between levels 5 and 1, there is also considerable difference in average size. In Fig. 4.8 this factor is exploited for reverting from the shape-coordinates to correctly scaled vectors of relative displacement. There is no resulting separation. The size differences are expressed by the nesting of sample 1 (the youngest one) within the scatter of sample 5 (the oldest one). Like the shape coordinates underpinning it, the mean differences of this vector, which is in the direction of "height" of the final chamber between levels, lies perpendicular to the average direction of uncertainty of location of point *D*.

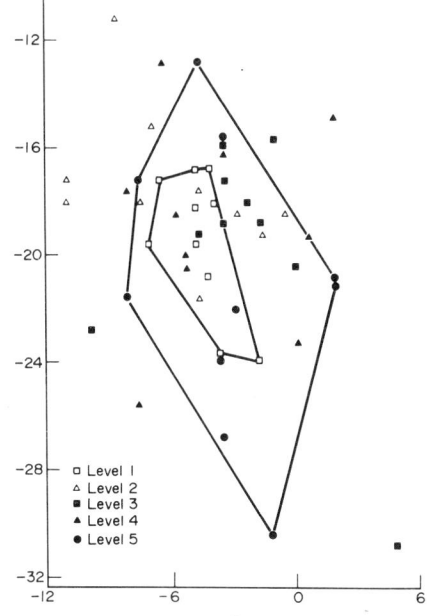

FIG. 4.8. Separation achieved by triangle *ABD* for levels 1 and 5 for *Brizalina*. The convex hull for level 1 lies entirely within the field traced out by the convex hull for level 5. The marked difference in dispersions for the two sets of data is obvious. Redrawn from Bookstein and Reyment (1989).

4.5.6 *The Biorthogonal Grid*

The method of biorthogonal grids applied to *Brizalina* yields interesting results of palaeobiological relevance. However, before reporting on the results of my analysis, I shall briefly outline the concepts underlying the grids. Note that this technique has been largely surpassed by that of principal warps (Bookstein, 1991).

4.5.7 *Note on Tensors in Morphometry*

Bookstein's (1978) biorthogonal grids represent growth as two **strains**. Growth is portrayed as a symmetric tensor field over one whole image which can be compared with others on the same image (cf. section on transformtions in Chapter 2). Dilatation (strain) patterns (cf. Boresi and Chong, 1987, p. 123) can be studied in some organisms without too much trouble in cases where homologous surface structures, such as various kinds of ornamental features, can be identified in phylogenetically homogeneous groups. The kind of data amassed by Benson (1972) for ostracods are an example of a typical palaeobiologically interesting situation in which figurative strain is given an areal meaning rather than being conceived in terms of volume.

Normally, the *dilatation* or expansion θ is defined as the increase in volume per unit volume. That is

$$\theta = (\delta V^* - V)/\delta V \qquad (4.15)$$

where δV^* stands for the strained volume corresponding to volume δV. At first sight, it might seem inappropriate to describe a statistical problem in terms of elasticity, curvilinear coordinates and hence tensors. However, the method of analysis can be reduced to standard multivariate form under certain assumptions (Bookstein, 1983).

Any analysis involving strain, real or figurative, can be described in terms of the tensor calculus (Watson, 1970) and such an approach is familiar to structural engineers, physicists and structural geologists. The algebra of multivariate statistics can be reformulated in tensor form, although such a step is seldom to be seen in statistical literature. For example, the Quadratic Form (see Chapter 2, p. 16) constituting the Mahalanobis generalized statistical distance is, in the terminology of the tensor calculus, a *metric tensor*. We can express this as follows. The distance between two infinitely nearby points P and Q in a coordinate manifold of k dimensions is given by

$$ds^2 = g_{ij}dx^i dx^j \qquad (4.16)$$

which we recognize as the Mahalanobis squared generalized distance. The coefficients g_{ij} in (4.16) are symmetric in the indices i and j. A coordinate manifold over which such a quadratic form is defined is called a k-dimensional Riemann-space. This quadratic form is invariant under allowable coordinate transformations. The tensor having the quantities g_{ij} as its components is called the fundamental or metric tensor of the Riemann-space. The foregoing formulation of the generalized distance was realized by Indian statisticians more than 50 years ago (cf. references in Blackith and Reyment, 1971). Scalars and vectors can be construed as being the first level of complexity for *tensor fields* that represent the interactions of multiple geometric parameters at a point. The next level up is the second order symmetric tensor which expresses two positive strains, or other rates, in directions at right angles to each other. In diagrams, this tensor is characterized by a small right-angled cross at each point. Figure 4.9 is an example of this mode of representation.

Thus, a symmetric tensor field offers a useful means of describing deformation. In a point-by-point shape change connotation for our two-dimensional model, the deformation is expressed as two rates of linear extension in two directions at $90°$. As was pointed out in Chapter 2, the usual discussion is in terms of three dimensions.

You may be interested to learn that the transformation producing the latent roots and vectors of familiar Principal Component Analysis can also be expressed in terms of tensors. The usual covariance matrix is a symmetric covariant tensor A_{ij}. The familiar determinantal equation of multivariate statistics is written in tensor symbolism as

$$|A_{ij} - \beta g_{ij}| = 0 \qquad (4.17)$$

of degree k in β and where g_{ij} forms the unit matrix in Cartesian space. The roots are latent roots of A_{ij} and the principal directions are those of the latent vectors. Analogous representation can be made for all other methods of multivariate statistical analysis.

As stated earlier in this section, these axes are oriented along the directions of maximum and minimum rates of change of length. The curves and dilatations that can be superimposed on the mesh of crosses describe the change in shape of the average outline by a tensor field which expresses deformation of the interior by particular rates in particular directions. The results summarized in Fig. 4.9 bring out the biologically significant fact that the deformational intensity is not evenly spread over the entire surface of the test and that the difference in the two rates tends to be greater along the medial

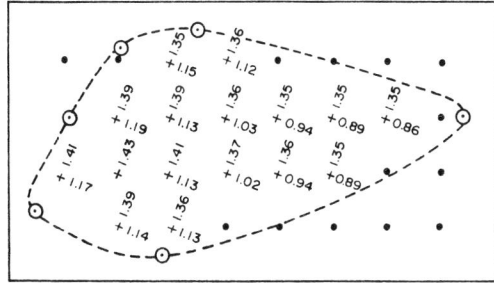

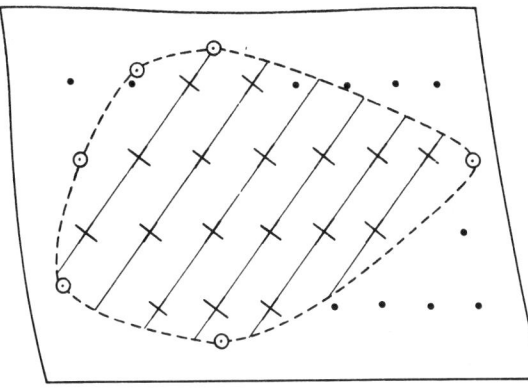

FIG. 4.9. Biothogonal grids for a comparison between levels 1 and 5 for the *Brizalina* data. The figure shows selected strains along principal axes of deformation (dilatations) superimposed on reference mesh points. The lower figure is "deformed" in relation to the upper one and indicates the average difference in shape between the two levels.

axes and proximal to the proloculus. This is a practical illustration of Huxley's (1932) interpretation of differential growth.

Example 4.6. Ecophenotypy in Two Species of Cretaceous Ostracods

We shall now review a second illustration of the shape-coordinates technique, using work by Abe *et al.* (1988). In this paper, variability in two species of Santonian (Cretaceous) ostracods from Shiloah, Israel was analysed. I have already had occasion to point out that ostracods often exhibit marked variability in some features of the carapace. Some of these variations can be identified as being due to genetic variability in ornamental details. Other sources of variability lie with the reaction of the organism to environmental factors, that is, ecophenotypic variation, which is the kind most commonly observed in fossils.

The Material: In the illustration reviewed here, *Veenia fawwarensis HONIGSTEIN* and *Oertliella cretaria* (*VAN DEN BOLD*) were shown to have been strongly affected by the lime-rich environment in which they lived and, in particular, by the content of magnesium in the seawater.

The landmarks measured are indicated schematically in Fig. 4.10. You will see that these are strategically located around the carapace and at three lateral positions, namely, the anterior furrow, the adductorial tubercle, and the termination of the lateral ornamental field. Spinose carapaces occur quite commonly in the material of both species from sampling level 7 out of the nine

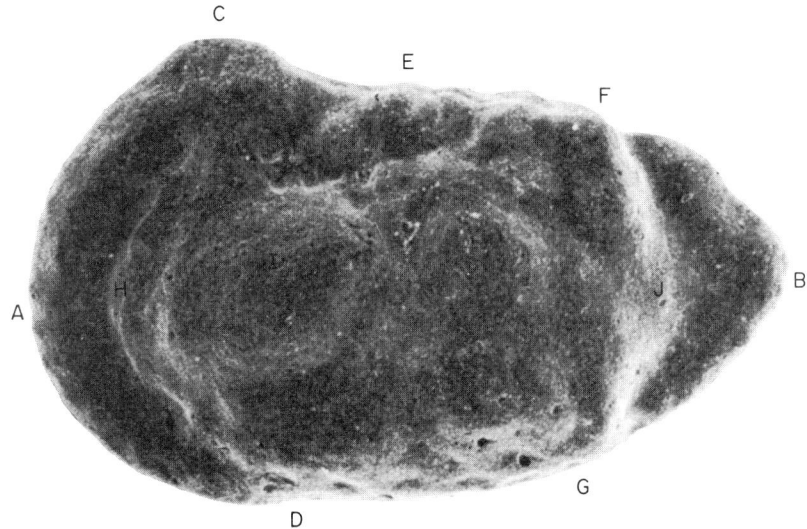

FIG. 4.10. The location of landmarks and pseudo-landmarks on the carapace of the Cretaceous ostracods from Israel. True landmarks are points *C*, *E*, *F*, *B*, *G*, and *D*, which all have well defined, homologous geometrical locations in relation to sharply defined shell features. Point *A* is a pseudo-landmark. Redrawn from Abe *et al.* (1988).

sampled in Shiloah quarry. All specimens observed of *Oertliella cretaria* bear spinosities and approximately 95% of the individuals of *Veenia fawwarensis* are spinose. In addition, the background bulbosities of most specimens of *Veenia* bear spined outgrowths in the sample from level 7. This is a remarkable feature and its morphological effects are probed in the following analysis.

The findings are presented firstly with respect to each of the species and then for the two species considered in the same connexion.

ANALYSIS OF *Veenia*: The first problem to be addressed is whether the ostracod carapace has undergone determinable changes in shape over the interval sampled at Shiloah.

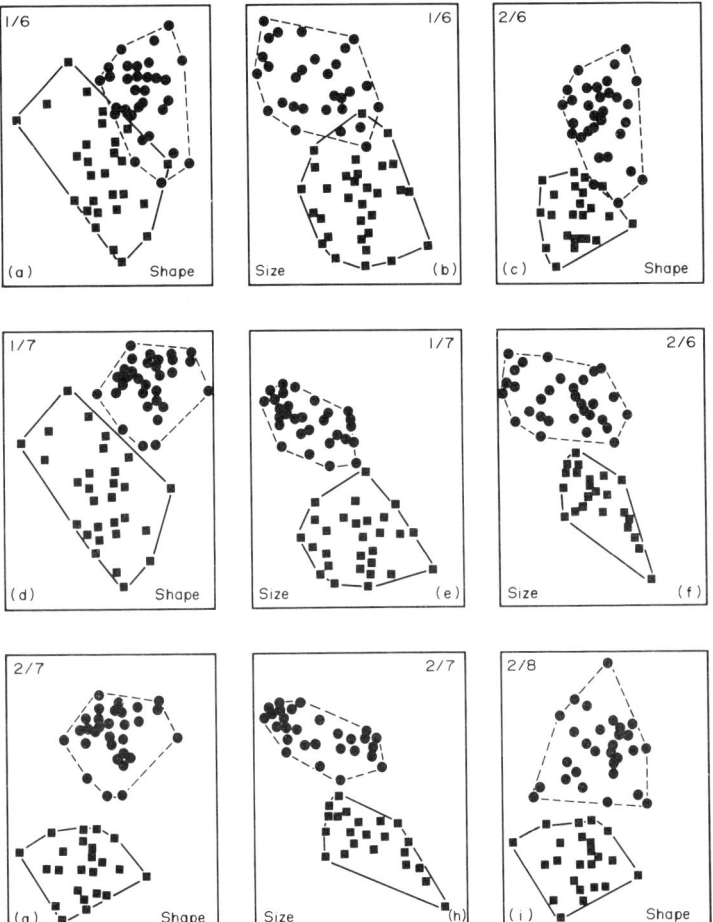

FIG. 4.11. Relationships for triangle *ABG*, displaying selected size and shape comparisons, for *Veenia fawwarensis*. Comparisons between samples for eight stratigraphical levels are represented in this figure. Redrawn from Abe *et al.* (1988).

Considering triangle *ABG*, which displays the most marked differences (Fig. 4.11), it is seen that the sample from level 7 is set off from all other samples irrespective of whether size is included or excluded from the calculations (see Figs 4.11d, e, g, h). Triangle *ABD*, embracing the anteroventral rib-intersection, also gives good separations for pairwise comparisons of samples. Sexual dimorphism is clearly manifested in the triangular displays of Abe *et al.* (1988).

ANALYSIS OF *Oertliella*: For the most part, the results for *Oertliella* echo those obtained for the other species, except that contrasts are less strongly

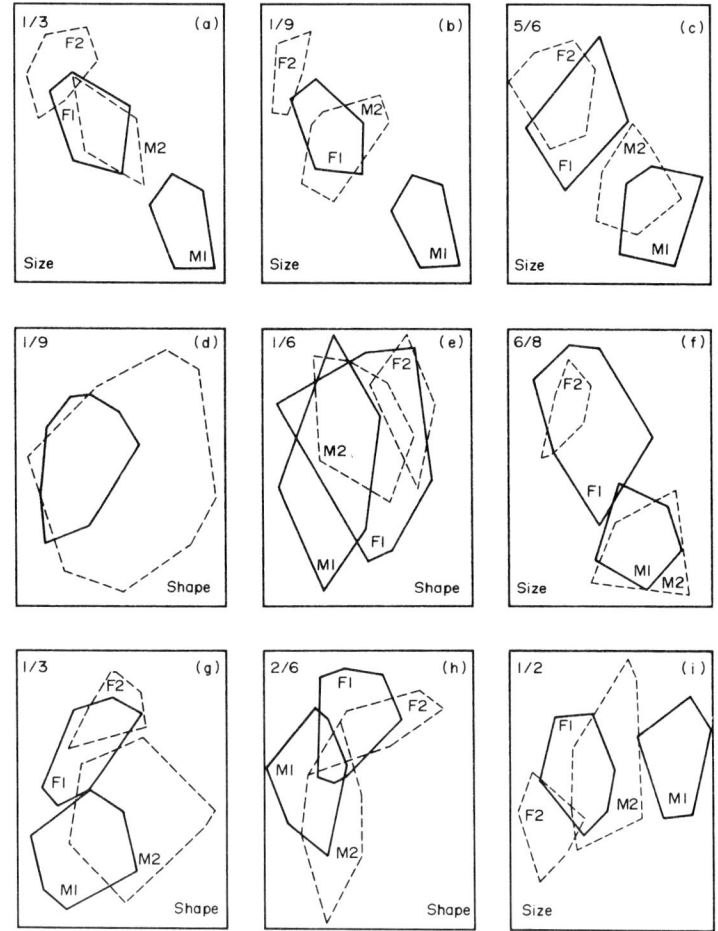

FIG. 4.12. Shape-coordinate analysis of sexual dimorphic relationships in shape and size in *Oertliella cretaria*. M1 and F1 are males and females for the first sample of a comparison and M2 and F2 denote females for the second sample of the comparison. The numbers in the top left-hand corner of each figure. Redrawn from Abe *et al.* (1988).

uttered. Once again, the best differentiation occurs for comparisons with sample 7.

The geometrical relationships for some of the comparisons are displayed in Fig. 4.12. The effects of sexual dimorphism are especially neat. Thus, with size retained, there is an informative relationship between the distributions of males and females of both samples located adjacent to each other. Two sexually undifferentiated samples (Fig. 4.12d) indicate that the convex hull for the sample from level 1 lies entirely within the convex hull for sample 9, attesting to substantial differences in variability for the two levels. An analogous situation occurs for *Brizalina* (see p. 131). This is an enlightening pointer to the power of geometric morphometrics.

Pooled Data: We turn now to an examination of all levels simultaneously. In order to avoid the interpretational muddles associated with overcrowded plots, each sample was limited to ten specimens. One result for *Veenia* is shown in Fig. 4.13. There is no clear clinal relationship between samples owing to overprinting due to oscillations in the succession of convex hulls. When the length of the baseline is multiplied back, there is a tendency to form two fields (Fig. 4.13) with samples 1 to 5 forming one field and samples 6 to 9 a second field. The results for *Oertliella* are essentially the same.

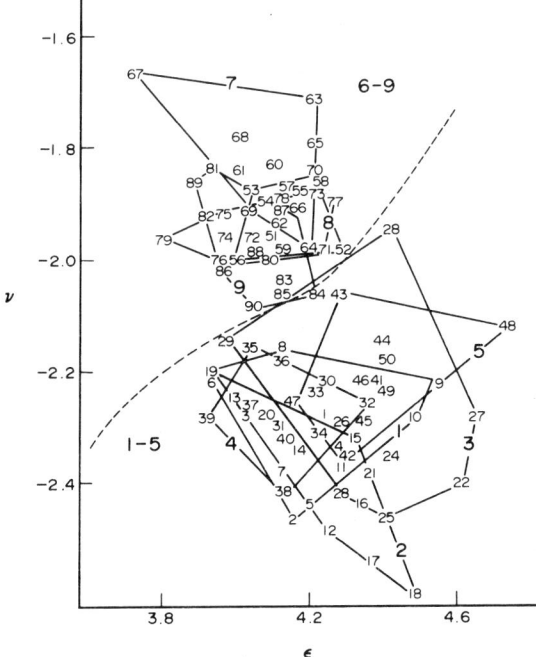

FIG. 4.13. Analysis by shape coordinates for all samples of *Veenia fawwarensis*. There is a clear division into younger and older samples. Redrawn from Abe *et al.* (1988).

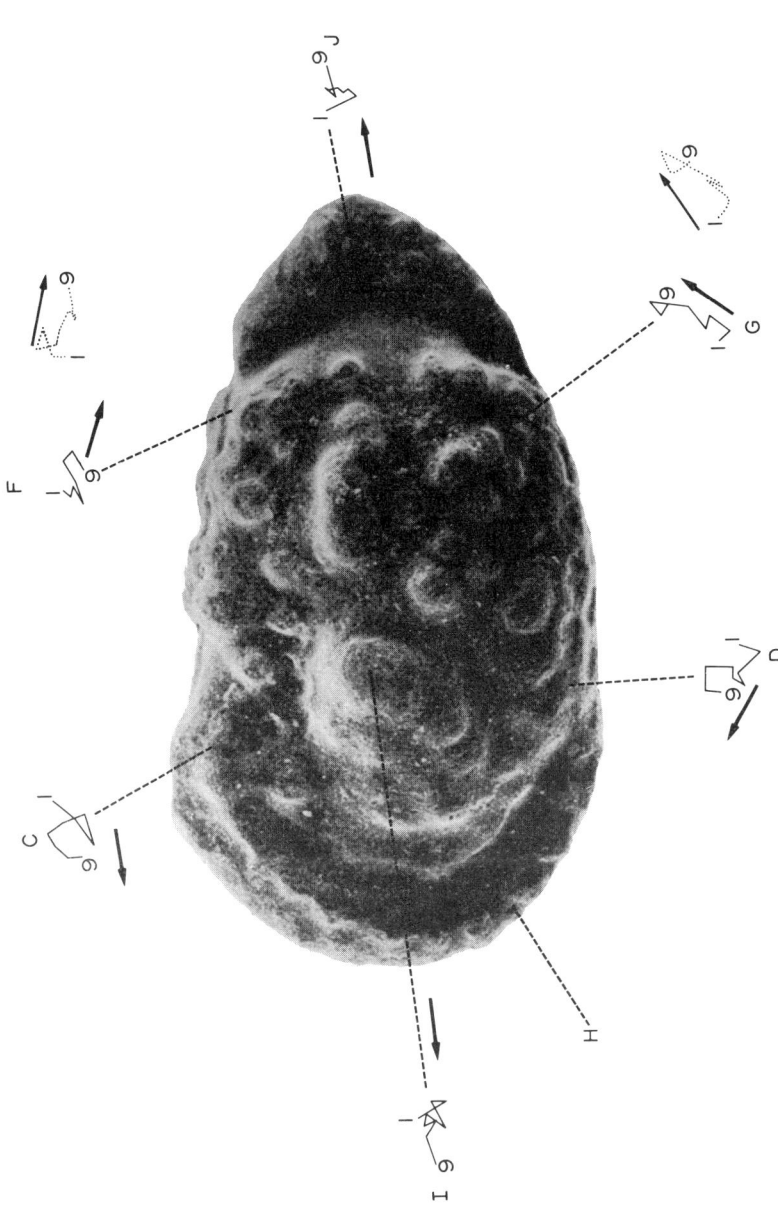

FIG. 4.14. Nine point trajectories for six shape coordinate means referred to the usual baseline for *Veenia fawwarensis* and two trajectories for *Oertliella cretaria* (dotted path). Redrawn from Abe *et al.* (1988).

Hence, although no smooth cline is to be seen, there is a difference between the first half of the sequence and the second half. This difference seems to be due to ecophenotypic effects caused possibly by an elevated content of magnesium in the seawater.

Analysis of landmark trajectories: Another way of studying secular trend in landmarks is by plotting the trajectories traced out by the set of shape coordinates over time. The three internally located landmarks are included for this part of the analysis (i.e. landmarks *H* to *J* in Fig. 4.14).

All landmarks show significant variability in the mean shape-coordinates in relations to the baseline *AB*. All MANOVA results are significant at the 0.005% level, or better and if all eight landmarks are considered at the same time (i.e. landmarks *C* through *J*), the significance of the MANOVA is overwhelming. In addition, there is strong size allometry with $R^2 = 0.62$ for size, measured as the summed square distance of all the landmarks from their centroid (Bookstein, 1986) as required on the sixteen coordinates for *Veenia*. The inclusion of size has little effect on the outcome of the MANOVA.

The graphical summary of the analysis for trajectories is shown in Fig. 4.14 which illustrates 9-point paths (for the nine sampling levels) for six shape coordinate means to an *AB* baseline for *Veenia* and two for *Oertliella*. Only those trajectories with a time-correlated displacement are shown in the figure. The information contained in Fig. 4.14 suggests that the situation may be more intricate than can be explained by a simple model of ecologically driven variation in size and shape. This is elegantly exposed in the spatially divergent pattern displayed by the ventral intercepts of the ornamental ridges. The points of intersection, which are landmarks *D* and *G*, are proportionately closest at sampling level 1, after which they gradually diverge. The ornamental features seem to be undergoing structural rearrangement involving upward and outward shifts. A posteriorly directed component is also manifested in the rear of the shells at landmarks *G* and *J*. The net result is expressed in a displacement of most anterior landmarks in a posterior direction. The same patterns occur for *Oertliella* with respect to the posterior configuration of the carapace (cf. Huxley (1932) for observations on crabs).

Discussion: The general impression conveyed by the study of the landmark trajectories is that we are observing a gradual microevolutionary shift in the lateral ornament of the carapace. The effects of fluctuations in the magnesium content identifiable in the earlier analysis do not make themselves felt here and it is possible that we are confronted by intrinsic directions of change in the organization of the animals and not only morphological reactions to ecological variability.

The observed changes are of very small magnitude, no more than 2.5 to 3% of the length of the carapace, but they seem to bear the imprint of permanency in the case of *Veenia*. It was found that about two-thirds of the shape history of the two species is shared; this is not unexpected in view of the susceptibility of the posterior region of some crustaceans to ecological factors. Agreements between other landmarks for the two species are, however, slight. Trajectories for landmarks *F* and *G* for *Oertliella* are shown by dotted lines in Fig. 4.14.

Example 4.7. Ecophenotypic Variation in Recent *Mutilus*

The example reviewed in this section was excerpted from work by Reyment *et al.* (1988) on the Recent ostracod species *Mutilus pumilus* from southern Australian waters. Ornamental details in some marine ostracods are known to be influenced by the chemistry of seawater. This may also extent to the outline of the shell and the size attained by adults. A palaeontological example was offered in the foregoing section. Salinity is doubtlessly one of the major chemical factors involved in this variation. Further aspects of variability in size, shape and ornament are taken up in Chapter 6.

Morphs that moult at periods of relatively low temperature, say, at night, tend to develop less heavily calcified shells than those that moult during the day when temperatures in tropical and temperate tidal pools can reach, and exceed, 40 degrees, thereby providing carbonate in the ambient medium and hence causing the secretion of more heavily calcified shells. A sample of fossil shells can therefore contain an inseparable mixture of thinner and thicker individuals.

The specimens studied in the present example come from nearshore sites around the coasts of South Australia, Victoria and West Australia. For the details of the full analysis I refer you to Reyment *et al.* (1988).

Findings: Using the characters illustrated in Fig. 4.15a,b, we have two kinds of landmarks and pseudolandmarks available for analysis. One set, Fig. 4.15a, consists of sites located along the median rib obtained by Blum's method of median-axial determination (Blum, 1973). The second set of landmarks, Fig. 4.15b, are of the kind with which we are now familiar; these are located around the outline of the shell. For the purposes of the present illustration, I consider

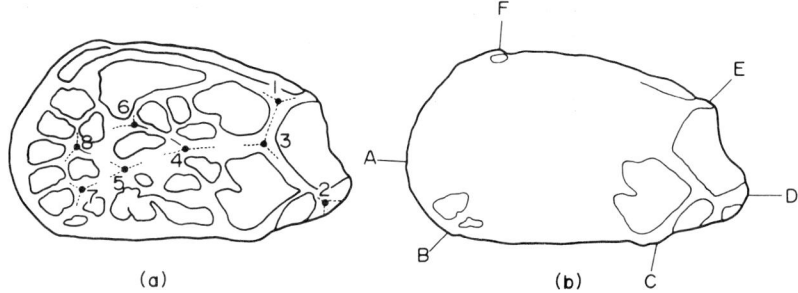

FIG. 4.15. Diagrams indicating landmarks measured on left valves of living *Mutilus pumilus* from several sites along the southern Australian coastline. (a) Shows eight pseudo-landmarks located along the central rib structure, obtained by the median axis method of Blum (1973). Illustration (b) shows the locations of points located around the outline of the shell. Redrawn from Reyment *et al.* (1988).

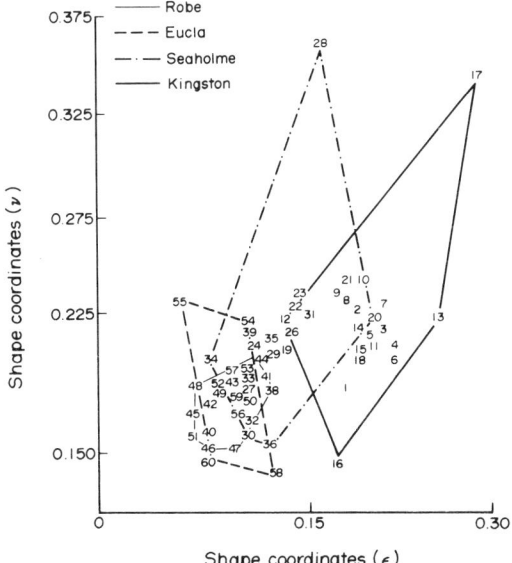

FIG. 4.16. Shape-coordinate analysis for *Mutilus pumilus* from the southern coast of Australia, from Eucla (Western Australia) Kingston (South Australia), and Seaholme (Victoria). Redrawn from Reyment *et al.* (1988).

only two samples for *Mutilus pumilus*, collected off Kingston in South Australia and off Seaholme, Port Philip Bay, Victoria, for the Blum variables. The analysis based on the variability of the ribbing was shown by Reyment *et al.* (1988) to be interpretable as an almost homogeneous shearing although with some slight rotation in a part of the field.

Using the outline landmarks, the triangles *ABD* formed by the anteroventral intersection, *AFD* made with the eye-tubercle, and *AED* with the posterodorsal angle, the following conclusions were arrived at. With *AD* as baseline, Fig. 4.16 was obtained for triangle *ABD* for the samples from Eucla (West Australia), Robe and Kingston (South Australia) and Seaholme (Victoria).

Discussion: The fields for Kingston and Seaholme (which are two geographically well separated locations) are remarkably similar in shape but overlap but slightly. The fields for Eucla and Robe overlap each other in the far left of the graph and they cut into the zone staked out by the points for Seaholme. They are completely separated from the Kingston field. Although Robe and Kingston are geographically close, the fields delineated by their points are differently shaped. These results could be indicating that seasonal variation in some ecological factor (or diurnal variation) such as temperature, is the primary cause of variability in the shell of *Mutilus pumilus*, including the polymorphic condition of the ribbing recorded by Hartmann (1982).

An interesting laboratory project suggested by the results obtained by the *Mutilus* study would be to rear the species in aquaria under controlled physical and chemical conditions to see what measurable effects on the ornament developed. By observing changes in the carapace produced by different settings of the extraneous variables, valuable information on the size of ecophenotypic effects could be obtained. Hartmann's (1982) study intimates that such a project would be well worth trying.

4.6 Thin-plate Splines and Principal Warps

Introduction: A fervent and long-entertained wish of morphometricians has been to find an intuitively attractive way of carrying out the kind of shape studies inherent in D'Arcy Thompson's grids, but for which no implicit structures for computing were provided. Bookstein's biorthogonal grids give part of the solution, but this method does not furnish a fully expressed graphical display of the point-to-point deformation. Bookstein (1989) has developed an ingenious way of realizing this goal by his method of Principal Warps. This is again a rather complicated subject and I shall only outline the main features of the method and give an example. I am indebted to Fred Bookstein for help in preparing this section.

Bookstein (1988, 1989a, 1989b) shows how the technique of the **thin-plate spline** can provide an elegant means of visualizing statistical analyses of deformational data by the agency of a general-purpose *Interpolation Function*. This function expresses one of the mappings that models a particular biological homology sampled by pairs of points (e.g. coordinate pairs). A purely uniform transformation leaves parallel lines parallel (see Chapter 2). In the two-point registration we have already met, all landmarks are displaced by multiples of a single vector whereby each multiplier is proportional to the distance of the landmark from the baseline. To any transformation of landmarks, there is a **Bending Energy**, which may be defined as the net energy required to bend an infinite, infinitely thin metal plate over a set of landmarks so that its height over each landmark is equal to first the x-coordinate and then the y-coordinate of the corresponding landmark in another set of landmarks. **Uniform translations** involve tilting and re-rolling this plate but not **bending** (see Section 2.9). In the terms of the theory, these manoeuvres generate zero bending energy.

We need some basic ideas now. For data in the form of landmark locations we have that:

(a) Any change in shape for a set of landmarks has (1) a *uniform* part and (2), a *non-uniform* part. Any sample of landmark coordinates encompasses variation in both of these parts about a mean configuration.

(b) The uniform and non-uniform parts of any change or scatter represent complementary subspaces of the full vector space of shape variation.

(c) The uniform part may be imagined as variation in a typical triangle, rigorously interpolated or extrapolated so as to apply to every landmark triangle in the same way. The uniform part is calibrated by anisotropy, which is the log-ratio of the principal strains (cf. Chapter 2).

(d) Deformations of equal anisotropy can be made to lie on ordinary circles. This permits anisotropy to be viewed in terms of distance.

(e) An important property of shape change in the present model is that the uniform and non-uniform parts are *incommensurate*. The uniform part, an anisotropy, and the non-uniform part, an energy, cannot be united into a single metric, nor can the third factor required for specifying the magnitude of a change in the configuration of landmarks, size, be merged with them into a metric.

Any single non-uniform transformation may be expressed as a finite sum of **Principal Warps**, which are characteristic functions (or eigenfunctions if you like) of the bending energy corresponding to Procrustean orthogonal displacements of the conceptual metal plate at the landmarks. (The term "Procrustean", as originally proposed in statistics as a fitting technique in multivariate analysis, is a fanciful borrowing from Greek mythology: *Prokroustes*, literally "stretcher", the bandit who fitted dupes to his bed by stretching or lopping.) These "warps" emerge in descending order of a latent root, bending energy per unit summed squared Procrustean displacement, which can be identified with an inverse geometrical scale; the sum of the products **over** the landmarks of the displacements in any particular direction is **zero**. Procrustean distance is the sum of the squares of these displacements.

A sample of shape changes, or their residuals, after subtraction of an estimated uniform part, may be decomposed into a series of *Relative Warps* which are the latent vectors of the covariance matrix of shape-coordinates with respect to bending energy. These are analogous to principal components as defined in Chapter 3 and therefore apply to the configuration of landmarks for a sample of specimens. In the usual manner, these latent vectors are calibrated by the corresponding latent roots which represent variance in shape per unit of bending energy.

The bending energy may be imagined as a metric (a distance-measure) on shape space. In mathematical terminology, it is a "deficient metric". Landmark configurations which differ by a uniform transformation are at distance zero from each other in this metric. Bending energy zeroes out all transformations that exactly fit any combination of those models. In most multivariate statistical theory, the appropriate geometrical representation of a distance measure is a generalized ellipsoid. Statistical distance in all directions is variously proportional to the Euclidean distances of the descriptor space: we have met this mode of depicting relationships in the Mahalanobis generalized distance, for example.

A different solid-geometrical situation pertains for the bending energy. Here, the correct picture is not an ellipsoid but a *cylinder* in a landmark configuration in space. The generators of the cylinder, i.e. the straight lines on it, are the sets of all transformations derived from a given non-uniform warping by application of any additional uniform transformation. All such additional transforms are at the same "distance" (bending energy) from the starting form.

The axes of this cylinder are the Principal Warps of the landmark configuration. These are, as mentioned above, the latent vectors of the bending energy with respect to the summed square displacement of landmarks in their original system of coordinates. Each principal warp specifies the displacement of each landmark by a particular distance (positive or negative, summing to nought) in an unspecified direction that is the same for all landmarks (Bookstein, 1989b). The cylinder pairs these axes into circles of equivalent bending in any direction of the plane. We note that:

1. The first principal warp represents the pattern of displacements having greatest bending energy per unit root-mean-square landmark displacement. Usually, it is the relative displacement of the two landmarks closest together with but small contributions from the others, which are effectively "at infinity".
2. At the other extreme, the last principal warp is the largest-scale non-linearity that can be considered to leave landmarks at infinity fixed. Its appearance is rather like what you get when you deform a square to take on the appearance of a kite.

How does the method of principal warps relate to Mosimann's Theorem (p. 108)? We have seen that the principal warps are a decomposition of size and shape space. The role of size is quite clear: if you double the size of the second form in a comparison, you will double the warps. We saw earlier on what in terms of Mosimann's Theorem, size must be construed as being independent of shape, given a probability distribution. For the Booksteinian Distribution (i.e. circular "noise" at each landmark), *Centroid Size* is the size variable in Mosimann's solution. On the null model (of no allometry), transformations of the principal warps (i.e. their ratios) are *independent* of Centroid Size analogously to any other measurement of shape, as also ratios of their bending energies, etc. Otherwise, the principal warps have no special relationship to Mosimann's Theorem, nor do the shape coordinates. However, the method of shape coordinates provide a convenient way of proving that Centroid Size is the right choice (Bookstein, 1991).

Needless to say, the mathematical details of principal warps are quite complicated and I refer the reader with a desire to know what Bookstein has done to two papers of his on the subject, Bookstein (1989a, 1990). Another illuminating reference is the discussion appended to the paper by Kendall (1989).

4.6.1 *The Thin-plate Spline*: *Practical Details of the Method*

The analysis of shape by the concept of the thin-plate spline is based on a function which describes the surface

$$z(x, y) = -U(r) = -r^2 \log r^2 \tag{4.18}$$

where r^2 is the distance $(x^2 + y^2)^{1/2}$ from the Cartesian origin. The function U satisfies the equation:

$$\Delta^2 U = \left(\frac{\partial^2}{\partial x^2} + \frac{\partial^2}{\partial y^2} \right)^2 U \propto \delta_{(0,0)}$$

where the right-hand term denotes proportionality to the "generalized function" $\delta_{(0,0)}$. It is zero everywhere except at the origin and its integral is 1. The appropriate mathematical statement is that U is a *fundamental solution of the biharmonic equation*, $\Delta^2 U = 0$, which is the equation for the shape of a thin metal plate levitating as a function $z(x, y)$ above the (x, y)-plane (see Bookstein, 1991, Fig. 2.2.2).

To bend the metal plate requires energy. The sharper the degree of bending, the greater is the energy required. This provides a logical foundation upon which to base a measurement of shape difference, if we construe this in terms of physical deformation.

If we have landmarks for two organisms, each regarded as being fixed to a thin metal plate, fitting one form to the other will necessitate bending. The degree of distortion involved can be expressed as energy generated by bending one plate so as to make its points conform with those of the second plate.

Bookstein (1988) defines a *Bending Energy Matrix*, $\mathbf{L}_k^{-1} \mathbf{P} \mathbf{L}_k^{-1}$ (see equations 4.18 and 4.19 below) the latent roots and vectors of which are interpretable as descriptors of deformation. These are a canonical descriptor of the modes according to which points may be displaced, irrespective of global affine transformations. The latent vectors of the bending energy matrix called by Bookstein PRINCIPAL WARPS.

Let $z_1 = (x_1, y_1), \ldots, z_k = (x_k, y_k)$ be k points in the Euclidean plane in a Cartesian system.

If $r_{ij} = |z_i - z_j|$

$$\mathbf{P}_k = \begin{bmatrix} 0 & U(r_{12}) & \cdots & U(r_{1k}) \\ U(r_{21}) & 0 & \cdots & U(r_{2k}) \\ \cdots & \cdots & \cdots & \cdots \\ U(r_{k1}) & & \cdots & 0 \end{bmatrix}$$

of order $k \times k$.

We have also a matrix of order $k \times 3$ to consider

$$
\mathbf{Q} = \begin{bmatrix} 1 & x_1 & y_1 \\ 1 & x_2 & y_2 \\ . & \ldots & \ldots \\ 1 & x_k & y_k \end{bmatrix}
\tag{4.19}
$$

These two matrices defined in (4.18) and (4.19) combine to form

$$
\mathbf{L} = \left[\begin{array}{c|c} \mathbf{P}_k & \mathbf{Q} \\ \hline \mathbf{Q}^T & \mathbf{O} \end{array} \right]
\tag{4.20}
$$

where $\mathbf{0}$ denotes the null matrix. This matrix is of order $(k + 3) \times (k + 3)$. Writing $\mathbf{L}_k^{-1}$ for the upper left sub-matrix of $\mathbf{L}^{-1}$, the latent vectors of $\mathbf{L}_k^{-1} \mathbf{P} \mathbf{L}_k^{-1}$ can be interpreted as the coefficients of a thin-plate spline attached to the base-plane since they are coefficients of the functions U based at the k landmarks. The latent vectors, which are the Principal Warps, furnish the bases for figures for each of the non-zero latent roots (normally $(k - 3)$ in number because $\mathbf{0}$ in equation (4.20) is a 3×3 matrix of noughts. The latent roots can be interpreted as measures of the **bending energy** required to produce the warp corresponding to each of them, to wit, the associated latent vector. Each principal warp is a geometrically independent mode of affine-free deformation at its own geometrical scale. This is the inverse of the corresponding latent root.

Bookstein's results indicate that for more than four landmarks, the bending energy matrix has a non-trivial spectrum which can encourage interesting biometrical speculations. In general terms then, the method of Principal Warps can be seen to localize **within-groups variation** in shape as distinct from **between-group differences** in shape. In physical terms, the principal warps express features of bending at successively higher levels of bending energy. In this representation, they correspond to qualities of deformation on *successively smaller physical scales*. Bookstein (1989b, pp. 573–574) illustrates this with an example that shows how the first latent root corresponds to a small feature, to wit, simple displacement in to closely located landmarks. The last significant latent root corresponds to deviation in large-scale features.

In order to illustrate the main features of the method, I have carried out a phylogenetic study on two species of ostracods (data extracted from Benson, 1972). The results of the calculations using the program TPSPLINE by F. James Rohlf (version of June, 1990) are listed in Table 4.3.

TABLE 4.3. *Principal warps for two species of* Agrenocythere

---

Number of landmarks $= 9$
Bending Energy $= 0.06774$
Affine Matrix (the uniform part)

1.1246	0.0382
-0.0877	1.1365

Latent roots		Energy contribution
1.	0.1917873	0.005863
2.	0.0991501	0.085903
3.	0.0606656	0.004142
4.	0.0346544	0.000041
5.	0.0147592	0.000869
6.	0.0094780	0.000926

There are $(k-3)$ warps $= 6$: see text.

Latent Vectors, the normalized Principal Warps. I list
the first two vectors here only: they were used to produce
Figs 4.18 and 4.19. They constitute the coefficients of the
first three of the six warps. There is very little bending
energy associated with principal warps four through six.

Landmark	Warp 1	Warp 2	Warp 3
1	0.0231	-0.1391	0.0337
2	-0.0749	0.0817	-0.1843
3	0.4339	-0.1450	0.6326
4	-0.7878	0.1016	0.0395
5	0.3935	0.0112	-0.6264
6	-0.0668	-0.1292	0.3164
7	0.1408	0.5867	-0.0685
8	-0.0756	-0.6909	-0.2384
9	0.0137	0.3229	0.0954

The second principal warp contains most of the
information on the shape difference between the two
species.

Example 4.8. Principal Warps for Two Species of *Agrenocythere*

The observations were made on nine landmarks on the shell of each of the
species *Agrenocythere hazelae* and *Agrenocythere radula* so as to display the
deformation involved in fitting one species to the other. The landmarks were
selected among those indicated by Benson (1972).

The findings are briefly reported in the figures accompanying the example
(Table 4.3). The two shapes can be superimposed whereby the grid indicates
the distortions and where they are located on the plate. The bending energy to
produce this fit is 0.0677. The appearance of the non-affine part of the
transformation is shown in Fig. 4.17; recall, that it is this deformation which
is associated with bending energy. Of the total bending energy, more than 83%
lies precisely along the second principal warp (in the terms of Chapter 3, this
is a percentage of the trace of the matrix for bending energy, defined above.
The effect is the difference in relative position of landmark 8 between
landmarks 7 and 9, an almost horizontal, that is, antero-posteriorly directed

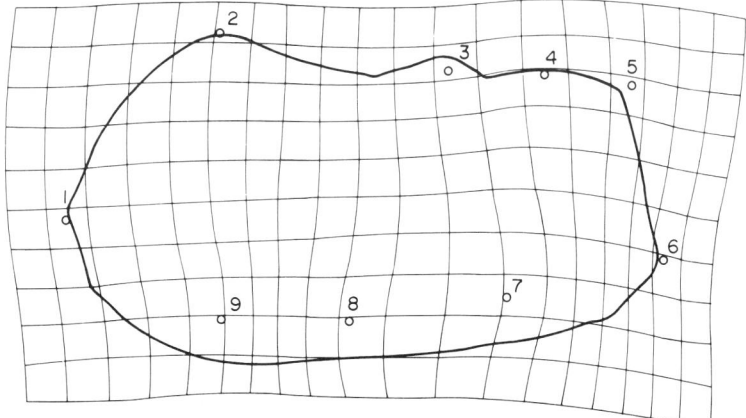

FIG. 4.17. Non-affine transformation for the species of *Agrenocythere*.

change. Two other warps have non-negligible amounts of bending. The first principal warp describes the rearrangement of landmark 4 with respect to landmarks 3 and 5. The third principal warp describes the relation of landmark 3 to landmark 5. Figure 4.18 indicates the result for the first principal warp with the distortion again appearing through warping of the grid and with warping energy = 0.00886. Figure 4.19 depicts the figure for the second principal warp, with energy = 0.05590. In this representation, the main warping effects lie in a different location from that in Fig. 4.18.

Remarks: This is a simple example, but I think it serves its purpose well enough in that the subtle power of the thin-plate spline method comes to full

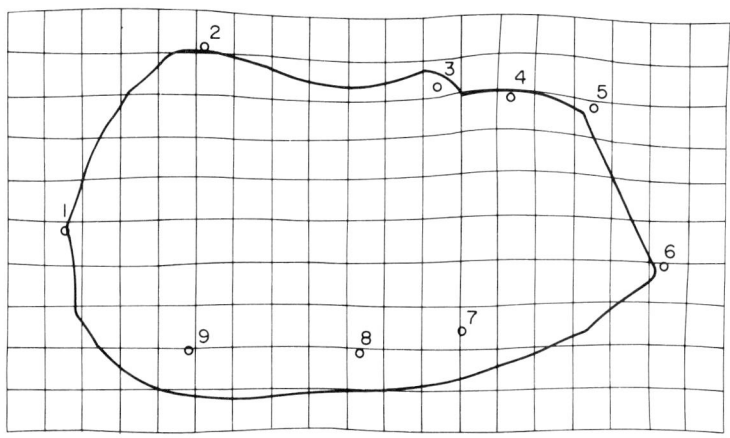

FIG. 4.18. First principal warp for the species of *Agrenocythere*.

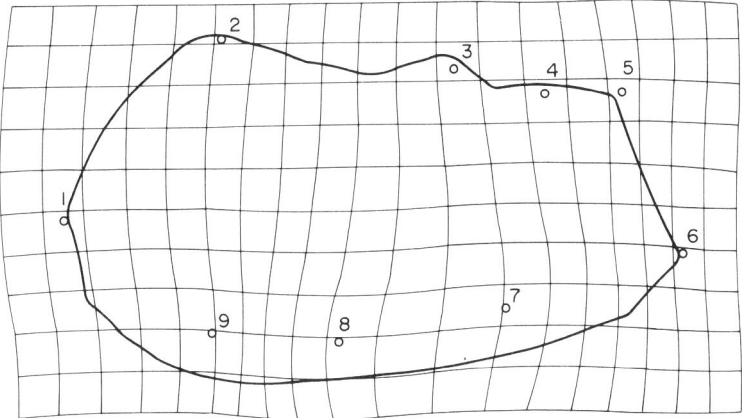

Fig. 4.19. Second principal warp for the species of *Agrenocythere*.

expression. The direct connexion to the grids of D'Arcy Thompson should be obvious.

Bookstein and Sampson (1987) have given a completely worked example for lateral cephalograms for pure-bred rats. Here, variability in form at a single age is studied. The classical model of growth gradients (cf. Huxley, 1932; Bookstein, 1991) is compatible with the highest order relative warps. The analysis of shape variation in the rat skulls disclosed that two relative warps are explicitly identifiable with rigid motion or rescaling of the vault of the cranium in relation to its base. Bookstein (1990) points out that Abe *et al.* (1988) found, in effect, that a cubic growth gradient accounts precisely for some of the changes in a lineage of ostracods.

4.6.4 *Summary of Main Concepts*

1. Differences in shapes, expressed by homologous LANDMARKS may be decomposed into two parts: a UNIFORM PART and a NON-UNIFORM PART.
2. Hence, complementary to the sub-space of uniform (affine) translations, there is a sub-space of NON-UNIFORM TRANSFORMATIONS.
3. Sample variability of the non-uniform part can be examined in the light of the concept of the **bending energy matrix** of the sample. The latent vectors of this matrix are a canonical description of the modes according to which points may be displaced irrespective of global affine transformations. These are the **Principal Warps** of the configuration of landmarks.
4. The uniform transformation portrays a change without bending.

4.7 A Possible New Frontier for Shape Analysis: Fractal Geometry

4.7.1 *Introduction*

Up to this point in my narrative, I have been almost exclusively concerned with shapes that are regular, more or less, and with smooth, or almost smooth, outlines. We all know that there are many organisms we meet in our daily work that are not regular in outline. Take merely the problem posed by a leaf. How do you go about quantifying that? Objects with irregular outlines can be very difficult to reduce to manageable form. Fourier methods and eigenshapes offer one possibility, as I have indicated earlier in this chapter. The rapidly expanding field of *Fractal Geometry* is another.

The concept of fractals derives principally from the work of the French mathematician Benoit Mandelbrot, who in 1982 introduced what he defined as the "Geometry of Nature". Fractal geometry provides a new scientific way of thinking about the edges of clouds, the shape of the distribution of a patch of plankton in the ocean, the quantification of a coastline as viewed from a satellite, and ornamental patterns in ostracods, in short, physical structures from dendritic patterns to galaxies. The advent of fractals has revolutionized some aspects of Physics, as has been brought home to me by listening to knowledgeable colleagues in the Royal Swedish Academy of Sciences. Fractal geometry is an extension of classical geometry which permits one to describe the shape of an object, no matter how irregular, in precise terms. The book by Barnsley (1988) is a useful introduction to the subject. Dennis Slice (1989) markets an easily accessible computer program, Fractal*D*, for the fractal dimension. Fractal geometry has proven to be a very valuable tool in computer graphics.

4.7.2 *Fractal Dimension*

The fractal dimension of a set is a number which indicates how densely that set occupies the metric space in which it lies. Among its properties, are:

1. It is invariant under stretching and squeezing of the underlying space and hence is meaningful as an experimental observable.
2. It has a certain robustness and is independent of the unit of measurement; this is a property of practical importance.

How can I justify the inclusion of a note on fractals in a book on multidimensional topics. I think the answer should be self-evident. Fractals are concerned with dimensionality, although the dimensions are fractioned. In the present connexion, our interest lies with the fractal dimension of two-dimensional data, the outlines of objects. The areas of application are many: I have already mentioned outlines, but other kinds of data are curves such as evolutionary series and ammonite sutures, digitized environmental boundaries, etc.

How does a fractal dimension differ from a topological one? Mandelbrot's conception of this question is that the "effective" dimension of a structure can depend on the scale at which it is observed and can differ from the more topological dimension. Take a simple example. Viewed from Venus, the Earth looks like a zero-dimensional point. From the Moon, the fact that the Earth is a three-dimensional body is clearly apparent. The nearer a hypothetical space traveller gets to the Earth, the more clearly can the continents can be made out, until they can be seen to be outlined by complex, winding unidimensional coastlines. Hence, the effective dimension of the Earth depends on the scale at which it is observed. And this dimension can even have fractional values, hence the origin of the term fractal.

A line has a topological dimension of one. If it is divided into N equal parts, each part is no more than a scaled version of the original. If the scaling factor is r, it is defined as

$$r = 1/N$$

or

$$Nr = 1.$$

Likewise for two dimensions, the scaling factor is

$$Nr^2 = 1$$

(because r is now expressed in terms of an area for two-dimensional topological space). The general expression is

$$Nr^D = 1.$$

The estimate for the fractal dimension is a simple ratio (Barnsley, 1988, pp. 174, 190) as expressed in the ensuing formula.

$$D = \frac{\log N}{\log \dfrac{1}{r}}$$

Thus, the fractal dimension of an object utilizes the relationship between the length and scale of observation. A commonly used method of estimation is to proceed by means of the *step-length*, i.e. an arbitrarily selected number of Euclidean distances around the perimeter. This is the equivalent of measuring the length of the outline by the familiar method of pacing with a pair of dividers.

It turns out that the choice of step-length is crucial. There is no absolute way of choosing the "right" distance and the best solution must be arrived at by a succession of trials. As step size is increased, the length estimate for complicated data, such as an ammonite suture, will drop sharply owing to changes in perceived complexity. Pronounced non-linearity in a part of the log (length)-by-log (step size) plot may lead to confounding in the fractal dimension due to a change in resolution from smooth to complex.

Example 4.9. The Anterior Vestibule of an Ostracod

Presentation of the Problem: As a simple illustration of the concepts involved in determining the fractal dimension, I have selected some data on the anterior vestibule of *Krithe*. (The vestibule is the pocket formed by the calcified part of the inner lamella and the outer lamella.) The extent of the vestibule in this genus of deep-sea ostracods is thought to be under the control of environmental factors, particularly the oxygen content of the water (Peypouquet, 1977). Without going into the details of the ideas underlying Peypouquet's conclusions, the morphometrical problem posed is interesting and one that seems to be amenable to treatment by fractal geometry. Four examples of the vestibular zone were digitized and then analysed by the program *FractalD*, using a series of exponentially increasing steps. The outline of the vestibule is, there is no doubt, a unidimensional contour, but it does fill the plane in which it lies to a certain extent. This extent will depend on the complexity of the contour. In all events, we should not be surprised if the dimension turns out to be somewhat greater than unity.

Findings: The four vestibules are illustrated in Fig. 4.20, A–D together with a sample for each specimen of the regression for the log (length) against log (step-size) and the appropriate fitted regression line. As you will see from the figures, the slope of this regression line provides the estimate of the fractal dimension. The greater the angle of slope, the greater is the fractal dimension for a data set. These diagrams make it clear that care must be taken in the choice of step-size. Moreover, the location of the starting point can influence the estimate. It is therefore recommended procedure to select starting points at random and examine the output for consistency (see Exercise 10 at the end of this chapter).

The estimates of the fractal dimension for the vestibules of the four morphotypes studied here are as follows (randomized initiations):

First specimen (Fig. 4.20A)
The average of ten estimates is 1.047, with a range of 1.004 to 1.126.

Second specimen (Fig. 4.20B)
The average value for ten estimates is 1.059, with a range of 0.943 to 1.227.

Third specimen (Fig. 4.20C)
The average value for ten iterations is 1.211, with a range of 1.118 to 1.284.

Fourth specimen (Fig. 4.20D)
The average for ten iterations is 1.047, with a range in values of 0.968 to 1.231.

For a completely rounded analysis in a situation such as the present one, where the estimates of the fractal dimension are not very stable and depend on where the starting point is located on the circumference, one would normally want to compute, say, 20 to 30 values and then select the mode of them as being the most interesting estimate.

What conclusions can be drawn from the foregoing analysis? Firstly we need to know what the right question is. The fractal dimension is *not* a measure of complexity; it is a summary measure of complexity. In the present illustration, it seems to indicate that the first, second and fourth vestibules *may have a common source* and that the third vestibule may have a different origin (the fact the *D* turns out to be the same for samples 1 and 4 could be fortuitous). I think this makes sense. Inspection of the sketches in Fig. 20 suggests a fair degree of similarity in shape for specimens 1, 2 and 4, while the

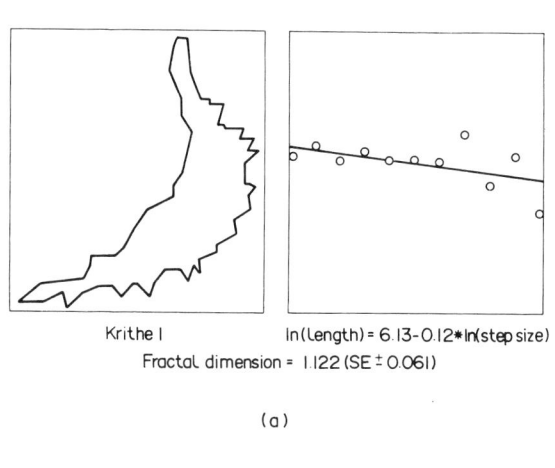

Krithe I   In(Length) = 6.13 - 0.12*In(step size)
Fractal dimension = 1.122 (SE ± 0.061)

(a)

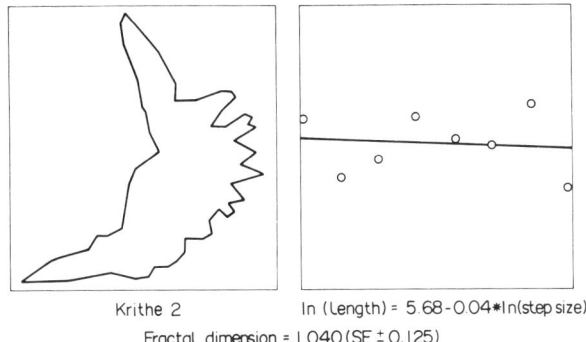

Krithe 2   In (Length) = 5.68 - 0.04*In(step size)
Fractal dimension = 1.040 (SE ± 0.125)

(b)

FIG. 4.20(a,b)

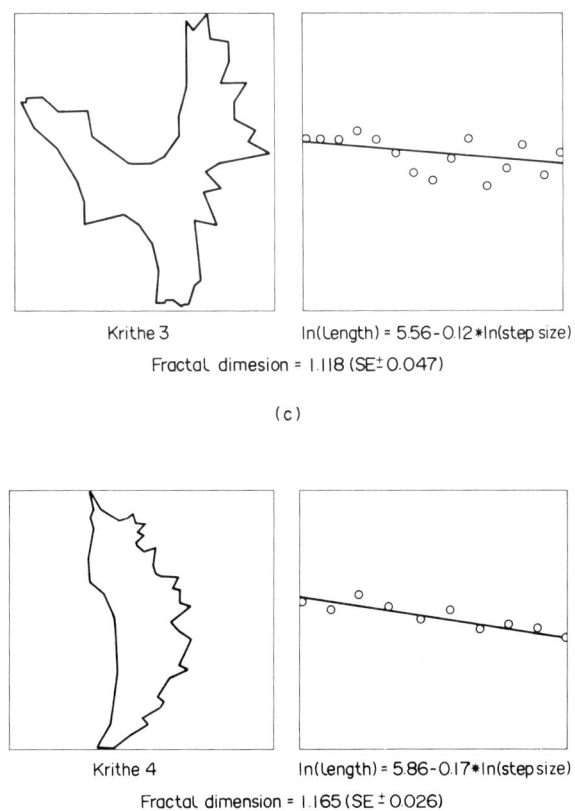

Krithe 3   ln(Length) = 5.56 - 0.12*ln(step size)
Fractal dimesion = 1.118 (SE± 0.047)

(c)

Krithe 4   ln(Length) = 5.86 - 0.17*ln(step size)
Fractal dimension = 1.165 (SE ± 0.026)

(d)

FIG. 4.20(c,d)

FIG. 4.20. Sample results for the determination of fractal dimension for four ecomorphic vestibules in living *Krithe*, *A*, *B*, *C*, and *D*. The left-hand vignettes contain sketches of the anterior vestibule. The right-hand vignettes display sample regressions of log (length) against log (step-size). The fractal dimensions indicated are provided with a (biassed) estimate of each standard deviation. Note that the slope of the fitted regression line provides the estimate of *D*. (N.B. In three cases, the average estimate for ten iterations happens to be less than the estimate displayed in the diagram.)

outline for individual 3 differs in two ways; firstly, there is a deeply indented concave zone and the convex marginal zone is far from regular. If the suggestion of a common ecological origin for the vestibules of individuals 1, 2 and 4 be true, then the result would imply a simplification of the model used by Peypouquet (1977) for ecologically monitored variation in this genus. A separate origin for the type represented by the third specimen is supported by Peypouquet's model.

There is as yet no broadly accepted unequivocal way of associating a fractal dimension with a set of data, but some points with respect to estimational procedures are clear. In a study of the present kind, it is necessary that the structures be observed at the same magnification. If not, then recourse must be made to standardization of the data to make them compatible. Secondly, the sets of data being compared should have compatible step-lengths. Estimates of the lengths of the contour are obtained successively for each element of the set of step-lengths. As step size increases, the estimate of D is less and less affected by intricacies of the contour. For this reason, it is customary to cite a value of the fractal dimension for a given experimental range.

Fractal geometry is a subject that we are going to see develop very rapidly and importantly over the coming decades. I have little doubt that the results for morphometry will be as far reaching and profound as they are proving to be for the physical sciences and technology. So keep your eye on the ball.

Exercises

1. In the method of shape coordinates, size is extracted from the analysis by standardizing the triangular shapes to baseline-length. This introduces a statistical constraint. Do you think that this is likely to have undesirable effects on the analysis? For a discussion of the effects of baseline standardization, see Bookstein (1986).

2. If you have measurements on a set of landmarks, how would you go about obtaining distance measures from them? Hint: Recall what you used to do in school geometry.

3. Devise sets of landmarks likely to be useful for analysing variation in shape in:
    a. trilobites
    b. ammonites
    c. fish-skulls
    d. mammal bones
    e. radiolarians
    f. human skulls

4. The examples given in Chapter 4 to illustrate geometric morphometrics only consider changes at and below the specific level. Can you devise procedures for studying relationships at a higher taxonomical level, say, within a family? Use the technique underlying the construction of the biorthogonal grid as a starting point and proceed to the principal warp method.

5. Look at the following piece of tensor notation:

$$\cos \theta = g_{ij} A^i B^j = g^{jk} A_j B_k = A^k B_k$$

If you were told that A^i and B^j are unit vectors, what would you expect that the above equations are saying in statistical terms. Hint: use equation (2.5).

If you would like to know more about the beauty of tensors, some good old classics are:

Tensor Calculus by Spain, Oliver and Boyd (1956).
Coordinate Geometry with Vectors and Tensors by E. A. Maxwell. Oxford University Press (1958).
Applications of Tensor Analysis by A. J. McConnell (1931), reprinted by Dover Publications in 1957.

Another good reference is *Space through Ages* by Cornelius Lanczos, Academic Press (1970).

6. The following data were used for constructing Figs 4.17–4.19 (measurements to a tenth of a millimetre). Plot the two sets of coordinate pairs on the same piece of graph-paper to produce the original shapes for the two species in relation to each other. Check this figure against Fig. 4.1.

$N = 9$ species $=$ *Agrenocythere hazelae*
Coordinates of landmarks

2541	1777	3830	1699
2837	2163	3392	1571
3340	2091	3175	1523
3508	2034	2807	1530
3679	2039		

27 coordinates around the outline

2544	1795	3334	2124	3355	1462
2574	1884	3410	2051	3191	1444
2612	1977	3533	2057	3046	1435
2674	2063	3652	2030	2908	1432
2776	2153	3687	1944	2804	1435
2843	2162	3752	1812	2704	1466
2925	2140	3815	1685	2619	1530
3055	2100	3660	1561	2564	1605
3190	2084	3512	1505	2544	1798

$N = 9$ species $=$ *Agrenocythere radula*

2458	0781	3937	0610
2832	1219	3529	0536
3398	1107	3154	0484
3632	1077	2822	0514
3846	1047		

28 coordinates around the outline

2451	0782	3563	1076	3453	0385
2472	0904	3684	1131	3197	0375
2516	1024	3835	1047	3076	0382
2582	1108	3886	0955	2936	0346
2725	1222	3946	0854	2650	0429
2829	1219	3996	0758	2564	0483
3057	1135	3969	0657	2459	0672
3196	1112	3917	0546	2451	0788
3323	1093	3810	0495		
3466	1116	3578	0412		

N.B. The set of nine coordinate pairs for each of the species refers to the landmarks. The second of coordinate pairs is for the outline of the valve and was not used in the calculations to produce the Principal Warps. The computations were made using a program by Professor F. James Rohlf, available as part of the publications from the Workshop on Morphometry, University of Michigan, Ann Arbor, U.S.A. (May 1988), published by the University of Michigan (1990). Note: If you have the volume and diskettes of the Michigan Workshop on Morphometry (Eds Rohlf and Bookstein, 1990), the format above will permit you to run the data directly in the program TPSPLINE.

7. Take two unequal sheets of square paper. Try to make the larger sheet fit the smaller sheet (you cannot do this legitimately by one or more linear folds). To do this you will have to crinkle the larger sheet. If you mark both sheets with a "homologous points", and try to fit these point-by-point, what happens around the most disjunct fits?

8. To what extent can the method of Procrustean Superposition be connected to the method of the Thin-plate Spline? Use the paper by Rohlf and Slice (1990) to get started. If you have the disk-package of the Ann Arbor workshop, you might like to use program GRF by Rohlf and Slice on the landmarks for ostracods in Benson (1972).

9. In estimating the fractal dimension of a closed contour, particularly one that is very irregular in shape, it is advisable to make several determinations, using randomly selected starting points. Can you find a plausible reason for this assertion?

10. A fractal dimension should be underpinned by adequate experimentation. Can you qualify this statement? What value would you consider to be an adequate representation of D on the basis of the following suite of randomly obtained estimates (for the anterior vestibule of *Krithe*)? Argue in terms of the mean *and* the mode.

1.041, 1.020, 1.046, 1.126, 1.028, 1.004, 1.032, 1.080, 1.072, 1.024, 1.083, 0.968, 0.984, 1.037, 0.997, 1.061, 1.027, 1.231, 1.053, 1.044.

CHAPTER 5

Application of Quantitative Genetics to Evolutionary Biology

Contents

5.1 Multivariate Analysis in Evolutionary Biology

5.1.1 *Introduction*

The rapidly expanding field of Evolutionary Biology encompasses various syntheses using results and methods of palaeontology, genetics, applied mathematics, and even new theories of its own. Notwithstanding that most of the deliberations are, and perforce must be, in qualitative terms, new avenues for quantitative analysis based on exciting new mathematical models are continually being opened. These are taken up in detail in a later section with particular reference to those that can be applied to fossils. In order to be able to profit from Chapter 5, I suggest you begin by consulting Falconer (1981) and Wallace (1981) in which the basic ideas underlying quantitative genetics are summarized and supported by numerous examples. If you wish to gain deeper insight into the mathematics of the subject, then I recommend the books by Bulmer (1980) and Manly (1985).

There are several concepts that need to be defined, some of which may be unfamiliar to some. I have therefore chosen to present a more general view of the field than for the topics taken up in other chapters. If you think that what follows is burdensome, pretentious, out-of-place or otherwise irrelevant, I suggest you sashay thru to Section 5.2. I shall first endeavour to define the concept of phenotypic plasticity in relation to quantitative work. Secondly, I shall consider the question of series in time in relation to evolutionary sequences in palaeontology and the notion of evolutionary rates. In connexion herewith, the idea of evolutionary stasis will be examined. The most extensive part of the chapter is concerned with the applicability to palaeontology of multivariate models in quantitative genetics.

5.1.2 *Phenotypic Plasticity*

Phenotypic plasticity, or ecophenotypy, is defined as the reaction of the organism with the environment. An important fundamental work on the subject is the book by Ivan Ivanovitch Schmalhausen (1949). A useful, quantitatively oriented introduction to the subject of phenotypic plasticity is to be found in an article by Scheiner and Lyman (1989). An informative way of entering into the topic is by reviewing some recent work on variability in landsnails.

Goodfriend (1986) has drawn attention to the fact that many species of landsnails display a remarkable breadth of morphological variability. Variation in size seems to encompass a large genetic component, but just how large this is remains to be assessed. The occurrence of smallish adults in dense populations seems to be due to the influence of pheromones. Another morphological feature of palaeontological relevance, particularly for quantitative studies on ammonites and other fossil molluscs, concerns the relative area occupied by the aperture. This tends to be smaller where snails live under dry conditions. Goodfriend suggested that this state could be the outcome of a reaction directed towards diminishing the loss of water from the shell.

Another finding of palaeontological interest concerns the fact that larger snails tend to have higher expansion rates of their whorls. Moreover, in the Mediterranean region keeled versus rounded peripheries occur. All of these features occur in fossil material of many species of gastropods and it is therefore important to keep in mind the far-reaching effects of ecological factors when interpreting a multivariate morphometric analysis. Many geneticists are directing attention towards studying how species evolve in response to complex environments (Scheiner and Lyman, 1989, p. 95). It is well known that one way for a species to adapt to complex environmental conditions is to be phenotypically plastic. Schmalhausen (1949) provides several examples of this kind of reaction to ecological conditions.

Smith and Patton (1988) were concerned with analysing geographical variation in gophers. They find size to have a strong environmentally controlled component and shape to be influenced by underlying genetic factors. The morphometric methods used by Smith and Patton include Burnaby's (1966) growth-invariant discrimination, canonical discrimination (by canonical variate analysis) and principal component analyses for each locality. The results obtained are, perhaps, not unexpected and in general, it was found that there is good agreement in the pattern of variation shown by gophers with Jolicoeur's (1963) principal-component size-shape model.

Another study I should like to mention here is that by Tissot (1988). He analysed geographical variation and heterochrony in two species of cowries in connexion with which descriptors of size and shape were employed for examining relationships between temperature and the growth of callosity.

5.1.3 *Evolutionary Models*

There is much discussion at the present time of models for describing morphological evolution. Few, if any, readers of this book will have failed to have run into the terms "punctuated equilibria" and "evolutionary gradualism". I have no intention of being sucked into the trap set by involvement with the polemics surrounding these terms. It is my opinion that the arguments presented in the controversies surrounding these ideas are often expounded *ex cathedra* in a form lacking scientific stringency and with many of the hallmarks of a doctrine and an impressive doxology for the innovators thereof. One's thoughts flee, unaided, to the fuzzy half-world of political debating and the kind of reasoning known to logicians as *ignoratio elenchi.*

Levinton (1988) gives a reasoned account, in my opinion, of many of the pertinent aspects of evolutionary modelling and Milligan (1986) viewed the field in terms of the quantitative genetic background required in support of the various arguments. We do well to recall that the idea of "punctuation" arose in palaeontological connexions with Simpson (1944), heralded by Alcide d'Orbigny in 1852 (cf. Laurent, 1987, p. 130) and is by no means new nor particularly frightening. Milligan (1986, p. 530) concluded that the punctuational pattern is a consequence of ecological interactions, an opinion he gained from the results of simulations. Charlesworth *et al.* (1982), in reviewing the subject of morphological change and speciation, remarked on the lack of resolution available from the fossil record and thought it unlikely indeed that interpretable morphological shifts could be recognized in the majority of fossil lineages. The degree of accuracy required can just not be obtained from fossil data was their conclusion. This is a problem to which we return repeatedly for it is a very real one which cannot be made to go away by the simple expedient of ignoring it. This leads to a consideration of averaging effects (cf. Abe, 1983): i.e. can significant shifts in size through time be expressed graphically in a meaningful manner, bearing in mind that any sample must perforce consist of a jumble of individuals. A further important argument to note in the article by Charlesworth *et al.* (1982, p. 490) is that adherents of punctualism concentrate on measures of body-size (they are not alone in doing that, I note). In a subsequent section, the vexing complexity of this problem is weighed. A more catholic selection of traits is required for a nuanced examination of the multitude of problems involved in analysing morphological evolution.

5.1.3.1 *Macroevolution and Microevolution*

Levinton (1988) has tried to tidy up in the attic of evolutionary nomenclature. He defined **macroevolution** as the sum of those processes that explain the character-state transitions that diagnose major taxonomical differences. A whole chain of microevolutionary changes may be entailed in an evolutionary transition which, when viewed collectively, constitute a macro-evolutionary event. By now you will have seen enough of the way in which evolutionary statistical analyses are constructed to understand that in many situations, it is the arbitrary choice of variables that dictate whether or not an observed evolutionary progression is of microevolutionary or macroevolutionary significance. An example I use in this book is the transition from *Echinocythereis isabenana* to *E. aragonensis*, two species of ostracods, in the Lower Eocene of Aragon, Spain (Reyment, 1985). One of the imprints of this transition is a marked reduction in volume, i.e. size, of the carapace. It might be argued that the admittedly permanent reduction in size is no more than the outcome of some long-term ecological whimsicality, however unlikely this may seem. But when this shift in volume is coupled with fundamental changes in ornament, the authenticity of the event would appear to be validated.

Statistical modelling of evolution in shape can be conveniently approached by way of the concept of *Bauplan* (Ger. structural plan), a structural entity that can be conceived of as evolving as a whole. Although perhaps not generally realized, it is this concept that underpins the Thompsonian grid constructions and aspects of the methodology of Principal Warps. It is recommended that once the pattern of the entity has been recognized taxonomically, and the dominant variables identified, multivariate procedures can be introduced for testing constancy in shape through time. One might also vary the choice of variables to see whether congruent results are obtained. In general, this can be expected since architectural constraints tend to determine form irrespective of function.

Benson (1972) attempted to portray relationships between taxonomic entities by graphical displays centred on and around prominent ornamental characteristics. This is often a useful and instructive way of expressing ornamental variability and Siegel and Benson (1982) have quantified the procedure (see Section 4.5.3). Benson's diagrams are an excellent way of setting up a chart of landmarks for ostracod carapaces.

5.1.3.2 *On Evolutionary Rates*

The challenge levelled by the problem of assessing evolutionary rates is a difficult nut to crack. The main source of the difficulties lies with geological factors and how is one to relate time to change in fossil material? Most studies of evolutionary rates employ labile distance measures (lengths, breadths, heights, etc.) with all the risks these entail of confounding ecophenotypic variability with the recognition of significant evolutionary change. I am no less guilty than any other practitioner of this analytical blandness.

Bookstein (1987, 1988) addressed the issues in two papers of fundamental significance for evolutionary biology. He showed that series in time of fossil data (for example, a sequence of univariate means for some measure of distance) cannot be interpreted, without the risk of courting disaster, in any of the simple fashions with which we have become accustomed, alas! He demonstrated that anagenesis and equilibrium in evolutionary sequences can be interpreted as deviations from Random Walk in opposite directions. Evolutionary rates only exist when the hypothesis of symmetric random walk can be rejected. Thus, the correct form for the null model is *not* that an evolutionary rate is nought but rather that it is a symmetric random walk. Moreover, time is not a suitable independent variable and one should strive to find a valid covariate, such as an ecological factor. This factor can then be used as an accessory for estimating an evolutionary rate.

In his second article on the subject, Bookstein (1988) points out that histograms of random walks in one or several dimensions almost always display what seem to be clusters of events, separated by gaps. Before such structures in a sequence of observations can be used as evidence of a systematic component in the data, one must reject that the possibility that they are consistent with a random walk having no measurable features relevant to location or separation. Bookstein tested his method on sticklebacks using data of Bell *et al.* (1985). Those authors had reported that they were unable to establish stasis in a sequence of these fish, but they did believe that they had confirmation of periods of rapid evolutionary change. Bookstein proved that time-trends do not exist in these data at all, and that evidence of stasis in some of the characters measured was, on the contrary, forthcoming. The general conclusion yielded by the analysis is that whenever there is a possibility that data conform with a random walk, any interpretation of an evolutionary time series in terms of an adaptive landscape can be misleading. This is a point of quite some importance for it necessitates much rethinking of conventional evolutionary studies. How many of these are spurious?

Kitchell *et al.* (1987) have also looked at spurious evolutionary rates, albeit with less mathematical sophistication than Bookstein. Another relevant contribution is that of Kirkpatrick (1982) who took up the question of quantal transitions in evolutionary sequences. One of Kirkpatrick's major conclusions is that the mean of a character (presumably a distance measure) can change by several standard deviations in only a few hundred generations; clearly, this makes a quantal shift "invisible" in most fossil sequences. Bookstein (1988) demonstrated by computer simulation that saltations interspersed with clumping can occur naturally in a sequence of random events. Hence, unless you have a specific statistical process to work on (Cox and Lewis, 1966), the interpretation of quanta in a sequence needs to be treated with great care.

5.2 Time Series in Palaeontology

The work reviewed in the preceding section can hardly be thought encouraging for the possibility of analysing series in time of fossils. Research done over the last few years demonstrates unequivocally that sequences of morphological means cannot be used for evolutionary reconstructions without an adequate statistical appraisal. Up to now we have only considered rather uncomplicated biological situations. Things in real life are far more complex. Let us now look at some further studies.

Hartl and Cook (1974) reported on autocorrelated (= serially correlated) random environments for which they used a model in which generations overlap and the environment fluctuates continuously. Positive environmental serial correlations were interpreted as having the same effect as increasing the variance in fitness of the organisms. Therefore, positive serial correlations intensify the possibility that the gene frequency will be near 0 or 1 and hence the population will be less polymorphic than in the absence of serial correlations. The palaeobiological significance of this result lies with the fact that polymorphisms are not infrequently pleiotropically connected to morphological differences as, for example, in the genus *Leguminocythereis*. It is therefore quite on the cards that serial correlations in a sequence can influence means in an arbitrary manner without significant evolutionary change having occurred. That is, the random course of positive autocorrelations influences the pattern traced out by some morphological character over time. The method of treating this situation statistically should lie within the framework of the results of Bookstein (1987), the only difference being that we are now required to examine the random course of serial correlations rather than that of the means.

I am afraid that it is not possible to get around the difficulty by saying that the means of morphological traits observed on a sequence of fossil organisms can be taken as uncorrelated. This is not true and the means are, and must be, correlated serially if they are the expression of evolutionary change in morphological traits. If the observed changes are being driven by some ecological force, then the series of ecological measures should be the primary target for testing. Charlesworth (1984) has pondered over several of the problems connected with this question while studying evolutionary patterns in morphological characters. We shall see later in this chapter that serial correlation can be induced as an artefact.

Thus, by their very nature, fossils generate series in time. There is a very natural desire to attempt to apply the means and goals of classical econometric time-series analysis to such data. The danger inherent in a straightforward assault by these methods should, however, by now be apparent to you, as has been brought home by Bookstein (1987, 1988) but, as we shall see (p. 204), the enlightened hierarchical application of the methods of time-series analysis to morphological sequences in time can yield interesting results.

What constitutes a statistically significant step from one morphological mean to the next? Kirkpatrick (1982) showed that a statistically significant difference between adjacent means in a time series is not necessarily evidence of significant evolutionary change. We need then to become clear on one vital point. It is important to be aware of the scale at which a discussion is being held. One of the cornerstones of classical palaeontology is known as Cope's Law. This expresses phylogenetic increases in terms of evolutionary advance. There are several favourite examples in the textbooks: the pachyderms, the horses, etc. and there is no doubt that a reasonable case can be made in respect of these lineages. In the context of the discussion in the present chapter, these changes are at quite a different scale from the series studied by Bookstein and many others in that they are concerned with taxonomically higher entities and not sequences within a species or a transition between two species. Charlesworth (1984) brought home the issues involved when he pointed out that a low rate of evolution in a very sparsely sampled sequence can well hide periods of rapid evolution in opposite directions. The role of the serial correlation coefficient in interpreting evolutionary series was well illustrated by Charlesworth. Connor (1986) has examined the question of serial correlation in palaeontological data in a detailed and stimulating paper.

5.2.1 *The Role of Covariates in Evolutionary Studies*

We have just learned that an initial consideration in a statistical appraisal of a sequence of morphological means measured on some fossil species must be to ask how valid these means are? Not even the most starry eyed practitioner of statistical palaeontology would like to suggest that the thanatocoeneses with which we must work can even be moderately representative of data obtained from living populations. Even sampling from Recent sediments can have its vicissitudes, notably the fact that a very thin layer of sediment may contain the remains of dozens and dozens of generations. Abe (1983) documented this convincingly for ostracods in his work on the associations of the Tokyo Bay environment.

What does all this tell us? Firstly, we cannot normally hope to recognize minor morphological fluctuations in the history of a species and hence seasonal ecophenotypic reactions. A notable exception is the chance offered by varved sediments where the varving is directly relatable to the seasons. Secondly, the short-term fluctuations we observe in sequences of means for fossils must be the expression of fortuitous compositions of the samples and cannot represent genuine morphological effects unless these are very strongly manifested over many generations. Thirdly, the hills and dales we actually measure are long-term reactions, be these genuine evolutionary trends or be they reflections of an environment under tardigrade displacement.

In order to stabilize the investigation of a series in time of fossil data, it is vital that *covariates* be sought out, such that observed variations in distance measures can be correlated with ecological factors and/or *identical covariational patterns* are proved to occur among other species sampled from the same bed. This confronts us with the difficult problem of carrying out valid correlations between series in time. The statistical aspects are far more complex than you might expect and special techniques are required. Gordon (1982) has given a review of the main methods for measuring the relationship between two variables which have been observed at specified depths down a single borehole. An example of the method is presented on p. 247. Gordon and Reyment (1979) examined several situations for assessing agreement in adjacent boreholes for multiple variables.

5.2.2 *Types of Sequences*

Types of sequences in time that occur in palaeontological studies are:

1. Univariate means of a distance measure.
2. Multivariate scores of means of distance measures.
3. Measures of outline such as the eigenshapes of Lohmann and Fourier shapes.
4. Frequencies of species, relative displacements in frequencies in relation to environmental influences.
5. Presence-absence data for species.

The statistical methods applicable to the first three categories are usually also suitable for the last two, as we shall see in Chapter 7.

In all five categories above, the role of the environment can be included by means of suitably constructed analytical procedures. Via and Lande (1985) underscore the kinds of questions that should be put in evaluating interaction between genotype and environment, although without an adequate feeling for points such as those raised by Bookstein (1987, 1988). It is, however, relevant in any study laying claims to reasonable completeness that due consideration be rendered to the expressions of evolution in the mean phenotype in different environments. In this way, the *norm of reaction* of the species can be charted. This vital factor is often glossed over lightly, or just ignored, in palaeobiological studies of evolutionary trends. Yet, it is a very real part of the mode of expression of any species. There is a weak point for item 2 in the above list. For many organisms, the multivariate mean scores have such a wide range of variation that it is hardly possible to construct a valid sequence in time from them.

Let us take the ostracods again as a useful means of illustrating the concepts involved. It has been shown that the ambient chemistry of the seawater during the developmental cycle of most ornamented forms plays a substantial role in determining the habitus of the adult individual. Thus, each set of chemo-environmental conditions (in turn coupled to temperature) is related to a particular degree of calcification and elaboration of the ornament—a *character state*. That is, the expression of a character in a given environment. The morphological features we observe, and measure, in an evolutionary study are often the expressions of character states, with all the statistical indeterminacy that this can engender (Schmalhausen, 1949). Consider, for example, the effects of ecologically cued polymorphism (discussed in Chapter 6). Usually, the coordinates of the environment will not be stable. In terms of quantitative genetics, the additive genetic correlation estimates the degree to which the phenotypes expressed in two or more environments have the same genetic basis, due either to pleiotropic effects of genes, or to linkage disequilibrium (Falconer, 1981).

5.3 Multivariate Applications in Quantitative Genetics

In this section, I review the role played by standard methods of multivariate statistical analysis in the quantitative genetics of the phenotype. The discussions are couched in terms of distance measures. However, there is nothing to stop the expansion to determinants of shape, such as considered in Chapter 4.

5.3.1 *Introductory Comments*

The field of quantitative genetics is one that has only recently begun to play a significant rôle in scientific research as opposed to its classical applications in agriculture. One would be foolhardy to deny that this is a difficult topic to treat in terms of fossils—it is hard enough to resolve for living organisms—but this is not a sufficient reason to give up without trying.

Quantitative genetics as we know it at present is largely the brainchild of one man, the distinguished geneticist D. S. Falconer, who adapted the theoretical developments of Sewall Wright and Kempthorne to the analysis of the phenotype. An early contribution by J. L. Lush was made in 1957. The subject can most conveniently be entered via Falconer's (1981) book, notwithstanding that the main field he treats is animal husbandry. Quantitative genetics can be conveniently defined as those differences between individuals that are of *degree* rather than of kind, **quantitative** rather than **qualitative**. The step to palaeontological applications was taken by Lande (1976) in a paper of fundamental importance, later expanded to the multivariate case (Lande, 1979). It is important to realize that these results were obtained with evolution in fossil organisms particularly in mind; they are not some half-baked amateurish extrapolation from standard genetics. Phillips and Arnold (1989) have recently given a synthesis of the multivariate relationships between fitness and phenotype in depicting natural selection. The relationship is expressed as a multidimensional surface relating fitness as a function of phenotypic traits. The statistical methodology is based on standard procedures of multivariate analysis in which the coefficients of (multivariate) selection are estimated by multiple regression.

Virtually every organ of any species shows individual differences—the differences in size in our own species, or of our domestic animals are familiar examples. Individuals form a continuously graded series from one extreme to the other and do not fall naturally into sharply demarcated types. Micropalaeontologists are very aware of this, faced as they usually are with a superabundance of material. The usual situation with which we are concerned in palaeontology is that of continuous variation. It is then different from what applies, for example, in the case of human serological characters in which classical Mendelian genetics pertain and the mechanism of inheritance can be observed. Inherited quantitative differences depend on genes the effects of which are slight in relation to variation from other causes. In the circumstances applying in the study of the inheritance of continuously varying characters, such as measures of size, quantitative differences are usually, but not invariably, influenced by gene differences at many loci; consequently, the individual genes cannot be identified by their segregation.

The unit of study in quantitative genetics then cannot be the individual but must be the *population*. As I have already pointed out, the foundations of the subject lie with the genetics of animal husbandry. There is a vast reservoir of information on experience with living animals from which to draw. Unfortunately, the data available for naturally occurring creatures borders on the scanty. If we go back to the very origins of work leading to what we now term quantitative genetics, we arrive at the days of the early biometricians, Galton (1887) and Pearson (1901). Later important developments are those of Fisher (1918) and Wright (1931). Some of the basic ideas of modern mathematical statistics, correlation, regression and the analysis of variance, were produced in answer to questions concerning the inheritance and evolution of continuously varying characters.

Although the theoretical advances of biometrics have evolved separately from quantitative genetics, the latter depends on the former for its development, this being true all the more of late, with the increased use of multivariate methods in applications to evolutionary biology. Here, quantitative genetics has proven eminently valuable for couching analyses in terms of Natural Selection and Phenotypic Evolution in continuously varying characters. Studies based on fossils cannot be expected to yield unequivocal information on selection and selective mechanisms. One difficulty lies with the possibility that a selection intensity of a few percent (which is quite strong) can go unrecognized owing to statistical difficulties, spatially varying selection, and differing genotype-environment interactions in various habitats. I referred to some of the work on this subject in the first section of this chapter. If you wish to become adept at the computational methods of quantitative genetics, I suggest you obtain the manual by Walter Becker (1987).

5.4 Evolution in the Phenotype

In simple terms, the theory introduced by Lande (1976, 1979) describes the evolution of the mean values of a set of quantitative phenotypic variables denoted by the column vector:

$$\mathbf{z} = (z_1, z_2, \ldots, z_p)^T. \tag{5.1}$$

For our purposes, the z_i will usually be "distance measures" on the carapace of some fossil—we have defined this concept on in Chapters 1 and 4. Suffice it to recall that such traits are the "common" things we measure on fossil shells. We could, of course, also consider some specific measures of shape, such as we can obtain from eigenshape analysis, Fourier analysis, the features of the average shape coordinates of Bookstein (1986), etc.

As intimated by equation (5.13), further on, the phenotype of an individual is taken to be composed of the sum of a vector of additive genetic effects (breeding values) and an independent vector of "environmental effects"; the latter can be conveniently made to encompass "developmental noise", dominance, epistatic genetic variance, as well as genotype-micro-environmental interaction. The theory requires the postulation that the breeding values and phenotypes have a multivariate normal distribution in the population before selection each generation. Using these rather mild assumptions, Lande (1979) derived a means of analysing a general form of natural selection that assigns **expected fitness**, $\bar{W}(z)$ to individuals with phenotype z.

Let $p(z)$ denote the distribution of the phenotype in any generation before selection. Then, the mean fitness in a population with mean phenotype z can be written as:

$$\bar{W} = p(z)W(z)dz. \tag{5.2}$$

In a panmictic population (equal chances of mating in the population) with discrete, non-overlapping generations, the change per generation in the mean phenotypic vector, measured each generation before selection is given by:

$$\Delta \bar{\mathbf{z}} = \mathbf{G}\nabla \ln \bar{W}$$

$$= \mathbf{G}\,\boldsymbol{\beta}. \tag{5.3}$$

Here, $\mathbf{G}$ is the *additive genetic covariance matrix* of the morphological characters, and $\boldsymbol{\beta} = \nabla \ln \bar{W}$ is the **gradient** of the surface of log-mean fitness with respect to changes in the mean phenotype, with ∇ defined as the differential operator. If you want to know more about the differential operator, I can refer you to Schwartz *et al.* (1960, pp. 80–86). This is a very clear exposition in which the del (or nabla) operator, ∇, is defined and its use illustrated.

Thus, we can draw the conclusion that the evolution of a particular character depends not only on selection acting directly on it, but also on indirect selection acting directly on genetically correlated characters.

$$\Delta \bar{z}_i = \sum_{j=1}^{p} \mathbf{G}_{ij}\boldsymbol{\beta}_j. \tag{5.4}$$

Here, $\mathbf{G}_{ij}$ is the *additive genetic covariance* between the characters z_i and z_j, and

$$\boldsymbol{\beta}_j = \frac{\partial \ln \bar{W}}{\partial \bar{z}_j}$$

represents the first element of **directional selection** acting directly on character z_j. This latter quantity is the relative change in mean fitness in the population produced by a small change in the mean of character z_j while holding the variance of that character, and the distributions of all other characters, constant.

Lande (1979) gave another formulation of *Natural Selection* in terms of a vector of *selection differentials* **s** on a set of characters (see (5.7) below)

$$\Delta \bar{z} = GP^{-1}s.$$

Using (5.4), the *selection gradient* can be written as:

$$\beta = P^{-1}s \tag{5.5}$$

where **s** is defined as the vector of differences for the characters before and after selection:

$$s = \bar{z}^* - \bar{z},$$

which then relates to the formulation of Lande (1988). You will observe that equation (5.5), is in general terms the statistical equivalent of a linear discriminant function (cf. formula 3.7). For many purposes, the above expression is to be preferred, as has been shown empirically by Cheverud (1988). The foregoing ideas are the mathematical expression of the concept of adaptive zones of Simpson (1944, 1953), in its turn based on an analogy with the notion of *adaptive topography* of Wright (1931) for evolution in gene-frequency. Simpson's ideas were partly based on the belief that natural selection increases adaptation.

Now, if we allow the population to be represented as a point on a surface of mean frequency $\bar{W}$ as a function of the mean-values of phenotypic characters in the population, we can say that the dynamic significance of the foregoing equations is that the evolution of a population does not follow the steepest uphill direction, as expressed by the selection gradient, β, but in a direction, partly determined by **G**, which also partially determines the rate.

We continue now to a consideration of how the theory attempts to arrive at measuring natural selection.

5.4.1 *Measurement of Natural Selection*

Evolutionary biological work runs into a difficulty with respect to determining which of a set of correlated characters are actually being acted upon by selection. In the case of *directional selection*, which functions so as to change the mean phenotype in a population, one can propose a kind of solution, as has been done by Lande (1988).

Let us define individual relative fitness as

$$w = \frac{W}{\bar{W}}. \tag{5.6}$$

We need $\mathbf{P}$, the *phenotypic covariance matrix* of the p characters of interest, and the difference vector

$$\mathbf{s} = \bar{\mathbf{z}}^* - \bar{\mathbf{z}}.$$

Here, the asterisk denotes a quantity *after selection* and within one generation. The required solution can be summarized in the following terms. As already stated above (equation 5.5) the selection gradient can be written as in equation (5.7)

$$\boldsymbol{\beta} = \mathbf{P}^{-1}\mathbf{s} = \mathbf{P}^{-1}\,\mathrm{cov}(\mathbf{w}, \mathbf{z}). \tag{5.7}$$

I have already pointed out that (5.7) has the general form of a statistical distance. It can also be seen to be a vector of partial regression coefficients in a linear multiple regression in the population within a generation (Lande and Arnold, 1983).

5.4.2 *Stabilizing or Disruptive Selection*

Stabilizing or disruptive selection on single characters, and the tendency of selection to correlate sets of characters, has been studied by Lande (1988), who notes that estimates can be provided by analysing the change in the covariance matrix of breeding values caused by selection within a generation. This is determined by a symmetric matrix of coefficients in a quadratic regression of individual relative fitness on the individual phenotypes.

$$\boldsymbol{\Gamma} = \mathbf{P}^{-1}\mathrm{cov}(\mathbf{w},(\mathbf{z} - \bar{\mathbf{z}})(\mathbf{z} - \bar{\mathbf{z}})^{\mathrm{T}})\mathbf{P}^{-1}. \tag{5.8}$$

For the phenotypic distribution before selection multivariate normal, $\boldsymbol{\Gamma}$ can be interpreted as a matrix of coefficients of average curvature of the surface of relative fitness, $w(\mathbf{z})$, weighted by the distribution of the phenotype (Lande and Arnold, 1983).

Directional and stabilizing (disruptive) selection may act together on a single character. A positive (negative) value of γ_{ii} indicates directional selection to increase (decrease) the mean of character z_i. A negative (positive) γ_{ii} indicates stabilizing (disruptive) selection favouring intermediate (extreme) values of z_i.

Although the foregoing results may seem abstruse and likely to be little to the point in palaeontological studies using real data, they are nevertheless important in that they establish the form of multivariate statistical and hence morphometrical analysis of evolution in the multidimensional phenotype. It is necessary to know what the theory says if useful interpretations of actual data are to be made. This may seem to be a trite statement, but it is my experience that a great many studies in evolutionary biology are made without due regard to the theoretical requirements of the methods used.

5.4.3 *Heterogeneous Environments and Phenotypic Plasticity*

Falconer's (1960) genotype-environment interaction in terms of correlations between the same character differently expressed in different environments is an important way of modelling a commonly occurring situation in the natural environment. Via and Lande (1985) modelled the evolutionary dynamics of *phenotypic plasticity*, or "norms of reaction" in a population occupying a discrete set of environments, a simplification of true life but none the less an important first step in model-building.

An *optimal* mean phenotype is produced in each environment by phenotypic plasticity, unless *there is no genotype-environment* interaction (Lande, 1988, p. 80). This is a basic premise of models that involve ecophenotypic reactions, as was indicated in the first part of this chapter. Phenotypic plasticity may lie at the root of the great variability displayed by some ammonites (see Example 8.8 in Chapter 8). Reyment (1988) thought that variation in Turonian (Cretaceous) vascoceratids in North and West Africa was in part, at least, to be placed at the door of phenotypic plasticity. If the plasticity of a character, or characters, has a greater heritability than the trait itself, it is not unreasonable to expect the primary evolutionary reaction to be an increase in the plasticity of individuals. Palmer (1985), referring to variation in the shells of gastropods, pointed out that intraspecific shell variation often correlates with modifications in environmental conditions. He reported a dimorphic state that seems to be inherited in a Mendelian fashion but in nature, there are intermediates between the two states. He also showed that the ecological conditions experienced as a juvenile can partly influence the adult phenotype. The important work of Scheiner and Lyman (1989, p. 96) addresses the question of quantifying the genetic component of phenotypic plasticity. At this point, you would do well to delve into the book by Manly (1985) on the statistics of natural selection.

5.4.4 *Genotypic and Phenotypic Correlations*

It will doubtless have occurred to you that all is not plain sailing in the palaeontological application of quantitative genetics. What do you do about genotypic and phenotypic correlations? Let matters become even murkier; it is also a difficult point to deal with for most living organisms. There are absolutely no data on genetic correlations for most animals of interest to palaeontologists and the best one can hope for is to borrow determinations from some presumed related organism. This was done by Lande (1976) in the vertebrate examples supporting his theoretical results. But there are rays of hope. Cheverud (1988) has reported on genotypic and phenotypic correlations in a study supported by empirical evidence. Cheverud asks whether phenotypic parameters can be used in place of genetic counterparts. This question has been asked by many others, but without the support of data to back up the arguments. The reason for putting the question should be immediately clear. If we can use phenotypic values, which are easily obtained, in place of genotypic, then applications to fossil data become palatable in that we can make direct applications of theory without the need for arbitrary approximations and guesses.

Cheverud's empirical study showed that on first sight, $R_g^2 > R_p^2$ and that their patterns are only broadly similar. The data providing the basis for this conclusion were obtained from the literature and, for the most part, derive from quite small samples. He found, however, that when large sample-sizes are used, the difference between genetic and phenotypic correlations is slight indeed and the general pattern of correlation turns out to be similar. Cheverud concluded that much of the divergence between genetic and phenotypic correlations reported in the literature is the outcome of experimental imprecision in determining the former. It seems logical to conclude that genetic and environmental causes of variation tend to act on growth and development in a similar manner. The phenotypic correlation coefficient can therefore be partitioned in the following way:

$$r_p = h_x h_y r_g + e_x e_y r_e \tag{5.9}$$

where e is the square root of the proportion of phenotypic variance due to environmental factors and h is the heritability (h_x for variable x and h_y for variable y). Cheverud's empirical results seem to suggest that there is a constant relationship between r_p and r_G. There is ample scope for further research in this field.

Cheverud (1988) uses the following relationship to express the connexion between genetic and phenotypic variance:

$$\mathbf{V}_G \mathbf{V}_G^T = k \mathbf{V}_p \mathbf{V}_p^T \tag{5.10}$$

where the $\mathbf{V}$ are standard deviations.

Scheiner and Lyman (1989, p. 5) define the *heritable component* of plastic variation as

$$h_{p1}^2 = \sigma_{GE}^2 / \sigma_P^2$$

where the suffix $P1$ denotes "plasticity", σ_{GE}^2 is the variance for the interaction between genotype and environment (actually a covariance), and σ_p^2 is the total phenotypic variance (Becker, 1987). The heritable component is thus expressible as a simple ratio of variances. The determination of σ_{GE}^2 can only be achieved by careful experimentation for a broad range of simulated environments.

Other thoughts on much the same subject, to wit, the constancy of relationship between the phenotypic and genetic covariance matrices have been expressed by Turelli (1988), Lofsvold (1986), and Kohn and Atchley

(1988). Mantel's permutation test is sometimes used for checking the equality of covariance matrices (cf. Manly, 1985), a procedure that is not without its weaknesses (Turelli, 1988, p. 1345). The main thrust of these investigations has, however, been directed towards assessing whether covariance matrices remain constant through time. All the various rather indecisive testing indicates that this probably is not true and that covariance matrices vary significantly as time proceeds. Surprised?

There is also the question of **Geographical Variation**. Riska (1985) found that correlations vary geographically in a random manner in local populations of the aphid *Pemphigus populicaulis*. Riska also found that significant differences in correlations among populations were mainly due to differences in the overall magnitude of correlation represented at the localities. He concluded that short-term selection could be the cause of apparent random geographical distances in this species. The study also contains observations on phenotypic and genotypic covariance matrices which might eventually prove valuable as research on the topic proceeds and can doubtless come to be of value in palaeobiological work. It must be emphasized that precision of the kind required for such analyses necessitates the use of quite large samples if stable, unequivocal results are to be obtained. I consider that samples of 200–300 may be necessary in many cases.

5.4.4.1 *Practical Estimation for Fossils*

In the case of some species of freshwater ostracods, quite reasonable estimations of the quantities discussed in the present section can be made. Estimates of the environmental correlation coefficient between two traits measured on an organism and the additive genetic correlation coefficient can be made in a realistic manner for parthenogenetic forms. Parthenogens permit us to gauge the order of magnitude of the reaction of the traits to environmental influences. Hegmann and DeFries (1970) pointed out that the phenotypic correlation coefficient r_p is a direct estimate of r_E, the environmental correlation coefficient, for individuals from an isogonic population (cf. Noordwijk *et al.*, 1980; Reyment, 1982e).

As an example, I use the results of Reyment and Brännström (1962) for the freshwater parthenogen *Cypridopsis vidua* (Müller), a species that displays obligatory parthenogenesis. The correlation between length and height of the carapace is 0.6. If this value is inserted in equation (5.11) we obtain

$$r_A = (r_P - (1 - h^2)r_E)/h^2 \qquad (5.11)$$

$r_A = 0.8$. Thus, the additive genetic correlation coefficient is only slightly different from the phenotypic correlation coefficient, which agrees well with the observations of Bailey (1956), Hashiguchi and Morishima (1969), Hegmann and DeFries (1970), Leamy (1975) and Cheverud (1988). There is, however, still a dearth of research on the relationship between the additive genetic correlation coefficient and the phenotypic correlation coefficient.

5.4.5 *Drift and Selection*

This section becomes rather technical and if you have not done much Physics, it can give a little trouble, for which I apologize. I suggest you prepare yourself by consulting some such book as Defares and Sneddon (1960, Chapter 8) for an introduction to partial derivatives. For the genetical aspects, you would do well to read Maynard Smith (1989). To quote John Maynard Smith in that book, "if you can't stand algebra, keep out of evolutionary biology". I think you should know that modern genetic theory is largely the work of mathematicians. Ronald Fisher, the renowned applied mathematician and who was for many years the incumbent of the Chair of Genetics at Cambridge, was not a biologist.

I now return to Lande's fundamental palaeontological developments of the theory of quantitative genetics. Among the many things taken up by him (Lande, 1976), random genetic drift and selection were the most important. Almost all genetic theory is concerned with the genetic system, embracing such topics as gene frequencies and recombination rates that cannot be observed or observed from observations on *polygenic characters* (Falconer, 1981).

The analysis of phenotypic data requires models that are expressed as far as possible in terms of the phenotype. Lande's contribution to this difficult and obviously controversial subject was to take Simpson's (1953) qualitative synthesis and, by means of the basic formula of quantitative genetics (Bulmer, 1980, pp. 144ff.) develop a theory of evolution in the average phenotype by Natural Selection or Random Genetic Drift, or a combination of these. In doing this, he employed relevant areas of the theory of population genetics. Lande's model shows that with constant fitnesses, the average phenotype evolves towards the nearest adaptive zone in the space of the phenotype. I am going to be so fey as to suggest you make the effort to work through Lande's (1976) paper. It is one of the gems of evolutionary biology.

The main idea underlying the Lande model can be formulated in terms of two considerations:

1. Firstly, we have the problem of estimating the *Minimum Selective Mortality* needed to produce an observed rate of evolution.
2. Secondly, how does one evaluate an hypothesis of evolution by random genetic drift in relation to the *Effective Size of the Population*?

The limiting case of drift in the absence of selection is important as a test of the power of random genetic drift in a particular situation. If drift cannot be ruled out, there is no basis in fact for assuming that phenotypic evolution in a specific situation is the outcome of natural selection by itself. The perspicacious will no doubt have discovered the soft belly of the drift-selection model as outlined above. We are required to direct attention to an "observed rate of evolution", but we have just been told in the foregoing section that this concept is fraught with doubtful assumptions. Thus, as long as we confine ourselves to thinking about the drift-selection model in a theoretical frame, we can proceed without qualms. It is first when we attempt practical applications that all sorts of complications can engulf us unless due care is paid to avoiding the snares. Charlesworth (1984) has tried to sort out some of the difficulties inherent in determining rates of evolution but these are mainly statistical (Bookstein, 1987, 1988). Long and well controlled series are needed, say about 1000 sampling levels (Lewis, 1964), if the statistical properties of a sequence of events are to be established with any degree of certainty.

We must therefore bear in mind that random genetic drift can never be proven for fossil data, only inferred, since any pattern of migrational episodes can have brought about the observed pattern of morphological change. Fluctuating selection could also have done this, and also aleatoric variation. Computer simulation experiments demonstrate that a monotonic trend can be produced for a character by chance alone (Raup, 1977). But remember when you evaluate these results that such a trend is an exception, not a rule In order to distinguish between such chance trends and evolutionarily significant shifts, it is necessary that the regional validity of the phenotype be assessed and established (Lande, 1979a). This amounts to testing the orientation of estimates of the selection gradient for collinearity at several well separated sites.

There are several time-bombs built into the theoretical development of the drift-selection model. I have already underlined the fact that the estimation of evolutionary rates can be something of a chimaera. Moreover, the model of selection used in the theory is not the most realistic one available, its advantages lying with the fact that it produces a tractable mathematical treatment. There is also the vexed problem of giving a rational definition of what constitutes a reasonable population size for drift to be invoked as an active evolutionary mechanism. Hartl (1980) and Falconer (1981) seem to opt for a few tens to hundreds of individuals. Some modern workers talk in terms of hundreds of thousands and serologists accept populations of that size in their analyses of the effects of random drift on the distributions of blood groups (Mourant, 1983).

Can we recognize the processes of evolutionary change in fossils? Petry (1982) casts a pessimistic shadow over our chances of being successful. Bell *et al.* (1985) sound more positive tones, at least for material deriving from episodes of mass mortality such as occur among fish and ammonites.

5.4.5.1 *Note on Heritability*

A fundamental equation of quantitative genetics describes the deterministic changes in the average value of a phenotypic character z in response to selection for a population with discrete generations and with constant fitness and infinite size (Falconer, 1981; Bulmer, 1980). We can write this as follows:

$$\Delta \bar{z}(t) = \bar{z}(t + 1) - \bar{z}(t)$$
$$= (\bar{z}_\omega(t) - \bar{z}(t))h^2. \tag{5.12}$$

Here, $\bar{z}(t)$ denotes the mean value of the character in generation t, before selection, and $\bar{z}_\omega(t)$ is the mean after selection, but before reproduction. The *realized heritability* of the character is h^2.

Let us briefly consider the heritability and its role in equation (5.12). The heritability is determined by the genetic system, the breeding structure of the population, and the environment. Presumably, it is not likely to change during the course of evolution (although some recent work may point in that direction). For living animals, the heritability is estimated by assessing an offspring–parent relationship or a sibling correlation. Obviously, you cannot do this for fossils. One way around the difficulty is to "borrow" a value of h^2 from some living organism, related to the fossil species under study. This can be a useful approximation for large mammals, since most available determinations of heritability are for farm animals. Another way is to construct empirical confidence intervals, a method made possible by the relative inertia displayed by equations (5.19) and (5.20) in the following. This can be done over a realistic range of values of h^2. Note that the heritability is defined for the range $0 \leq h^2 \leq 1$.

The phenotypic variance V_P can be partitioned into two components, the genotypic variance V_G and the variance due to environmental effects, V_E. We have already considered an aspect of this partition on p. 174, equation (5.11). The equation for the partition may be written as follows; note the presence of a covariance term.

$$V_P = V_G + V_E + 2\text{cov}_{\text{GE}}. \tag{5.13}$$

It has been customary to regard the covariance term as insignificant (Searle, 1961). In an alternative mode of partitioning, the genetic variance is broken down into a term V_A, the additive genetic variance, V_D, the variance due to dominance, and V_I, the variance arising from deviations due to interaction. In some studies, the covariance term is absorbed into V_A. However, if the additive genetic variance cannot be estimated consistently, there will be instability in the heritability. Hallauer and Miranda (1981) have demonstrated for maize that there may be considerable interaction between V_A and the environment. The additive genetic variance is the main determinant of response of a population to selection. Making use of the foregoing set of definitions, we can express the heritability as the quotient

$$h^2 = V_A/V_P \tag{5.14}$$

that is the additive genetic variance divided by the phenotypic variance. In the case of offspring–parent relationships, this is equivalent to the regression of offspring on parent (Falconer, 1981, p. 148).

Consequently, the heritability is a property not only of a character but also of the population and the environmental conditions surrounding it. I should perhaps mention that it has been shown that there is an association between the size of the population and the heritability. Generally, the heritability displays low values for characters of reproductive, hence evolutionary, significance (and economic significance in farming), but high values for characters of size and shape. Such characters tend to be less important as a determinant of natural fitness.

5.4.5.2 *Rate of Evolution and the Adaptive Landscape*

Analogous to Wright's (1932) expression for change in gene-frequency, Lande (1976) derived the following equation (5.15), which shows that the rate of evolution is controlled by the local topography of the adaptive landscape. With average fitness $\bar{W}$

then

$$\Delta \bar{z}(t) = h^2 \sigma^2 \frac{\partial \ln \bar{W}}{\partial \bar{z}(t)}. \tag{5.15}$$

The logarithmic slope expressed by

$$\frac{\partial \ln \bar{W}}{\partial \bar{z}(t)}$$

indicates the above postulation to be in accordance with the model. The other determining factor is provided by the amount of heritable variation $h^2 \sigma^2$. Here, σ^2 denotes the phenotypic variance and $\ln \bar{W}$ is the natural logarithm of the mean fitness of the individuals in the population.

5.4.5.3 *Natural Selection and Phenotypic Evolution*

We shall now examine a model for assessing the part played by natural selection in phenotypic evolution. This model is based on a measure of the rate of evolution proposed by Haldane (1949), namely, that a meaningful yardstick is the *rate of change in units of the phenotypic standard deviation*. This measure is tractable for mathematical modelling, although it is not the only way of representing the rate of phenotypic evolution, nor necessarily the best; recall what we said in the first part of this chapter about the ever-present danger of spurious conclusions concerning rates of evolution in the phenotype.

Lande's model uses a quantitative measure of the minimum intensity of natural selection required in order to explain an observed evolutionary change **in the absence of random genetic drift**. The model is based on *truncation selection*, well known from animal husbandry, in which a fixed proportion of the most extreme deviants are culled every generation. This is, as I have already mentioned, not the most realistic approach for naturally occurring animals and plants and some other forms of selection could well be more relevant (cf. Charlesworth, 1984). All phenotypes are said to have a fitness of one, unless they lie beyond the truncation point, in which case they have a fitness of nought.

Let b denote the number of phenotypic standard deviations between the average phenotype and the truncation point (Haldane, 1949); the latter lies at

$$\bar{z}(t) - b\sigma$$

$$\frac{\partial \bar{W}(z)}{\partial \bar{z}(t)} = \pm \delta[z - (\bar{z}(t) - b\sigma)]. \tag{5.16}$$

Where δ denotes the Dirac delta (Condon and Odeshaw, 1958, pp. 1–69). The function (5.16) is zero everywhere except at the point of truncation, where it is infinite.

For weak truncation selection on a normally distributed character z and with $\bar{W}$ approximately $= 1$, it can be shown that

$$\Delta\bar{z}(t) = \pm \frac{h^2\sigma}{2\pi^{1/2}} \exp\left(-\frac{b^2}{2}\right). \tag{5.17}$$

If the heritability and phenotypic variance are approximately constant, the total morphological change after t generations is

$$z = t\Delta\bar{z}(t)$$

which through equation (5.17) yields the relationship:

$$\frac{|z|}{\sigma} = \frac{h^2 t}{2\pi^{1/2}} \exp\left(-\frac{b^2}{2}\right). \tag{5.18}$$

This can be readily rearranged to give an expression in terms of b (equation 5.19):

$$b = \pm\{-2\ln[(2\pi)^{1/2}(|z|/\sigma)/(h^2 t)]\}^{1/2}. \tag{5.19}$$

The estimate of b yielded by (5.19), entered into a table of the standard normal integral, yields an estimate of the proportion of the population culled each generation required to result in the observed morphological change.

An approximate multivariate analogue of (5.18) and (5.19) can be obtained by using discriminant function scores instead of univariate z (cf. equation 5.23). Alternatively, the scores of the first principal component provide a reasonable multivariate approximation for p variables. In both cases, you can expect considerable variability in the values. Mention should be made of Charlesworth (1984) who has discussed various methods for providing an estimate of the "rate" of evolution.

5.4.5.4 *Random Genetic Drift and Phenotypic Evolution*

The discussion in the previous section was made in terms of selection in the absence of random genetic drift. In the present section, I shall be concerned with evaluating drift in the absence of selection. These are, of course, both unrealistic assumptions as far as they go, but this development is necessary since it supplies bounds that we shall need for the practical survey of various problems.

Wright (1931, 1932) seems to have been the first to realize that random genetic drift due to finite population size may be an important factor in evolution. Genetic drift causes random shifts in gene frequencies, which can cause a population to move out of an adaptive zone, against selection pressure, and possibly enter a new adaptive zone on a higher level of adaptation. The drift hypothesis is intellectually appealing and it forms a cornerstone in many an evolutionary model.

Lande (1976, p. 321) derived an approximate statistical test for ascertaining whether an evolutionary event could have been caused by drift, with the limiting case of **no selection** as a test of the power of drift. This test is constructed so as to determine how small the **effective population** size would have to be for there to have been a statistically significant chance of producing an observed morphological change by drift in the absence of selection. After t generations, in the absence of selection, the probability distribution of the average phenotype is normal at the starting point and with variance

$$\sigma_\phi^2 = h^2 \sigma^2 t / N.$$

The effective population size N^* at which there is a 5% chance of drifting a distance at least z in either direction in t generations is obtained when the observed magnitude of morphological change $|z| = 1.96\sigma_\phi(t)$. The solution with respect to N^* is

$$N^* = (1.96)^2 h^2 t / (z/\sigma)^2. \tag{5.20}$$

If $N > N^*$, i.e. the true effective population size is greater than N^*, then the shift in means is greater than can be accounted for by genetic drift alone. Estimates of the means inserted into (5.20) do not take account of sampling error, which naturally creates a source of uncertainty (Manly, 1985). Equation (5.20) is hardly a test in the true statistical meaning of the word and it is to be regarded more as an indicator of a likely situation than a proof. Care must therefore be exercised in the application of (5.20), since uncritical use can lead to fallacious conclusions.

Even for living organisms it is extremely difficult to estimate genetically effective sizes for naturally occurring populations (Futuyma, 1979, p. 277). Falconer (1981) seems to accept that drift can only be a real factor in populations of effective size less than 200 individuals (cf. Lande, 1979b, pp. 234, 244, 247). Lande's test (5.20) seems to indicate that drift can actually take place in relatively large populations. For oreodont mammals, he calculated a population size of $N^* = 200,000$.

I think you will have gathered by now that the methods reviewed in this chapter are stimulating and valuable. Their main importance lies with the possibility they offer of producing hypotheses for evolution in the phenotype. I have gone to considerable pains to point out the limitations of these deterministic to semi-statistical models (see the remarks by Crespi and Bookstein, 1989). Providing you keep these reservations in mind, much useful information can be obtained with their help. I shall now illustrate the methods with a simple example for a single size measure, drawn from the sphere of micropalaeontology.

Example 5.1. Morphological Change in a Species of Cretaceous Ostracods

Presentation of Problem: Two borehole levels, estimated to be separated by 100,000 generations, were available for the ostracod species *Veenia rotunda* REYMENT from the Cretaceous of southern Morocco. The average height of the carapace was ascertained to be the most diagnostic of all the characters measured on the carapace. The average difference in height between the two levels is 0.07 mm and the phenotypic standard deviation is 0.037.

Theoretical Requirements: The procedures used for the analysis are those summarized in formulae (5.19) and (5.20). We shall examine the hypothesis of selection in the absence of drift, then the hypothesis of random genetic drift in the absence of selection. This is biologically unrealistic but this approach is used here because it gives values for opposite poles. The final assessment should be made in the light of evolution proceeding mainly by one of the confronting models, with interaction from the other. This is discussed further on in this chapter.

The Analysis: The results of the calculations are reported in Table 5.1. The first column in the table, bearing the heading N^* tells us that the population sizes for the drift hypothesis would have to be quite large if the observed difference were to have been caused by random effects alone. However, noting that there are very few males in the sample, the effective population size would in actual fact have been much less. Hartl (1980) gives a good account of this factor. The column bearing the heading "culled" indicates the number of selective deaths per generation per million individuals that would be needed in order to bring about the observed morphological change. If we compare these results with the discussion given by Lande (1976) for his mammalian examples, we see that the figures presented in Table 5.1 are indicative of fairly

TABLE 5.1. *Results of the calculations for opposing hypotheses of random genetic drift and selection for* Veenia rotunda *REYMENT from the Moroccan Upper Cretaceous for* 100,000 *generations*

Difference in means = 0.07 mm; standard deviation = 0.037.

Variation in Heritability Test

h^2	N^*	Number culled
0.2	21,466	15
0.3	32,199	10
0.4	42,932	7
0.5	53,665	5
0.6	64,398	4
0.7	75,131	4
0.8	85,864	3

strong selection. Given that ornamental characters seem to have a heritability around 0.4, a culling value of about 7 seems quite reasonable.

Concluding Comments: What, you might ask, happens if I have wrongly estimated the number of generations? We shall see. Table 5.2 gives the same sequence of calculations for 200,000 generations. Note that the results could have been obtained without the aid of computer. The culling values are halved and the population sizes doubled. The culling values are clearly still quite high, in fact high enough to support an hypothesis of moderately strong selective effects.

The *Veenia* example shows that although the theories are all very nice and necessary, they can do no more than indicate the content of an idea. Obviously, random sequential variation cannot be accommodated by the procedures employed here, inasmuch as we are concerned with the phenotypic distance measured between two levels, and hence a deterministic representation of the evolutionary situation. The multivariate analogue of the present example could have been made using average scores for the first principal component based on a set of morphological traits and the square root of the

TABLE 5.2. *Results of the calculations for the same set of data as presented in Table* 5.1, *but with the number of generations set at* 200,000. *The difference in means is* 0.07 mm *and the standard deviation* 0.037

Variation in Heritability Test

h^2	N^*	Number culled
0.2	42,932	7
0.3	64,398	4
0.4	85,864	3
0.5	107,330	2
0.6	128,795	2
0.7	150,261	1
0.8	171,727	1

corresponding latent root as the estimate of the phenotypic variance. Heritability could have been estimated tentatively as an averaged value as a first approximation, since heritabilities for the dimensions of the carapace of ostracods can be expected to be roughly the same.

5.4.5.5 *Estimating Heritability for Fossils*

I have already broached this subject earlier in this chapter, but a few more words need to be expended thereon. The method I use is to compute curves over a meaningful range of heritabilities which for one-organ systems, such as occur in the study of fossil organisms, lie between 0.3 and 0.6. Morphological, including ornamental, traits tend to have heritabilities in the upper reaches of the scale, whereas those of reproductive significance are relatively low. This procedure can be used to produce empirical confidence intervals. The method has proven effective, particularly for obtaining culling estimates since the curve for b rises slowly over a wide range of critical heritabilities. The relationship between h^2, t, and x/σ is illustrated graphically in Fig. 5.1 for selection and in Fig. 5.2 for drift population-sizes for data on a Cretaceous benthic foraminifer. Figure 5.1 shows that the value of the heritability is far less important than the ratio z/σ, the quotient for expressing the change in the phenotype in relation to the phenotypic standard deviation, for influencing the course adopted by the curve.

5.4.5.6 *Selection and Drift Operating in Unison*

Random genetic drift interacting with selection is the fundamental motor of the evolutionary theories of Wright (1931, 1932), Simpson (1953) and Eldredge and Gould (1972). Lande (1976, p. 325) showed that the equilibrium distribution of the average phenotype, univariate or multivariate, is directly related to the adaptive topography and the effective population size. He has given charts of expected first-passage times in units of generations, which are useful for estimating the order of magnitude of effective population size required for crossing a threshold between adaptive zones by random genetic drift. These charts indicate that even very slight selective effects can bring about an overwhelming increase in the time taken to explore the adaptive zone by genetic drift and so to cross any threshold into some other adaptive zone. This conclusion was confirmed by later more explicit analysis of peak shifts (Lande, 1986).

Manly (1985, pp. 346–359) has reviewed aspects of the selection drift models discussed here and has pointed out that the results obtained depend greatly on the value of b in equation 5.21. Manly proves how easy it is to produce a nonsense-result if this fact is overlooked and a high value of b employed. Please keep this in mind if you intend experimenting with the methods of this chapter.

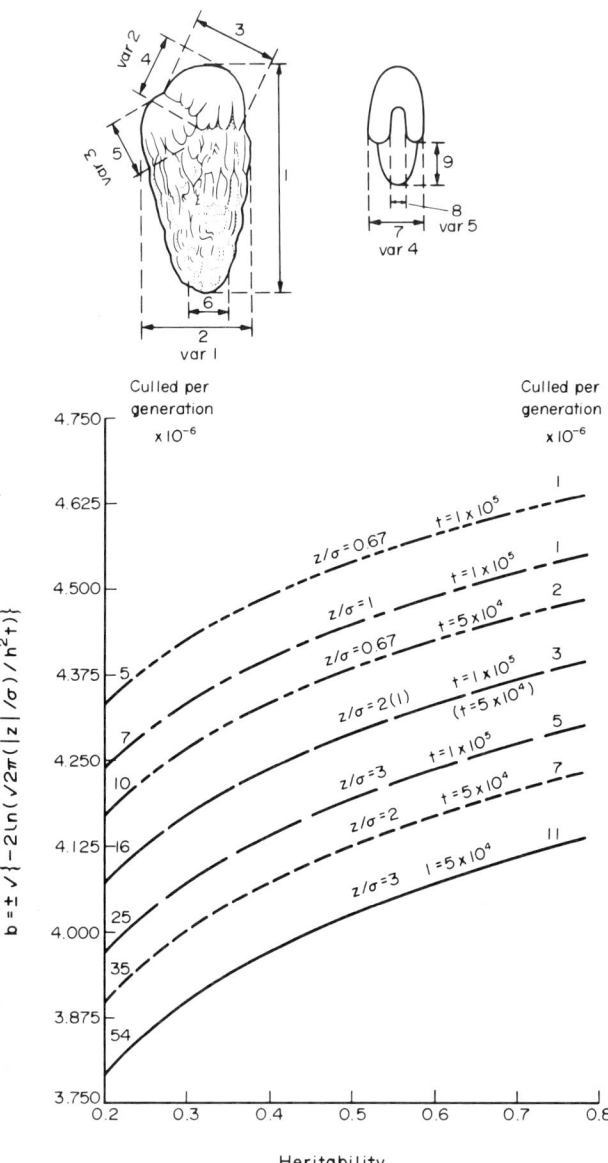

FIG. 5.1. The relationship between h^2, culling values, number of generations and the ratio between phenotypic change and phenotypic standard deviation for *Afrobolivina*. The graph gives an indication of how much b is influenced over a range of heritabilities. The vignettes show the variables measured on the test of the foraminifer.

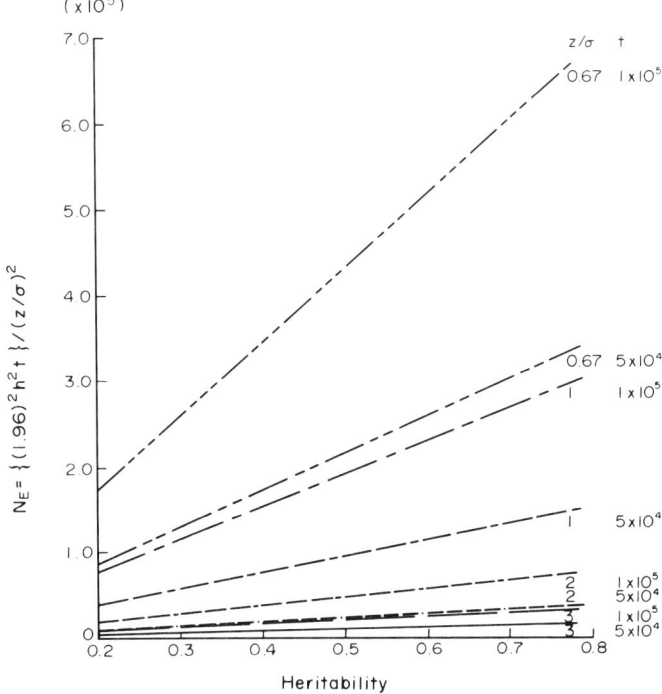

$(\times 10^5)$

$N_E = \{ (1.96)^2 h^2 t \} / (z/\sigma)^2$

Heritability

FIG. 5.2. Relationships between N_E and heritability, the ratio of phenotypic change to the phenotypic standard deviation and the number of generations for the data on *Afrobolivina*. This figure provides an indication of how much N_E is affected over a range of heritabilities.

5.5 Regional Validity of the Multivariate Phenotype

The selection differential and the selection gradient are essential tools for ascertaining whether the regional phenotype is stable or unstable. If the regional phenotype turns out to be unstable, it is doubtful whether the changes recorded for a particular evolutionary sequence are more than ecophenotypic in nature at best, or statistical artefacts, at worst. The temporal pattern of selection, that is, fluctuations in the "rate" and direction of evolution of p characters has no influence on the slope of the lines of change in these characters (Lande, 1979a, p. 404). The multivariate extension of Lande's models requires the assumption of a valid selection strategy and that the variables are multivariate Gaussian.

The vector of mean phenotypes in a population, **z**, can be partitioned into a vector of additive effects, **x**, and a vector of environmental influences, **e**, the latter also encompassing non-additive genetic effects. This is a simple generalization of the univariate situation (Falconer, 1981). Hence

$$\bar{\mathbf{z}} = \mathbf{x} + \mathbf{e}.$$

The corresponding partitioning of the phenotypic covariance matrix, $\mathbf{P}$, is then

$$\mathbf{P} = \mathbf{G} + \mathbf{E}.$$

Referring back to equation (5.5) we have that the vector of total selection differentials on adults, before reproduction, is:

$$\mathbf{s} = \mathbf{P}\mathbf{G}^{-1}\Delta\bar{\mathbf{z}} \tag{5.21}$$

and the selection gradient

$$\nabla \ln W = \mathbf{P}^{-1}\mathbf{s}$$

$$= \mathbf{G}^{-1}\Delta\bar{\mathbf{z}}. \tag{5.22}$$

The *net selection gradient* is found by summing (5.22) over 0 to $(t-1)$ generations. The resulting expression has the form of a linear discriminant function, to wit:

$$\mathbf{G}^{-1}(\bar{\mathbf{z}}(t) - \bar{\mathbf{z}}(0)). \tag{5.23}$$

This measure of selection is robust with regard to changes in the "rate" and direction of evolution since it is independent of the path followed between the initial and final multidimensional phenotypes. In practical applications, however, one would like to see

1. An adequate interval of time between the two sampling levels;
2. Reasonably large, multivariate normal samples, say $N > 60$, and,
3. Material sampled *in situ*.

The question of replacing $\mathbf{G}$ by $\mathbf{P}$ in (5.23) was discussed on p. 173. Charlesworth (1984) reviews some aspects of the concepts aired here, in particular the rather debatable point concerning the path between the initial and final observations.

5.5.1 *Multivariate Analogue of Minimum Selective Mortality*

The multivariate analogue of the minimum selective mortality can be obtained by computing a selection index, I, which is a linear compound of the original variables (Lande, 1979, p. 408):

$$I = (\nabla \ln \bar{W})^T \mathbf{z}$$

$$= \Delta\bar{\mathbf{z}}^T \mathbf{G}^{-1}\mathbf{z}. \tag{5.24}$$

Random samples from the populations in generations 0 and t provide the required sample mean vectors. Under the assumption that $\mathbf{G} \cong h^2\mathbf{P}$, where h^2

is an averaged value of the heritability (Manly, 1985, p. 359), the vector of coefficients in (5.25) can be estimated as

$$\mathbf{a} = (\bar{\mathbf{z}}(t) - \bar{\mathbf{z}}(0))^T \mathbf{P}^{-1}. \tag{5.25}$$

Index values can then be calculated for generations 0 and t and equation (5.19) can be invoked in order to find the minimum level of truncation selection needed to explain the observed change.

5.5.2 *Effective Population Size*

Assessing the question of whether or not changes in the population mean are due solely to random genetic drift in the multivariate case, the following procedure was put forward by Lande (1979). If drift is the only factor evolved (i.e. selective effects negligible), then the following statistic, which has the form of a generalized statistical distance, can be employed:

$$\chi^2 = (\bar{\mathbf{z}}(t) - \bar{\mathbf{z}}(0))^T \mathbf{G}^{-1}(\bar{\mathbf{z}}(t) - \bar{\mathbf{z}}(0))\frac{N}{t}. \tag{5.26}$$

This equation has a chi-squared distribution with p degrees of freedom. Here, N denotes the effective population size. You will see that equation (5.26) is in statistical terms, a weighted generalized statistical distance and the chi-square property is natural for it. Setting this statistic equal to the upper 5% point of the chi-squared distribution and solving for N gives

$$N^* = \frac{t_{\chi^2 0,0.05}}{(\bar{\mathbf{z}}(t) - \bar{\mathbf{z}}(0)^T \mathbf{G}^{-1}(\bar{\mathbf{z}}(t) - \bar{\mathbf{z}}(0))} \tag{5.27}$$

as the maximum effective population size consistent with an hypothesis of drift and in accordance with the assumptions underpinning equation (5.20). Also here the approximation $\mathbf{G} = h^2 \mathbf{P}$ is appropriate.

An example of the use of the multivariate analogues of the procedures for examining the likelihood of the drift/selection hypotheses will now be given.

Example 5.2. Application of the Multivariate Analogues of Lande's Univariate Procedures

This example considers the problems of testing hypotheses for drift in the absence of selection and, conversely, selection in the absence of genetic drift to multivariate phenotypic evolution in a lineage of Paleocene ostracods.

The relevant references for the example are Lande (1979a) and Manly (1985). By way of contrast, I can refer you to a study on the application of quantitative genetics to diatoms by Wood *et al.* (1987).

Statement of the Problem: The genus *Soudanella* occurs in the Paleocene of West Africa and Brazil. There is stratigraphical evidence to suggest that

Soudanella ioruba (REYMENT) was the immediate ancestral form to *S. laciniosa* APOSTOLESCU. The data available consist of the covariance matrices and corresponding mean vectors for three standard measures on the carapace: length, height and breadth. The samples comprise 24 male carapaces and 14 female carapaces.

Theoretical Considerations: Our aim is to apply Lande's results, in multivariate garb, to the analysis of presumed evolutionary changes in the carapace of *Soudanella*. Manly (1985) has outlined the steps in more detail than was done by Lande and my analysis is based on the presentation of the former author.

The first step is to carry out the calculations for formula (5.26) in accordance with the following procedure. We need to compute the Index of Selection for the beginning of the observed sequence, that is, the situation pertaining at zero generations. Then we compute the same Index for the end of the observed sequence, i.e. at t generations. The analogue of equation (5.21) requires an estimate of the variance of the sequence $I(t) - I(0)$ which we shall take, following, for example, Anderson (1984, p. 206) as:

$$(\Delta \bar{\mathbf{z}}^T \mathbf{P}^{-1} \Delta \bar{\mathbf{z}})^{1/2}. \tag{5.28}$$

The effective population size for the hypothesis of multivariate random genetic drift is obtained from equation (5.27). For the purposes of the present exercise, t was set at one million generations. As regards the method of analysis, the various parts thereof are quite straightforward, being a direct adaptation of the linear discriminant function and the Mahalanobis generalized statistical distance, as noted in the preceding section. The procedure is easy to program.

The Analysis: The Selection Gradient was computed to be

$$\mathbf{s}^T = (37.53 \quad 49.94 \quad 369.96).$$

This gradient finds special use for ascertaining regional validity in the phenotype. A second sequence, directly comparable with this sequence, should form a non-significant angle with the above vector.

The Selection Index Coefficients are

$$\mathbf{a}^T = (71.37 \quad 3.94 \quad 382.11).$$

In both of these vectors, the third trait, breadth of the carapace, is the dominant element and presumably the one most affected by evolutionary change.

The effective population size was computed to be 145,155 individuals, which is fairly large, and possibly far greater than would permit the observed phenotypic differences to have arisen by random genetic drift alone.

The minimum degree of truncation selection required to explain the morphological change is not big, being 3 selective deaths per generation per million individuals.

Concluding Comments: The analysis briefly sketched above can do no more than provide a rough scenario for the possible course of events. The multivariate generalizations I have used compound any inaccuracies inherent in the univariate cases, and this *caveat* we must keep in mind.

5.6 A Set of Case Histories

The field of applications of quantitative genetics in palaeobiology can best be illustrated by recourse to some actual investigations. I have included in the following pages several accounts drawn from my own work over the last few years. These are all micropalaeontological in origin and deal with foraminifers and ostracods. I should point out that the methods developed by Lande are more generally suitable for events on a large phylogenetic scale, larger than most of the problems with which I have been concerned. The entire subject of palaeontological applications of quantitative genetics is very much in the experimental stage and new developments appear all the time, many of which are ventilated in the pages of the journal *Evolution*.

If you intend experimenting with quantitative genetics, which I hope you will do, do not rush off and start drawing all sorts of conclusions on the basis of numerical results. If you want to use the methods summarized in this chapter, do so with due consideration of their tentative nature. With all these warnings, you are probably beginning to ask why I have included this chapter at all. The reason is that the biometrical analysis of evolution in the phenotype is a necessary advance, made at a high intellectual level, and it was bound to come sooner or later. As I have noticed already, theory and practice are developing rapidly and I have little doubt that the next few years will witness the advent of many new ideas and methods.

Example 5.3. Phenotypic Evolution and the Palaeoenvironment: the Cretaceous Foraminifer *Afrobolivina afra*

Introduction: The late Cretaceous to Paleocene benthic foraminiferal species *Afrobolivina afra* REYMENT crops up in many places in this book and, in a way, it has become the "fruit-fly" of multidimensional palaeobiology. The species occurs abundantly in boreholes over a wide geographical extent in West Africa, from Senegal to Angola and it has been well studied both stratigraphically and anatomically. It is endemic to the western coast of Africa.

The study summarized here is that of Reyment (1982). The material encompasses samples from 92 levels in one borehole, drilled at Gbekebo, western Nigeria. Thirteen levels are also available from a borehole located 70 km distant at Araromi, in the same sedimentary basin.

The characters measured on the test of the several thousand individuals analysed are illustrated in Fig. 5.1. These are all distance measures, although some of these variables also contain information on variability in shape. Secular variation in the phenotype of the foraminifer is often associated with ecological factors. In favourable circumstances, can such factors be determined in the sediment yielding the fossils. Foraminifers have well defined life-cycles which result from the alternation of the asexual schizont (= agamont) generation, which usually produces tests with a microspheric proloculus, with the sexual gamont generation, which produces megalospheric tests.

Statistical Analysis: Some of the statistical properties of the megalospheric individuals from the 92 samples are shown in Fig. 5.3. This figure displays secular variation in the diameter of the proloculus, the first canonical variate mean and the electrical resistivity log of the host sediment. There is good agreement in the pattern of swings to the right and the left of the latter two curves, except for the sandy zone indicated in the illustration, which is the only sedimentological heterogeneity in an otherwise homogeneous sedimentary sequence. Pirson (1977) tells us how electrical logs can be used for reconstructing the environment of deposition of marine sedimentary rocks in transgression-regression environments. Reyment (1982b) showed that the strength of association between the curves for Resistivity and the canonical variate means is highly significant and, also, that the diameter in the megalospheric proloculus is not significantly allied with variation in the environmental measures.

It is not infrequently intimated that evolutionary stasis implies genetic standstill, hence stagnation in the evolution of morphological characters. However, a species that is under the control of fluctuations in ecological factors will perforce display ecophenotypic variation and if the environmental changes are under the control of some major effect, such as eustatic changes of sea level, long-term shifts in its morphology can be expected to exhibit trending. This is the kind of situation envisaged by Bookstein (1987) when he called for an external determinant to be related to a morphological sequence. Small temporal adjustments in morphology by genetic tracking of ecological fluctuations will normally be without enduring effects—rather like the return of the tide erasing its earlier traces on the sand.

In a situation such as that just sketched out, there may be evolutionary stasis, but the sequence of means may trend. In Fig. 5.3, both long-term and short-term variations in the curves for the electrical log and canonical variate means are closely associated. The small oscillations probably reflect minor adjustments in the phenotype to small physical and chemical shifts, whereas the long-term trend in the sequence of canonical variate means may represent

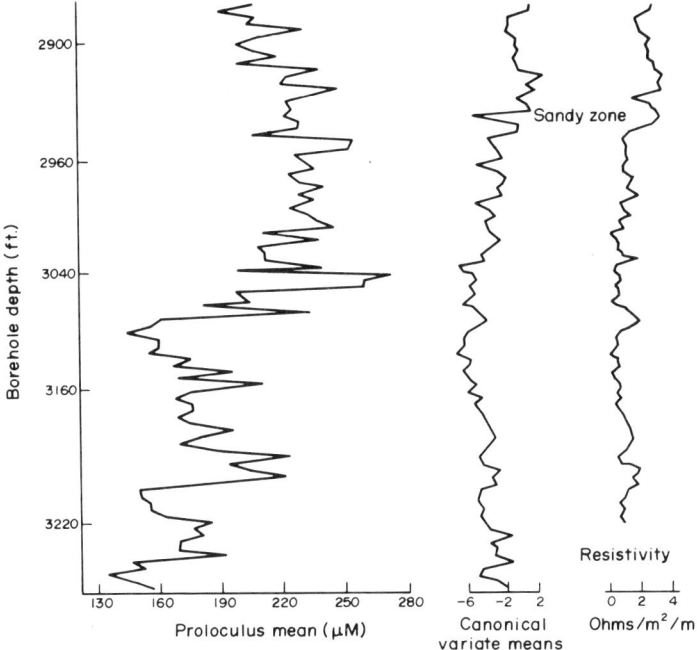

FIG. 5.3. Relationships between the means of the diameter of the proloculus, the first canonical variate means for the characters measured on *Afrobolivina afra* (cf. Fig. 5.1 for the variables) and the electrical resistivity log, which is here used as a palaeoenvironmental indicator. The data are for megalospheric individuals from the Gbekebo borehole, Nigeria. Modified from Reyment (1982b).

a more significant selective response to the eustatic change of sea level which so strongly marked the Maastrichtian of the South Atlantic.

Random Drift or Selection?: For the quantitative genetic calculations, I took 0.3 as an estimate of the heritability. I know of no determinations of h^2 for living protozoans. This a middle-of-the-road figure, since the heritability of morphological characters of the kind considered here lies between 0.2 and 0.6, with a preponderance of values for the upper limit (Falconer, 1981, Chapter 10; Van Noordwijk *et al.*, 1980; Strickberger, 1976). Although there is no necessary relationship between individual and populational heritability, I calculated this for gamont (= megalospheres) and agamont (= microspheres) generations in 47 samples. An estimate of 0.3 was yielded as an average value for eight characters. With respect to the question of estimating the number of generations, I used the mode of available determinations of lengths of the life-cycle for analogous living foraminifers, which is one cycle per year (references in Reyment, 1982b).

The proportion of the population culled each generation was estimated for three approximately equispaced intervals for which t was taken as 1×10^5

generations. The greatest value computed is 400 selective deaths per million individuals per generation (Reyment, 1982b). Two other cases gave values of 300 selective deaths per million individuals per generation. These figures may seem low, but they are many times higher than those obtained by Lande (1976) for fossil mammals. Equations (5.19) and (5.20) were used for the calculations.

Afrobolivina afra was extremely abundant and some pieces of core I have seen consist of more than 30% by volume of individuals of this species. Therefore, N, the deme-size, was taken arbitrarily as one million individuals, which is certainly an underestimate of the effective population size. Data in Sen Gupta and Strickert (1982) indicate that the numerical density of benthic foraminifers per 3 ml of sediment lies between 10 and 30,000 individuals.

With t set at 5×10^5 generations, equation (5.20) yields $N^* = 2.6 \times 10^5$, which is substantially less than the minimum estimated deme-size. Hence, random genetic drift may not have contributed importantly to the observed phenotypic changes.

Comments: It is appropriate to consider sources of inaccuracy liable to endanger the value of the analysis. In addition to the usually invoked bugbear of "geological causes", the most obvious of these concerns the different morphologies of microspheric and megalospheric individuals. This source of uncertainty was eliminated from the analysis by excluding all obviously microspheric individuals, which was done by inspection under the scanning electron microscope. (Eva Reyment and I have accounted for the SEM study of this material in papers published in the journal *Micropaleontology*.) A second source of inaccuracy arises naturally from the irregularity in form displayed by many foraminifers, which is largely the outcome of foraminifers lacking a terminal size. This leads to large variances and resulting heterogeneity in covariance matrices. The effects of this variability can be reduced by using logarithmically transformed observations.

Example 5.4. Analysis of Evolutionary Change in a Lineage of the Ostracod Genus *Echinocythereis*

Introduction: For the purposes of this illustration, I have chosen the clearly manifested morphological transition between *Echinocythereis isabenana* OERTLI and *E. aragonensis* OERTLI, illustrated graphically in Fig. 5.4. A useful indication of this transition between the two species of ostracods can be obtained from the lateral area of the specimens, as computed by the Zahn-Roskies algorithm. The material was studied in great detail by means of the SEM and variations in ornamental properties carefully noted. The morphometric work was supported by geochemical determinations on the host sediment (Reyment, 1985b).

The Analysis: The range of heritabilities over 0.2 to 0.82 is plotted in Fig. 5.5 against computed values of formula (5.19) in terms of the selective mortality per million individuals per generation and under the assumption that the

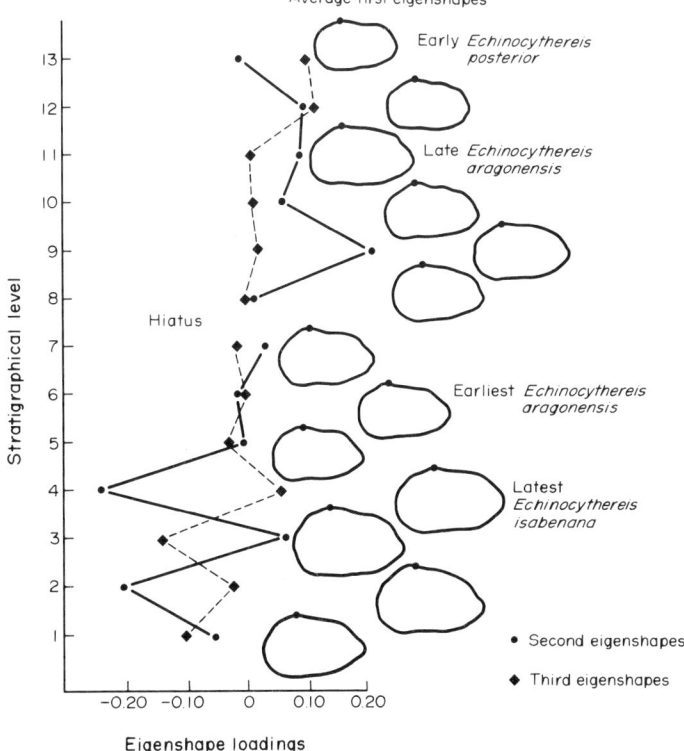

FIG. 5.4. Eigenshape-analysis for the transitions between three species of *Echino-cythereis* in a section in the Rio Isabena of northern Spain (Eocene). The data were selected for crucial ranges in the speciation events. The left-hand pair of plots shows the path traced out by the second and third eigenshapes. Redrawn from Reyment (1985b).

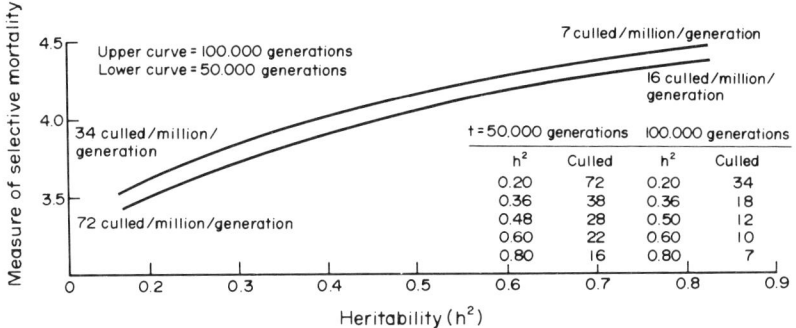

FIG. 5.5. Evolution in the phenotype. The estimated culling figures in relation to a range of heritabilities for the transition of *Echinocythereis isabenana* to *E. aragonensis* for the data of Fig. 5.4.

TABLE 5.3. *Computed minimum population sizes for a speciation event in the lineage of* Echinocythereis *under an hypothesis of genetic drift in the absence of selection*

h^2	Number of generations = 50,000 minimum population size	Number of generations = 100,000 minimum population size
0.2	2466	4933
0.3	3699	7399
0.4	4933	9865
0.5	6166	12,332
0.6	7400	14,800
0.7	8632	17,264
0.8	9865	19,731

The complete account is listed in Table 5 of Reyment (1985b, p. 189) in which variations in z/σ are taken into account.

transition between species could have taken 50,000 to 100,000 generations to accomplish. For a heritability of 0.4, the model requires 34 selective deaths per million individuals per generation over 50,000 generations. The alternative of 100,000 generations leads to 17 selective deaths. Both of these values support the hypothesis of relatively strong selective pressure and, in the absence of the effects of genetic drift, selection as the driving evolutionary force is an acceptable evolutionary mechanism. Figure 5.5 also includes a table of the culling values for a range of heritabilities for both of the models considered here.

Let us now examine what the situation is for the chance that the changes were brought about by random genetic drift in the absence of selection. Some of the values computed from equation (5.20) are listed in Table 5.3. These results are enlightening. If we take into account the fact that the effective population size for ostracods is something of the order of 5–6% of the values computed, which encompasses *all* individuals, juveniles, and males and females in grossly unequal proportions—see Hartl (1980), the chances of drift having been a factor in the speciation are high. For the arbitrarily selected heritability of 0.4, Table 5.3 indicates that we are dealing with effective population sizes of no more than 200 to 500 individuals. Hence, the transition between the two species could have been engineered in small demes. Granted that the population densities for these Aragonese Lutetian ostracods seems to have been small, it seems well within the bounds of possibility that the species transition could have been initiated by random drift. If we also allow selectional effects to join the model, then a biologically likely mechanism for the speciation event would be selection on small local populations in combination with genetic drift.

Comments: The data used for the present analysis are homogeneous with respect to variances and covariances, which is no doubt a result of (1) the relatively low variability usually displayed by the dimensions of the ostracod carapace, (2) the fact that the material shows no evidence of reworking, as indicated by detailed SEM scanning of all specimens in the collection, and (3) the ease with which adult shells can be identified and isolated for measuring (evidenced by the presence of adult hinge characters and marginal pore canals).

Example 5.5. Phenotypic Variation and the Palaeoenvironment for the Cretaceous Benthic Foraminifer *Gabonita elongata*

Introduction: A further case history treating secular variation in the phenotype of a benthic foraminifer is offered by *Gabonita elongata* (DEKLASZ and MEIJER) from the Gbekebo borehole in western Nigeria and studied by Ivert (1980). There are numerous ecological factors to which benthic foraminifers may respond morphologically, such as salinity, temperature, water depth, content of carbon dioxide, content of oxygen, sedimentological characteristics and redox conditions in the interstitial sedimentary environment. In rather general terms, Ivert recorded a tendency for megalospheric and microspheric individuals to increase in size and then to decrease in size over the interval studied. He also found that *G. elongata* tended to develop larger tests in a shale environment that was somewhat enriched in silica, whereas smaller tests were likely to be produced in carbonaceous shales, low in silica. The rôle of chemical factors in governing phenotypic variability in this species is unquestionably important.

Ivert's data were re-analysed using multivariate procedures. These analyses show that calcium and magnesium are correlated with smaller test sizes and silica with larger test sizes. The physiological significance of this situation is elusive.

The Analysis: Although genetic drift in small populations could have been a functioning mechanism for the history of at least part of the sequence (see Fig. 5.6), a selection model was thought to be more relevant for the evolutionary situation represented in this material (Reyment, 1983). Effectively, the observed sequence encompasses two major excursions (Fig. 5.6) which are marked by relatively pronounced morphological changes at both the univariate and multivariate levels, with a return to almost the same morphological status as manifested at the beginning of the observed sequence. Level-by-level comparisons indicate effective population sizes of between 1600 and 55,000 megalospheric individuals. However, in view of the significant association between the morphological status of the test and geochemical components of the palaeoenvironment, a model involving selectional responses seems to fit the situation better, though doubtlessly with an important phenotypic component.

As regards the slightly different traces made by megalospheres and microspheres in Fig. 5.6, a possible explanation for this is that asexually reproducing individuals could have reacted somewhat differently from those of the sexually reproducing phase, in particular with reference to the Ca–Mg

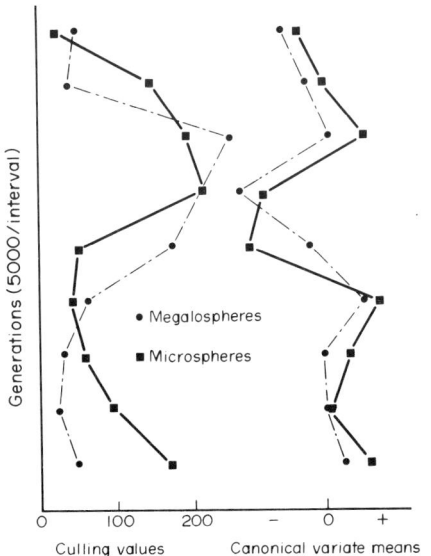

FIG. 5.6. Phenotypic evolution in megalospheric and microspheric individuals of *Gabonita elongata* for nine levels in the Gbekebo borehole (Nigeria). The two left curves show the range of culling values under an hypothesis of evolution by natural selection for microspheres (squares) and megalospheres (dots). The two right-hand curves display the first canonical variate means for microspheric tests (squares) and megalospheric tests (dots). Each division along the ordinate axis denotes approximately 5000 generations. A heritability of 0.4 was used for constructing the left-hand curves. Redrawn from Reyment (1983).

versus Si dipole. Only direct experimental work on living forms is likely to prove this point.

Example 5.6. Multivariate Selection Gradient for Two Species of the Ostracod Genus *Soudanella*

Introduction: The illustration offered here is an example of the application of the concept of the selection gradient to the problem of checking regional stability in the phenotype. I have already had occasion to underscore the fact that regional stability in the phenotype is an absolute basic necessity for any work hoping to enlighten us on the existence of genuine evolutionary trends in a sequence. Without confirmation of geographical stability in sequences, the risk that an observed morphological trend lacks other than the status of a graphical artifact is overriding. The two ostracod species reported on here have been studied earlier on in this chapter, but only with respect to changes recorded for a single sequence. The study will now be expanded.

Material: The data derive from two boreholes in western Nigeria, the one located at Gbekebo, the other at Araromi. These boreholes, already referenced in this chapter, were drilled in connexion with the search for oil in

TABLE 5.4. *Covariance matrices and mean vectors for species of the ostracod genus* Soudanella *at two sites in the Paleocene of western Nigeria*

Site 1			Site 2		
Mean Vectors variables					
length	height	breadth	length	height	breadth
0.644	0.345	0.271 (ioruba)	0.640	0.361	0.259 (ioruba)
0.816	0.456	0.384 (lacin.)	0.818	0.454	0.375 (lacin.)
Covariance Matrices					
Soudanella ioruba ($N = 36$)			*Soudanella ioruba* ($N = 45$)		
0.00196	0.00036	0.00018	0.00220	0.00050	0.00025
0.00036	0.00050	0.00029	0.00050	0.00069	0.00035
0.00018	0.00029	0.00030	0.00025	0.00035	0.00050
Soudanella laciniosa ($N = 36$)			*Soudanella laciniosa* ($N = 45$)		
0.00158	0.00023	0.00019	0.00115	0.00045	0.00025
0.00023	0.00034	0.00017	0.00045	0.00080	0.00030
0.00019	0.00017	0.00026	0.00025	0.00030	0.00050

Nigeria. The basic statistical data for the length, height and breadth of the carapaces of the ostracods are supplied in Table 5.4 in which the respective covariance matrices for samples from two comparable levels in the boreholes are listed, as well as the sample mean vectors.

The aim of the exercise is to see whether the selection gradient (equation 5.23) is stable over the range of the species transition. In both cases, the purported transition passes from a smaller species, on average, to a larger species.

The Analysis: The first step in the analysis is to compute equation (5.23) using Cheverud's (1988) approximation for matrix **G**. The "normalized" equations for the two sequences are:

$$\text{Site (1): } 0.0745x_1 + 0.2234x_2 + 0.9719x_3$$

$$\text{Site (2): } 0.1930x_1 + 0.2475x_2 + 0.9495x_3$$

The hypothesis of collinearity of the gradients can be explored by computing the angle between these two vectors. The cosine of this angle is 0.992, which corresponds to an angle of 7°. This difference is negligible and there is no reason to doubt, on the data available, that the selection gradients are collinear at the two sites sampled.

The above equations for the respective selection gradients indicate that most of the action lies with the breadth of the carapace, the third variable. I was at first surprised by this result, but after a moment's reflection realized that it was quite a logical result. The outlines of the two species hardly differ at all. It is in dorsal aspect that the differences in proportions become apparent. A third selection gradient was available for a pair of smaller samples, also from boreholes in western Nigeria. This vector is $(0.1000, 0.1331, 0.9860)^T$. This vector forms an angle of 3.8° with the gradient for Site (1) and an angle of 8.7°

with the gradient for Site (2). These differences are inconsequential and the conclusion can therefore be ventured that *S. ioruba* gave rise to *S. laciniosa* and that this event took place over a fairly broad geographical extent.

5.7 On Stasis

5.7.1 *Introduction*

A species may seem to be at an evolutionary standstill with respect to distance measures, that is, characters expressing size variability, but even shape variability, but, nevertheless, be undergoing significant and permanently manifested changes in other properties which, in the case of fossils, largely concern ornamental features. What does this tell us? It tells us that Stasis is in the Eye of the Beholder. To a large extent, this lack of change over time we believe we observe may be no more than the expression of what we have chosen to observe. Hence, any assessment of stasis needs to be made on as complete a range of traits as possible. Much of the current discussion, and dissension, in evolutionary biology is concerned with the identification of long, static periods in the history of a species, a condition formalized as stasis by Eldredge and Gould (1972). The question being put by many is, just what is stasis? Statistical arguments of varying complexity and validity are being advanced in order to provide a more comforting basis for these interpretations. The statistical methods employed are usually centred around some particular statistical test for elucidating the properties of some measure on a fossil organism. Thus, one tries to fit some suggested evolutionary model to a set of data, or one attempts to propose a model on the results yielded by some hopefully diagnostic statistical test.

This is all very well, and some valuable results have been accrued from such activities (cf. Bookstein, 1987, 1988). However, as far as I can make out, all the results published so far suffer from the classical malaise, well known to professional statisticians, namely, that extremely long series are required in order for success to attend any of the above endeavours. There is a fundamental mathematical paper on this subject by Lewis (1964). The definitive treatise on the subject is the book by Cox and Lewis (1966). Bearing in mind that morphologically defined evolutionary sequences are being treated as random in time by a growing corps of practitioners, the need for very long series of observations must be made clear.

From the biological viewpoint, the idea is often promoted that stasis in a character is the result of stabilizing selection on that character. In palaeontological studies, it is seldom enquired whether the selective processes can be supplied with a reasonable biological evaluation. If length is being selected for some reason or other, is the height of the organism being dragged along, and are other traits also enchained? If the length of a carapace is strongly bound to other morpho-characters, what is happening to the shape of the organism?

Perhaps, what we are really seeing is selection for shape, and length just happens to be a convenient feature to measure. Crespi and Bookstein (1989) have looked at this problem. If, on the other hand, the suite of changes being observed are the outcome of random variation in morphology, modifications in shape lack intrinsic interest apart from the fact that we should expect really bad products to be eliminated by selection.

There are many more aspects to the analysis of evolutionary change than just secular variation in distance measures analysed in a standard manner (Charlesworth *et al.* (1982), Cheetham (1987), Cronin (1985), Foote and Cowie (1988), Maynard Smith (1983), Milligan (1986), Stanley and Yang (1987), Wake *et al.* (1983), etc.). Clearly, in addition to searching for stasis in characters that represent variation in size, we should also consider characters that register changes in shape. Even more importantly, and far less frequently considered, evolutionary changes in ornament need to be brought to the fore. This seeming dearth of initiative with respect to the study of ornamental variation is doubtless due to the fact that it is difficult to quantify fugitive variations in surface sculpture. Image-analytical methods will no doubt soon appear that will allow the easy numeration of ornamental characters. In the meantime, however, we are largely restricted to using dichotomous variables, and counts on frequencies.

Geneticists encounter a variety of situations in their daily research that could lie at the root of stasis, if stasis is to be seen as the expression of reduced variation in the biology of a species. Wake *et al.* (1983) examined the oft-made observation that an organism may display surprisingly little variability in a varying environment. They coined the term "persistence" for categorizing this phenomenon, whereby is understood an active process embracing genetic, developmental, and physiological variation in an auto-regulatory system.

Maynard Smith (1983) reviewed possible causes for stasis and in so doing, accepted the biological reality of evolutionary standstill in some lineages. In terms of population genetics, stasis can be interpreted as the condition arising from stabilizing selection according to Charlesworth *et al.* (1982). Alternatively, stasis can arise as the action of developmental constraints (Maynard Smith *et al.*, 1985). Stanley and Yang (1987) recorded stasis in the morphological history of certain bivalves. They used multivariate statistical techniques to support their arguments, albeit in a rather naive manner. The distances they use are all constrained (cf. p. 225), hence the standard forms of multivariate analysis utilized by them are not appropriate (Aitchison, 1986a). A more fitting methodology for expressing generalized distances in relation to shape, and one that avoids the trap sprung by compositional variables, is Burnaby's (1966) method of growth-invariant discrimination. This procedure was, in fact, specifically designed by Burnaby for studying Carboniferous bivalves (cf. Gower, 1976; Reyment and Banfield, 1976; Rohlf and Bookstein, 1987).

A patent weakness of many quantitative palaeontological studies, including that of Stanley and Yang (1987), is that attempts are made to extract more statistical information from the data than it can possibly contain. I sound this note of warning here, because it is appropriate to do so in conjunction with a discussion of the statistical analysis of stasis and in view of what follows now. I have had occasion to mention on several occasions that special care is required for assessing the reliability of secular data for fossils.

In the articles referred to earlier on, Bookstein (1987, 1988) shows that according to his model, evolutionary rates exist only when the hypothesis of *symmetric random walk* can be rejected. It is not enough to identify the occurrence of a random walk, the mathematical properties of the process must be specified. Stasis, that is, evolutionary equilibrium in a morphological character, and anagenesis, may be shown to be deviations from a random walk in opposite directions. A crucial point is here that one must first be sure that the random-walk model is appropriate to the data, which may not be the case when the samples are composite (cf. Bookstein, 1988).

To summarize the foregoing discussion, the question of establishing stasis requires that the following questions be asked:

1. What is stasis, granted that stasis is of genuine biological significance?
2. Is stasis in morphological characters matched by stasis in other features, e.g. ornament?
3. How can stasis in ornament be defined?

5.7.2 *The Shifting Balance Theory and Stasis*

Lande (1986) reviewed the concepts in Wright's shifting balance theory and how it can be viewed in relation to "punctuated equilibria". Thus, random genetic drift and selection for multiple adaptive peaks may produce long periods of stasis according to Wright's model, interrupted by periods of rapid evolutionary change. After studying data on microfossils, Lande concluded that relative stasis and gradual fluctuating trends are the normal course of events for this model.

Directional selection in large populations in which more genetic variability can be maintained can result in faster evolutionary responses than are usually observed in small populations. Here we run into semantics. What is meant by "small populations". Hartl (1980) and Falconer (1981) have given this question close attention.

Stabilizing selection leads to a different situation from that just defined. Only small populations are likely to escape from its grasp by the expedient of random genetic drift. Developmental constraints are a valid factor to be considered in any effective analysis of stasis. They may be defined as a bias or limitation on phenotypic variability caused by the structure, composition, or dynamics of the developmental system. The effects of developmental constraints show up in patterns of variation and correlation of characters within populations. These can in turn restrict the possible rates and directions of phenotypic evolution.

Lande's (1986) evaluation indicates that long-term stasis requires stabilizing selection. What stimulates the onset of stabilizing selection? One factor is doubtlessly the choice of individual habitat. A second vital factor concerns the demography of populations. During global environmental change, populations tend to track the habitats to which they are adapted. This is usually achieved by dispersal with increases in the size of the population in favoured habitats.

Another interesting model relevant to the content of this chapter has been put forward by Slatkin (1982) in terms of pleiotropy and parapatric speciation. The interested reader is referred to Slatkin's paper for the details. Other pertinent references are Wake *et al.* (1983, pp. 211, 218), Maynard Smith (1983, p. 19), Turner (1986, p. 204) and Levinton (1986).

5.7.3 *Practical Considerations and Some Illustrations*

The Problem of Heterogeneity in Samples: The statistical treatment of evolutionary sequences poses a special problem for several organisms. There are pitfalls, the most obvious of which derives from the wide occurrence of biologically homogeneous, but statistically heterogeneous material. The most immediate root of such mixed samples is polymorphism in size and shape (see Chapter 6). If each sample comprises two or more morphs in unknown proportions, we shall experience difficulties in producing a meaningful variational curve for a polymorphic character. The successive means will depend on the accidental composition of morphs. The use of the mode can be a stabilizing procedure for graphical purposes, but this is not always crowned with success if the material consists of morphs in roughly equal proportions.

How can the problem of mixed samples of the foregoing kind be resolved? If there are only two opposing categories, such as angular outline versus ovoid (Ducasse and Rousselle (1979) consider this situation for Tertiary ostracods), one may arrive at a solution by visual inspection under the SEM and then manufacture a subsample for each morph. The more numerously occurring morph can then be used for making the required variational plot. Should the species possess more than two morphological states, difficulties will certainly arise in creating the sub-samples. Screening at both the univariate and multivariate levels may be necessary. A suitable approach to multivariate screening is offered by the method of cross-validation (Krzanowski, 1987a) which permits the operator to distinguish specimens having subtle atypicalities.

Example 5.7. Stasis in a Species of Foraminifers

The Campanian through early Paleocene species *Afrobolivina afra* REYMENT has been the object of numerous studies owing to its great abundance and the availability of external palaeoenvironmental controls (see also pp. 189, 289, 304). Recent papers dealing with evolution in the species are those of Kitchell *et al.* (1987), Turelli *et al.* (1988) and Reyment (1982b). In this latter article, I analysed nine measures of the test, including the diameter of the proloculus. I reached the conclusion that the species was in evolutionary stasis over the time encompassed by the available 92 samples in one borehole and 30 samples in a second borehole. The compound sequence of univariate and multivariate means fluctuates haphazardly in the short term, under the control of aimlessly varying ecological conditions. Kitchell *et al.* (1987) used the technique of bootstrapping to arrive at essentially the same conclusion for secular variation in the diameter of the megalospheric proloculus. They found that the fastest observed sub-set of changes in this trait is unexpectedly slow.

In a similar vein, Turelli *et al.* (1988) analysed the same set of data with respect to the length of the test (not the best choice of variable bearing in mind the great relative variability manifested by length in most benthic foraminifers), using information in Reyment (1982b). Those authors concluded that the observed changes in length of *A. afra* are too slow to be explained by a model of evolution by random genetic drift under an hypothesis of mutation-drift equilibrium. This conclusion harmonizes with modern opinions on genetic drift in large populations (Maynard-Smith, 1983; Mourant, 1983).

The record of *ornamental* variation in *Afrobolivina afra* betokens the existence of three morphs, to wit, (a) a smooth morph lacking longitudinal ornament apart from marked lobations in the distal sector of the test, (b) a morph characterized by *lateral costae*, and (c) a morph in which the lateral costae are united into an anastomizing pattern by transverse riblets to form a *regular* pattern of elongated hexagons. The regular morph was examined closely under the SEM for time-correlated changes in geometry over a period of time corresponding to late Maastrichtian to basal Paleocene. The general observation is that the ornamental pattern remains unchanged over this time interval. The hexagonal pattern of the youngest observed specimens is no different from that occurring at the beginning of the available sequence. Some excursions from this pattern occur *en route*, but they do not persist. This is a good example of conservative evolutionary behaviour.

Example 5.8. Stasis in the *Echinocythereis* lineage

We now pass to a completely different type of example. The ostracod lineage considered now was marked by two speciation events, without branching (Reyment, 1985b). One event is of a *quantal* nature, the other involved a gradual displacement of the ornamental pattern from one category to another. The reduction in average size attendant on the passage from *Echinocythereis isabenana* to *E. aragonensis* is clearly manifested in five standard distance measures on the carapace. The average decrease in size is 14–17% and it is a permanent displacement. None of the subsequent meandering back and forth cross the track staked out for *isabenana* by the canonical variate means. The use of multivariate mean scores for the present purpose is not without danger in that the error variance for these measures can be very large (Bookstein, 1988). Recommended procedure is to examine the coefficient of variation for univariate characters at the start of the study. Ostracods tend to have very low coefficients of variation for the carapace ($\cong 3.5$) and hence more definitely fixed multivariate mean scores.

Neither the initial part of the sequence of canonical variate means nor the series of means for *E. aragonensis*, nor the final species of the lineage, *E. posterior*, display statistically significant trend. It seems very likely that the meandering pattern of small-scale size variation over time represents the ecophenotypic response of the ostracods to fluctuations in the environment. This conclusion is supported by the almost identical variational history exhibited by an unrelated species of ostracods observed over part of the sequence. External chemical factors were proved to be significantly correlated with size traits. The driving force for the environmental shifts was thought to derive from the shallowing of the sea in late Lutetian (Eocene) time. The full biometrical analysis was reported in Reyment (1985b).

5.8 Time-Series Analysis in Multidimensional Palaeobiology

5.8.1 *Proem*

I have referred to time-series analysis in the foregoing pages on several occasions, and do so again in Chapter 8. I have inserted this section here because it has a direct connexion to the concept of stasis. It is now appropriate to consider the subject in more formal terms. Although time-series analysis is usually applied to single variables, the methods can be used for sequences of multivariate observations. Owing to the spread associated with such data, the mode of each sample of multivariate scores is likely to give more stable results than the multivariate mean, which, as has been noted on several occasions in the foregoing pages, may suffer from substantial error in estimation. This applies particularly to the canonical variate mean.

A little history to start proceedings. According to Kendall (1973), the modern era of time-series analysis was ushered in by work of Udny Yule who developed the subject in connexion with his work on sunspots. Yule discovered that amplitudes of series of sunspots and the distances between successive peaks and troughs are not regular. He likened this condition to what happens to a pendulum swinging under gravity when it is repeatedly subjected to small shocks at irregular intervals. The normal motion of the pendulum is harmonic and can therefore be represented by a sine or cosine wave. The swing of the pendulum when given minor shocks becomes irregular and the shocks will be absorbed into the future motion of the system.

I have quoted Yule's homely example at length because it contains a message for the palaeobiologist working with series in time, namely, that the "motion" of the observed system at one specified point in time reflects the previous history of shocks (=ecological events) experienced by the system. Translated into biological terms, this means that the major irregular fluctuations in morphology, or frequencies, displayed by the sequence derive from past events, thus implying that the events are serially correlated and hence not independently distributed.

5.8.2 *Summary of Methods of Statistical Analysis*

The first step in any statistical analysis of a time series is to test the material for trend. This may be obvious to the eye, but this is not always so. The data for *Echinocythereis* spp. seems to show trend on a broad scale. If significant trend occur in the set of means, the series is not stationary in the statistical sense. The simplest hypothesis for a series showing change fluctuations is, of course, that it is random.

Kendall (1973, p. 21) considered several approaches. Firstly, to examine the residuals after detrending in order to ascertain whether they show vestiges of systematization. Secondly, the hypothesis of trend should be examined against other alternatives. Thirdly, testing for trend must not be confused with testing for periodicity.

A useful way of testing for trend is to enter the analysis by means of an examination for a non-homogeneous or modulated Poisson process. We can examine the observed changes with respect to rate of occurrence, a simple model for which is

$$\lambda(t) = \exp(\alpha + \beta t). \tag{5.29}$$

A test for $\beta = 0$ against $\beta \neq 0$ is given by Cramér's statistic

$$U = \frac{\Sigma\left(\dfrac{T_i}{t_0}\right) - \dfrac{n}{2}}{\dfrac{n^{1/2}}{12}} \tag{5.30}$$

for a series observed for a fixed interval $(0, t_0)$ and associated with n events (observational levels) and T_i denotes the sequence of events. This statistic converges rapidly to a normal unit variable (Cox and Lewis, 1966, p. 47; Cramér, 1946, p. 245). For absolute values of U greater than 1.96 (which is the 5% level of significance), the hypothesis of $\beta \neq 0$ is rejected.

Although it is not common to find systematic fluctuations (=modulation) in $\lambda(t)$, it is as well to know how such data must be treated. (Note, that Slutzky and Yule were both amazed to find that random events can produce the appearance of systematic oscillation—according to Kendall, 1973.) If you think you have such data (which could conceivably arise from systematic geological influences), and you have sufficient observations, the recommended approach is to analyse contiguous sub-sets of the sequence.

A reasonable statistical approach to see whether a Poisson-distribution is applicable to the data is to see whether the Coefficient of Variation is about 1 in value. The significance of a value greater than unity is that there is bunching of the events representing a spate of concentrated activity, followed by a lull. This kind of situation could occur with rapid speciation.

An important quantity is the **serial correlation coefficient**. If we have shown trend to occur, then serially correlated intervals can be expected to occur. If trend has not been proven, it is of interest to check the sequence for serial independence of successive intervals. The serial correlation coefficient may be represented as:

$$\rho_j = \frac{Cov(X_i, X_{i+j})}{Var(x)} \tag{5.31}$$

($j = \ldots, -1, 0, 1, \ldots$). The serial correlation coefficient is computed for different lags. One "variable" is made up of all the observations running from 1 to $(n\text{-}1)$ and the second "variable" of the observations running from the second to the n-th. This is called the serial correlation coefficient of lag 1. Likewise for the second serial correlation coefficient with the first variable going to $(n\text{-}2)$ and the second running from the third measure to n. In this manner we obtain information on the degree of association between levels of increasing lags. Many stop at the third or fourth serial correlation for in a series showing independence between intervals, it is usually a waste of time hunting for what things can be like at lag 20. You can always turn up a fortuitously high correlation of no statistical relevance if you churn through long series. In the example following on this section you will however find a very long sequence of significant serial correlations. There are many more things you can do in a fully fledged time-series study but the simple introduction I have given here should suffice for getting you started.

The Correlogram: Pielou (1974) has discussed at length the value of the correlogram for probing the properties of a sequence of ecological observations. The original work was done by P. A. P. Moran. These ideas are of relevance for palaeobiological work and I shall now summarize them. We are all well aware of the fact that series in time fluctuate and that there are major hills and dales as well as ripples. Pielou asked if it is possible to ascertain whether the fluctuations are

(a) endogenous (= intrinsic)

or

(b) exogenous (= extrinsic)

Interest is also directed towards finding out whether the fluctuations are regular (periodic or oscillatory) or irregular (aperiodic). For exogenous data, a chance displacement is not *cumulative*; its effects are soon lost and the wave is undamped as the serial correlation does not become progressively less. For an endogenous series, a chance disturbance will persist. The curve is damped because as the time interval between two observations increases so does the dependence on an earlier event recede.

There is, nevertheless, one procedure I like very much which I have not yet introduced you to; this is the *logarithmic empirical survivor function curve*. For making this curve, one considers the *n* observed intervals between events, X_i, which may be ordered to give the observed-order statistics $0 < X_1 \le X_2 \le \ldots \le X_n$. The empirical survivor function may be defined as:

$$R_n(X) = \text{proportions of intervals longer than } X,$$

which is plotted against X itself. Here *n* denotes the number of distances, or intervals, in the sample. Where the distribution is approximately exponential in nature, the plot of the natural logarithm of the empirical survivor function is a useful graphical tool, since departures from linearity will indicate divergencies from the exponential distribution. The values used to make Fig. 5.8 were obtained from

$$\log_e\left(1 - \frac{i}{n+1}\right)$$

Example 5.9. Examination of Properties of a Sequence of Multivariate Means for *Echinocythereis*

Presentation of the Problem: The data consist of 103 multivariate means measured on three species of *Echinocythereis*. The five distance characters employed have been discussed in previous examples. For the purposes of the

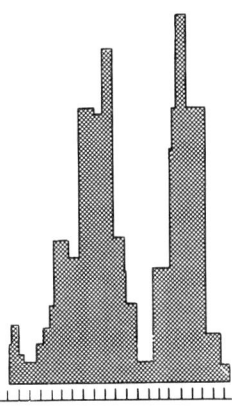

FIG. 5.7. Histogram of the time series for the first principal component scores for *Echinocythereis*.

present analysis, the squared lengths of the observational vectors were used. I am not recommending this as the ultimate technique for time series in Palaeontology, but for present objectives, this measure will do.

The Analysis: Although the plot of the canonical variate means in Reyment (1985b) suggests fairly strong tending in the sequence, Cramér's U-statistic gives a value of only 0.54, which is not significant. The histogram of the distribution of distances is pronouncedly bimodal. This is reflected clearly in the plot of the log-survivor function, which has a step in it. The serial correlations are significant up to a lag of 29, being highest for a lag of 1 and then falling successively. This seems to be an artifact induced by the displacement in size between the first and second species of the sequence, as evidenced in the bimodality of the histogram. An analysis of the first 46 means,

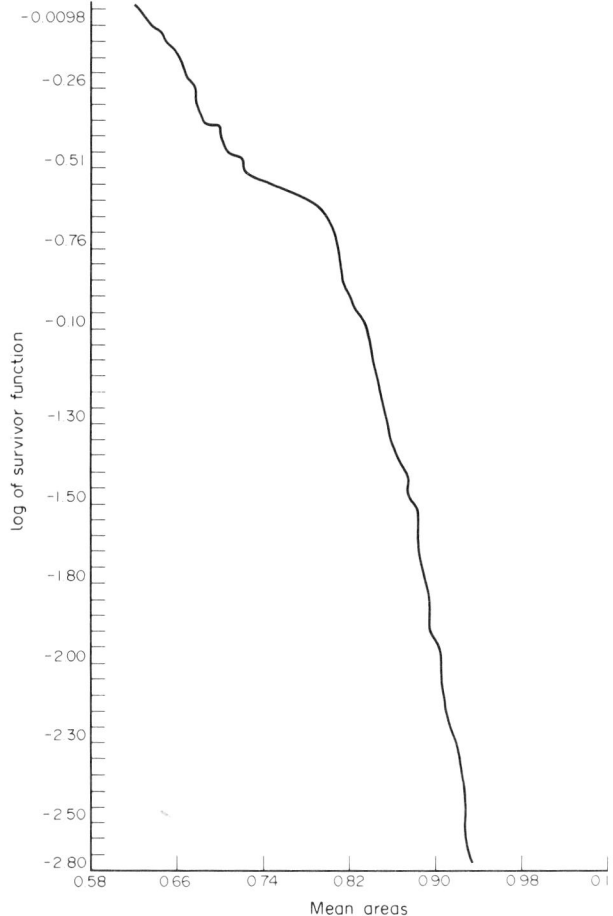

FIG. 5.8. The curve of the log-survivor function for the first principal component scores for three species of *Echinocythereis*.

which correspond to *Echinocythereis isabenana* yielded much more variation in successive serial correlation coefficients and with no significant values after a lag of 6. The value of U is not significant and the histogram is unimodal.

By way of contrast, I used the first vector of scores for a principal component analysis of the 103 quinquevariate mean vectors. The results obtained are closely similar. I obtained a value of U of 1.14, which is not significant, and the histogram of scores is bimodal, with a suggestion of trimodality (Fig. 5.7). The plot of the log-survivor function, shown in Fig. 5.8, displays the step which marks the zone of transition between *Echinocythereis isabenana* and *E. aragonensis*.

Example 5.10. The *Globorotalia tumida* Lineage of Neogene Planktic Foraminifers

Presentation of the Problem: Malmgren *et al.* (1983) analysed size and shape measures in a lineage of foraminifers. Briefly, the data consist of a 10 My-long sequence ranging from late Miocene to Recent from *G. plesiotumida* to *G. tumida*. The transition from one species to the other took place at the Miocene–Pliocene boundary. Sampling resolution was good being at a level of just two to five thousand years, occasionally 15 thousand years. The method of eigenshapes was employed to express the outline of the test, in edge-view. This orientation permitted observation on the important features of inflation and elongation of the test.

The general conclusions reached by this detailed study were

1. That periods of stasis occurred throughout the observed sequence, and,
2. Morphological change occurred rapidly, without branching. This mode of evolution was termed "punctuated gradualism".

Using the figures for area of the test in edge-view, I undertook an hierarchical analysis of the sequence of 104 observations. In passing, I note that Crespi and Bookstein (1989, p. 27) found that a determinant of shape gave a satisfactory stability to a measure of selection.

The Analysis

The complete suite of 105 observations yielded $U = 3.91$, which indicates highly significant trend in the points. This is also clear from the diagrams in Malmgren *et al.* (1983). The plot of the histogram shows the areas to be polymodal, with five peaks. This indicates the likelihood of several events in the sequence. The graph of the log-survivor function shows several distinct, though short, steps.

The correlogram displays exactly the same pattern as recorded for the ostracods above for Example 5.9. The serial correlation coefficients are significant up to a lag of 29 and in general, the curve traced out by the plot of the coefficients is monotonic decreasing. This is no doubt a reflection of the fact that successive values change very slightly and do not jump about wildly. There is a tendency to follow an increasing path, on a broad scale, and to decrease in the same manner. The same observation was made for the ostracod-data of Example 5.9. It is not reasonable to invoke the coin-tossing paradigm each time such a pattern is observed and such patterns tend to be common. The fact that test-size can be influenced by long-term ecological effects seems to be an important contributing factor to the motif displayed by these planktic foraminifers.

By way of contrast, the same analysis was applied to the first 72 observations. This time $U = 1.51$, which is not significant. However, the histogram of mean areas is bimodal and the log survivor curve contains some steps. The serial correlation coefficients are still significantly correlated, up to a lag of 15.

The last 45 observations (the oldest of the sequence) trend significantly with $U = 2.36$ (the cut-off point is 1.96). However, the histogram is almost unimodal, though skewed. The curve of the log-survivor function shows one very pronounced step. The serial correlation coefficients are, as to be expected, significantly correlated up to a lag of 11.

Remarks: What is all this about? I think the model of Malmgren *et al.* (1983) is reasonable as far as it goes. The successive values do not differ markedly, which leads to significant serial correlations. The meaning of these correlations may not be directly interpretable although the following points seem to be clear:

1. The further apart two levels are, the smaller are the correlations.
2. At the extremes of the sequence, the correlations become negative, which is the outcome of the gradual fall in size towards the past.
3. Statistically, the suite of values for the foraminifers as well as that for the ostracods of Example 5.9 are not independently distributed. This is not exclusively due to departure from serial normality, for the approximately normally distributed sub-sample of the oldest individuals also shows significant serial correlations.

The analysis reported on here is very much of a preliminary nature. A reasonable next step would be to examine the statistical properties of the residuals after detrending (cf. Kendall, 1973).

Exercises

1. Taxonomic work involving large samples of invertebrate species is often dogged by the problem of excessive variability. Explain the significance of this remark in terms of phenotypic plasticity. Can you suggest a statistical strategy for distinguishing between a sample of organisms showing continuous plasticity and a sample in which the variability is quasicontinuous (Grüneberg, 1952)?

2. What do you think about series of means of a morphological character as a useful way of establishing and tracing evolution in morphology in a lineage? In answering this question, you might like to examine articles in which "evolutionary trends" are presented in, for example, the journals *Paleobiology, Science* and *Nature*.

3. Lande (1979) generalized the concept of the selection gradient to

$$\beta = \mathbf{P}^{-1}\mathbf{s}$$

as was explained in equation (5.5). Here, **s** denotes the difference in phenotypic mean vectors before and after selection. Is this approach biologically realistic? Is the approach statistically realistic in the light of Fred Bookstein's (1987) results? Can you see any fundamental difference in interpreting the selection gradient for infra-specific variation as opposed to evolutionary relationships at the genetic level? In working out this problem, I suggest you consult a paper by Charlesworth (1984) and another one by Crespi and Bookstein (1989). In the latter reference, see what is said about $\mathbf{P}^{-1}$ and "noise" in relation to signal.

4. Equation (5.10) partitions the phenotype correlation coefficient. Can you suggest how this partitioning can be applied to distance traits on a parthenogenetic organism (fossil or living) to obtain an approximate estimate of r_G (see Reyment, 1982a, p. 167).

5. What advantage does the mode have over the mean in the construction of sequential plots of morphometric data? How would you go about making some kind of multivariate mode?

6. Use the information provided in Table 5.2 for *Veenia rotunda* to make a culling plot over a range of heritabilities from 0.2 to 0.8 like that displayed in Fig. 5.5.

7. Granted that the genetic covariance matrix **G** is square symmetric and can therefore be transformed into latent roots and vectors, rearrange equation (5.24) into an approximate, simplified form. Note, that if the original variables of **G** are multivariate normal so are any linear combination of these multivariate normal, such as the index of equation (5.24).

8. Can you see any reason for wanting to examine a series of observations in an hierarchical manner? What is wrong with "spot-testing" hypotheses?

9. When can you accept a trending sequence as being genuine (without the need for testing for random walk)?

10. Why have I been so "cagey" in my use of the word rate in this chapter?

11. Bookstein (1988) pointed out that histograms of symmetric random walks in one or several dimensions always show apparent clusters separated by apparent gaps. The locations of these features are not meaningful parameters of the underlying process and before clusters or gaps in evolutionary data can be used as evidence of systematic structure, one must reject the possibility that the data are consistent with a symmetric random walk. This is of fundamental importance for the interpretation of what has come to be called "phylogenetic gradualism". The means in Table 5.5 were gleaned from Matsuoka and Okada (1990), being stratigraphical changes in average coccolith size of a purported evolutionary sequence of *Gephyrocapsa*, Malmgren and Kennett (1982) for planktonic foraminifers, and Kellogg (1975) for radiolarians. In order to test the hypothesis of significant evolutionary trend (anagenesis), prepare a program along the following lines.

a. Compute the range for the series, $|S_k|$, that is the absolute value of the difference between the first and last values of the sequence.

b. Compute the root sum square of the differences between all steps in the sequence of means, to wit:

$$\text{est}(\sigma\sqrt{n}) = [\Sigma(S_{n_i} - S_{n_{i-1}})^2]^{1/2}$$

Compute then the ratio $|S_k|/\sigma\sqrt{n}$.

At the 95% level, the two-tailed test for judging hypotheses concerning an evolutionary series is that values less than 0.62 point to the likelihood of *stasis* in the sequence and values in excess of

TABLE 5.5. *Sets of sequential means for four samples* (*oldest observation first: measurements in µm*)

1. A constructed sequence showing anagenesis ($N = 20$)
1. 1.5.2. 2.2 3. 3.8 7. 7.2 7.9 9.3. 8.9 10.9 14.7 16. 19.5 23. 29.5 35. 40. 50.

2. Average coccolith size for purported evolutionary sequence of *Gephyrocapsa* ($N = 36$)
2.06 2.21 2.04 2.05 2.22 2.02 2.23 1.90 1.95 1.88 1.81 1.83 1.92 1.73 1.83 1.96 1.70 1.83 1.94 1.78 1.86 1.97 1.98 2.18 1.98 2.11 2.06 1.95 1.88 1.80 2.03 1.90 1.80 1.85 1.55 1.52

3. Circumference of planktonic foraminiferal lineage ($N = 70$)
34 35 38 31 23 29 24 23 29 24 29 27 22 27 24 28 31 23 28 30 33 34 29 28 26 32 27 29 23 25 24 20 37 30 31 32 38 40 42 43 41 40 48 49 63 57 54 55 51 58 52 53 62 58 67 61 67 64 77 59 70 86 79 65 81 110 87 102 93 126

4. Thoracic width in the radiolarian *Pseudocubus vema* ($N = 34$)
91. 89. 94. 86. 95. 88.5 92.5 90. 100. 105. 106. 103.5 105. 106. 107.5 105. 107.5 105. 107. 104. 108. 108.6 115. 117. 113. 119. 113. 109.5 114. 116. 119.5 119. 125. 134.

2.25 indicate the possibility of *anagenesis*. The broad zone in between incorporates the null model of random walk. Use the first data set of Table 5.5 to see what the result for anagenesis looks like. What do you conclude about the assertion of evolutionarily significant trend made by Matsuoka and Okada?

Using your program, what do you make of Fig. 2 for "chronospecies" in Malmgren and Kennett (1982, p. 291) and the concept of morphometric phylozonation? The data are the third set in Table 5.5. The fourth set of data was claimed by Kellog (1975) to display meaningful trend; what does your own analysis indicate?

CHAPTER 6

Polymorphism

Contents

6.1 The Nature of Polymorphism

6.1.1 *Introduction*

Let me start by telling you why I devoted an entire chapter to something that may, on first acquaintance, seem obscure and out of place in a palaeontological book. Despite diligent searches in standard palaeontological texts, I have not been successful in finding the subject mentioned, let alone given an authoritative evaluation. Even standard textbooks on genetics often fail to give the topic adequate consideration, particularly from the statistical viewpoint. Yet, polymorphism is of fundamental importance in taxonomy, as I hope to convince you in the ensuing pages. Pause a moment and try to recall the literature in your own field of specialization. How often have you run into a taxonomical analysis in which the possibilities of polymorphic complications have even been summarily considered? Serological work has usually a statistical basis, but how often do we find the appropriate statistical models being employed for analysis? Not often, I can assure you. Thus, the entire field of the quantitative treatment of polymorphism is one needing comprehensive reappraisal, re-education, reflection. By its very nature it is multidimensional.

The application of multivariate statistical techniques to data in the form of frequencies requires the special techniques needed for analysing compositions. This fundamental fact is understood by few; as I shall demonstrate, failure to use the appropriate mathematical procedures can lead to wrong conclusions. Most of the data on polymorphism in fossils comes from ostracods and the examples used to illustrate the methods for the statistical analysis of morph frequencies are drawn from this group. The meaning of the word polymorphism is "many-shaped". Polymorphism can take many forms, and we are familiar with the way in which this works from our own species. A very good example is provided by the serological systems of the blood. We have the ABO groups, Rhesus groups, MN and P groups, and many more. A good introduction to serological polymorphism can be found in Mourant (1983). The polymorphisms of the blood provide excellent didactic material and we shall have occasion to take up their analysis in more detail later on. In addition to Mourant's book, advised preparatory reading for this chapter is the statistical textbook by Aitchison (1986) and chapters 3, 12 and 14 of Manly (1985).

In order to understand how to construct a multivariate analysis of polymorphism, it is necessary to review and define concepts. In the present connexion, I have concentrated on features of analytical significance in multivariate analyses. I am well aware that the proem to this chapter can be written in a somewhat different manner.

Levinton (1988, p. 160) maintains the distinction that *Polymorphism* refers to Genetic Variants within a population and *Polytypism* applies for certain variants that are fixed in local populations. This is one way of looking at things; there are others. Clark (1976) provided a different set of definitions, to which we shall return shortly. In my opinion, it is necessary to have a firm grip on the subject of polymorphism in palaeontology, otherwise it is not difficult to produce a confused taxonomy. For example, polymorphism is rife among ostracods, but this is seldom realized. The result is that morphs of a single species end up by being classified as different species. Good references are Levinton (1988, pp. 109–113) and Lewontin (1974).

Almost all polymorphisms we observe in fossil material are the expression of polygenic states. We can seldom hope to identify, with certainty, Mendelian segregation such as can be done for the ABO blood groups. We do often observe what appears to be a two-state alternative polymorphic condition, such as the presence or absence of a posterior spine in a species of ostracods, but the genetic mechanism underlying this is likely to be continuous. *Quasicontinuous variation* has been well studied by workers in the field of

Quantitative Genetics. One observes what appears to be discontinuous (Mendelian) variation, but the mode of inheritance is continuous. The best, and perhaps only, way of getting to grips with the identification of polymorphism among fossils is to enter the field by way of studies on living organisms. The most important pioneering works on quasicontinuous variation are those of Grüneberg (1952) and Rendel (1967). There is also an informative review paper by Van Valen (1969). In these first two works, the problem was identified and given a satisfactory interpretation involving underlying continuous variation, threshold states and a plateau of developmental inertia. Many of the cases I have met in my own studies seem to belong to this category.

A genuine discontinuous relationship can, in palaeontology, only be tentatively inferred if it can be shown that there is a pleiotropic relationship in the material. A *pleiotropic* relationship is one in which of two quite unrelated characters are bound to each other. The example I usually quote is that of the West African Paleocene species *Leguminocythereis lagaghiroboensis* Apostolescu. The pleiotropic relationship affects the presence of a posterior process linked to regularly reticulated lateral ornament. The alternative state is expressed by a stubby posterior linked to a pattern of breached lateral reticulations.

6.2 Ecophenotypic Variability

So far we have only considered one type of morphological complication. A very common variational complexity is that known as **ecophenotypic variability**. Many organisms, including our own species, react morphologically with the environment, particularly with respect to size, but also to shape and ornament; we all know what happens to him that quaffs ale and scoffs crisps thereto. The chemistry and temperature of seawater can have a marked phenotypic effect on the ornament of ostracods. The remarkable variability exhibited by some ammonites seems to have an ecophenotypic background, albeit a complicated one. We shall be considering this in Chapter 8 in the example on *Knemiceras*.

Not all groups of animals are equally exposed to the whims of the environment but, none the less, those of importance in palaeontology do seem to be sensitive to fluctuations in chemical and physical ecological determinants. Unfortunately, little organized work has been devoted to this question and, apart from some groups of insects, a few lower vertebrates, etc., we have slight information on natural populations at our disposal. Animal husbandry has been responsible, however, for yielding a wealth of valuable details on the variability of domesticated species and it is to this field we shall have to turn for useful models. In fact, almost all we possess of an organized nature on the phenotypic plasticity of animals and plants comes from controlled breeding experiments.

A sufficiently intensively planned study of living organisms can sometimes succeed in disclosing whether two species are siblings or not. For fossils we can seldom be certain. More reliance has to be placed on the level of statistical analysis employed and the mathematical sophistication of the practitioner, the number of variables selected and just how diagnostic the chosen traits may be. Remember, many species of fiddler crabs can only be distinguished on male mating behaviour.

Some recent, interesting studies can yield useful guidelines for palaeontological work. Moran and Whitham (1988) studied evolutionary reduction in life-cycles in the genus *Pemphigus*. The suppression of a life-cycle leads to a quantum-type change, although gradual evolution is involved. This kind of evolutionary situation may be quite important in nature. The implications for palaeontological work are distressingly self-evident. Moran and Whitham express the opinion that phenotypic plasticity can be the functional mode permitting the existence of full and reduced life-cycles. They suggest that the transition to a reduced cycle is under selective control.

Another study of importance from the aspect of general principles concerns polymorphism in bill-size in an African finch, *Pyrenestes ostrinus* reported on by Smith (1987). This species displays non-sexual polymorphism in the size of the bill, but bill-size in that species and related species is extreme and bimodally distributed. The size of the bill is important from the viewpoint of feeding in that the morphs perform differently on distinct varieties of food. The reasons for this perplexing situation are not clear, but there may be a component of differential survivability involved in that strong variations in climatic conditions, such as occur in the areas in West Africa sampled, can lead to the partial suppression of one of the morphs on the basis of availability of a particular kind of seed. Breeding experiments indicate that large and small bills result polymorphically but not intermediate sizes.

6.3 The Panoply of Polymorphism

We shall now examine the panoply of polymorphism with particular reference to what we can hope to achieve for fossil material. Some of the points have already been touched upon but our present goal is to systematize our thinking on the subject.

6.3.1 *Coexisting Character States*

Genetically determined polymorphism is characterized by the ABO system of blood groups (Mourant, 1983), but also Mendel's sweet peas. The particulate nature of this class of polymorphism requires experimental confirmation in neontology. In palaeontology, our only hope of identifying Mendelian segregation is to recognize pleiotropism in our material. Levinton (1988) despairs of ever finding this kind of polymorphism in fossils. In the formal definition of discrete polymorphism we say that a polymorphic locus is one at which the most common allele has a frequency less than 0.99. *Disruptive selection* is a mechanism for preserving polymorphism in a population. It favours phenotypes at two or more modes in a continuous distribution and acts against phenotypes lying between the modes. It is an important operator in a set of discrete environments and is conceivably a catalyzing function in speciation. We have already met this selectional variant in the models used in the foregoing chapter.

Polygenic Coexisting Character States are the kind of polymorphic variability most commonly encountered in palaeontological connexions. It seems to bear all the hallmarks of Mendelian segregation, but is none the less polygenic and *quasicontinuous*. The underlying variability is continuous but the visual manifestation is discrete. The different character states appear when thresholds are passed. The classical monograph on threshold variation is that of Rendel (1967), who exemplified his model by the concept of "Make", an unspecified evolutionary force, applied to fruit-flies and mice. Reyment and Van Valen (1969) gave a practical description of threshold variation in fossil and recent ostracods, using Rendel's theory. Environmental control can be generally seen to be the driving force.

Quasicontinuous variation is actually quite common among living organisms. It has been very closely studied in *Drosophila*. In our own species, common diabetes might well be an ecologically regulated sickness with a threshold. Reyment (1982a) described a trimorphic ornamental condition in the Cretaceous foraminifer *Afrobolivina afra* in terms of threshold polymorphism.

6.3.2 *Discrete Morphological Categories*

I group here types of variation that can be referred to as within-state ecophenotypy. For example, the occurrence of vibrissae in the mouse, anterior denticulation in ostracods, and other ephemeral characteristics, many of which are amenable to analysis by multivariate statistical methods. These are basically multidimensional. Reyment *et al.* (1977) studied anteromarginal and posteromarginal denticulation in a species of the ostracod genus *Cytheridea* (Figs 6.2 and 6.3).

Discrete phenotypes may be mistaken for **discrete genotypes** explained by a single locus of large effect, as indeed I once did (Reyment, 1963a; Reyment and Van Valen, 1969). This is an important factor in the multivariate analysis or ornamental features in fossils as well as many living groups of animals and plants.

6.3.2.1 *Contiguous Polymorphism*

Contiguous polymorphism is a state in which the morphs of a species do not overlap but are separated almost entirely by ecological factors. A well-known case concerns polymorphism in barnacles provoked by a predator. The controlling ecological factor determines indirectly the geographical distribution of the morphs in that the predator itself is bound to some particular environment, usually related to depth and/or wave action. A specific example is the relationship between the acorn-barnacle *Chthamalus anisopoma* and the carnivorous gastropod *Acanthina angelica* (Lively, 1986). This gastropod preys on the acorn-barnacle by prising open the shell to gain access to the viscera. The level of predation is usually severe and results in the virtual extinction of the normal morph in the shore zone in which the gastropod operates. The alternative morph is protected morphologically in that it has a hooded shell shape which is very difficult for the gastropod to force. Outside of the immediate zone inhabited by the gastropod, the normal morph develops by default as it were. The hooded morph has a slower rate of growth and is less fecund. In the absence of the predator, it is at a selectional disadvantage. Another term applied to this kind of polymorphism is patch-bound monomorphism.

6.3.2.2 *Continuous Variability within Populations*

We have already discussed this important type of variability. Other terms for it are ecophenotypy and phenotypic plasticity. This is very strongly under the control of the environment. It may at times be difficult to distinguish ecophenotypy from, say, threshold variation, if the sample is small. The scarcity of intermediate morphologies can give the erroneous impression that the variation is discontinuous. Some examples of this sort of variability in a palaeontological context are: ornamental variability in the living ostracod species *Mutilus pumilus* discussed by Reyment *et al.* (1988), ornamental variability in species of Cretaceous ostracods, reported on by Abe *et al.* (1988). Shape variation in species of *Echinocythereis* (Reyment, 1985b) is also of this kind as is also the morphological transition in this lineage from the Eocene of northern Spain. A term used for ecophenotypically produced variants is *ecomorph*. Polymorphism was recorded for the Cretaceous ostracod genus *Oertliella* by Reyment (1982d). Levinton (1988, p. 362) suggests this could have been under the control of epidermal cells, possibly due to the effect of a single gene in accordance with the findings of Okada (1981) for a species of ostracods.

6.3.2.3 *Discussion*

One might well wonder whether polymorphism has some adaptive significance. It is difficult to see, for example, how blood groups can confer an evolutionary advantage. However, it has been shown over the last few decades, that some blood types are distinctly disadvantageous. The connexion between B and certain illnesses and the well-known Rhesus mortality factor on foetal mortality are just two cases. Differential mortality in morphs of the gastropod species *Nucella lapillus* has been studied by Etter (1988)—this has been shown to be related to temperature and the colouring of the morphs. No doubt, many more examples will come to light in the future. The palaeontological significance of the foregoing lies with the fact that morphological categories are often found to "disappear" in a sequence. In the usual mode of interpretation, this is seen as representing the migration of a species from the area and, conversely, the arrival of a new one on the scene. In actual fact, the situation may have a more complicated explanation. If we were following the palaeontological record of the acorn-barnacle, it might not be obvious that the periodic disappearances were the outcome of slight movements back and forth of the shoreline and hence small lateral migrations of predator and normal morph in response thereto.

We have had occasion to look at serological polymorphisms in another part of this book (p. 226) in conjunction with the analysis of compositional data and I have a simple example below (p. 224). We shall also be considering ecophenotypic variability in ammonites in Chapter 8 (p. 314), a quite different type of variation in that it falls under the rubric of Phenotypic Plasticity. Foraminifers are polymorphic, as demonstrated by Reyment (1982a), but here each situation must be judged separately. Ephemeral ecophenotypic utterances also occur such as in fugacious spinosity of the proloculus in bolivinid foraminifers (Reyment and Reyment, 1989).

As palaeontologists we are faced with the double-task of identifying, say, ornamental categories in our material and then trying to decide whether we are face to face with morphs of infra-specific status, or genuine species. Obviously, statistical analysis, particularly multivariate analysis, can and does provide an invaluable help, but it can decide no issues, only point you in the right direction.

Various attempts have been made to link polymorphism to environmental forces. Manly (1985) considered the statistical analysis of polymorphic variation in different habitats. He concluded that if there is significant correlation between the distribution of the morphs of a species and factors of the environment in which they live, this may be indirect evidence of selection. He proposed a battery of procedures for probing this hypothesis, including a randomization test for the habitat model and a negative binomial model for the distributions of morphs from random locations. The latter procedure applied to land snails indicated that variation in the proportion of morphs tends to be due to differences in location. Manly also applied a standard program for linear discriminant functions to data on *Cepaea* sp. (seemingly without taking cognizance of the fact that he was dealing with compositional data) to see whether morph-proportions would correctly allocate samples. A level of 85% of correct allocations was obtained for one of the species, but a somewhat lower level for the other. He also computed canonical correlations to check the proportions of morphs for one species against the proportions of the same morphs for the second species in order to ascertain if there was an overall statistical relationship between the two. Although a relatively high canonical correlation was obtained, to wit, 0.83, this is not statistically significant, no doubt due to the small sample size.

6.4 Polymorphism in Ostracods

I have chosen the ostracods to exemplify the palaeontological significance of polymorphism, not solely out of personal research-interests, but mainly because our knowledge of this group in this respect is greater than for any other fossils. Arthropods are often very highly polymorphic. This is obvious for insects and a significant portion of modern genetics has been constructed around the study of morphs of the fruit-fly. Crustacean polymorphism has been given much less attention.

Ornamental polymorphism in marine ostracods was first reported by Reyment (1963a) and further studied by Reyment (1966b) and Reyment and Van Valen (1969). The subject was taken up by Keen (1982) who recorded several morphs to occur in cytherettids. Further substantial contributions have been made by Ducasse and Rousselle (1978), Ducasse (1979), and Ducasse and Cirac (1981), all for Tertiary ostracods. Reyment (1988) reviewed the field and attempted to establish and stabilize concepts. Ikeya and Ueda (1988) demonstrated for living populations of the ostracod species *Cythero-morpha acupunctata* (BRADY) the occurrence of seasonally linked polymor-phism. Not only size is affected, as was indeed shown earlier by Abe (1983) for another species of ostracods, but also the lateral ornament reticulation was thought to be under the control of salinity and temperature, but with an underlying genetic component. Smaller individuals tend to have well-devel-oped reticulation, whereas larger carapaces tend to be weakly ornamented. Hence, in a single sample, spanning several seasons, one can find two classes of shells presenting a state that simulates pleiotropism. Ikeya and Ueda note that the ornamental variation is actually continuous, sex-linked in intensity and it is not possible to recognize naturally discrete classes. The lessons to be learned from this work are (1) that close attention must be paid to the occurrence of transitional ornamental categories in ascertaining the possible occurrence of pleiotropism; (2) the presence of several morphological variants in a sample is not necessarily a reflection of threshold polymorphism, but may also depend on seasonal sports. Unfortunately the work on *Cytheromorpha* lacks adequate statistical information and it is therefore not possible to comment further on eventual implications.

The polymorphism described by Reyment (1963a, 1966b) features mainly alternative ornamental states, but polymorphism in the outline of the shell, including pleiotropism in shape and ornament, was also recorded. (Note, that the ornamental classes here are discrete.) Keen (1982, p. 387) gave a succinct account of some of the kinds of polymorphism that may occur in marine ostracods. The features recognized in that article lend themselves to quantification and Manly (1985) has developed a quite general approach to the statistical treatment of polymorphism that would appear to be suitable for this purpose.

The recognition of ornamental polymorphism in living ostracods by zoologists is only a recent event and dates from Hartmann (1982). This perhaps is hardly to be wondered at owing to the relatively few people actively engaged in the study of living ostracods, coupled with traditionally differently stressed areas of interest. Doubtlessly future work will yield information of substantive evolutionary value. Perhaps the greatest stumbling block in the path of a general evolutionary model is the possibility that size, shape, and ornament are under the control of fundamentally different evolutionary processes (cf. Leman and Freeman, 1984). It is therefore essential that evolution in each of these three categories be examined in its own right and that they are not lumped indiscriminately without prior consideration. There is a growing body of evidence to support the conclusions reached by Leman and Freeman (1984) to the effect that size and shape are fundamentally different evolutionary directions. This may well be so in many instances, but I think that the examples assembled in this book proved that there are many cases when the two are, and must, be closely interrelated. It all boils down to one thing: the choice of variables.

Work on waterstriders (Gerridae, Hemiptera) by Zera (1984) provides a useful illustration of the concepts involved. Some species of waterstriders are wing polymorphic, whereby spatial or temporal variation in morph ratios can occur, both among populations of the *same* species and *between* species. Thus, there are varying proportions of fully winged, short-winged and/or wingless morphs. Observable polymorphic differences have been shown to be linked to a set of coordinated traits of another kind. As an example, short-winged morphs usually have a rapid rate of development, an early age of first reproduction, and high fecundity. This variation has been proved to be quasicontinuous. Zera also reports on a polymorphism in non-gerrids that has a bearing on crustaceans. There are differences in fitness among morphs and these differences tend to be located with particular morphs, even in distinct species.

Not very much is known about the genetics of ostracods, but things seem to be looking up a little. Havel *et al.* (1990) have reported on some freshwater ostracods from low-Arctic Canada and found, among other things, that in a sexual species, *Cyprinotus glaucus*, proportions are strongly biassed in favour of females. Sex-ratios are variable in adults but seem to be stable during ontogeny. Among six species studied, the proportion of polymorphic loci ranges from 0–33%, with an average of 20%; genotypic frequencies at these loci range were found to be conformable with Hardy-Weinberg expectations. One interesting side of the investigation concerns gene flow between sub-populations of *C. glaucus*. It was found that gene-frequency differences occur among local populations which was taken to indicate restricted gene flow in this passively dispersed species.

6.5 Environmentally Cued Polymorphism

Clark (1976, p. 256) expounded the semantic history of polymorphism arriving thereby at the logical conclusion that any feature that gains expression in the phenotype must perforce have a genetic basis. Clark therefore recommended that such terms as polyphasy and polyphenism should be abandoned. However, in genetically determined polymorphism, the environment may play little part in the determination of morphs. The term *environmentally cued polymorphism* was proposed by Clark to cover the situation where environmental stimuli interact with the genome so as to tease forth a particular morph. Thus, cueing can be responsible for producing seasonal forms in insects, such as butterflies. Clark noticed that there are intermediate states of polymorphism in which the characteristics of genetic and environmental polymorphism are manifested. Thus genotypes differ in their capacity to develop a particular morph under various ecological conditions. Reyment (1982c, d) applied Clark's theory to Cretaceous ostracods. The morphological variation in *Cytheromorpha* recorded by Ikeya and Ueda (1988) merits consideration in the light of Clark's results.

Harking back to Zera's waterstriders, it has been ascertained that a particular genotype produces the fully winged morph in summer and the wingless morph in the autumn. A second genotype produces only the unwinged morph during both of these seasons. The necessity to adapt to a particular season is met by the model of environmental cueing by requiring that the morph appearing at a particular time is largely determined by current, or immediately prior, ecological conditions. Clark considered that speciation might well result under conditions in which only one morph is environmentally cued whereby the ability to produce the alternative morph is ultimately lost. Several cases of this process seem to be known for New Zealand aphids.

6.6 Shape Polymorphism

The topic of shape polymorphism will be introduced by means of an illustration. Reyment (1985b) identified several shape-morphs in Eocene *Echinocythereis* from the Eocene of Aragon, Spain. The most commonly occurring shape-morphs are:

1. Asymmetrical rounding of the anterior margin.
2. A concave posterodorsal margin.
3. Underslung ventral region.
4. A concave dorsal margin.

It is not an easy task to identify evolutionarily significant shape polymorphisms, but in the present example, morphs (3) and (4) could be connected to evolution in the lineage. Morphs (1) and (2) are found throughout the history of the lineage and are not identifiable as evolutionary innovations. The main features uncovered by the multivariate analysis are treated in the next section.

6.6.1 *Multivariate Shape Analysis of the Lateral Silhouette*

I shall now consider two transitions in the ostracod lineage, of which certain aspects were taken up in Chapter 5. Analysis of the outline of the carapace by the method of eigenshapes was invoked for the present study.

The first of the evolutionary transitions concerns the passage from *Echinocythereis isabenana* to *E. aragonensis*. This took place rather rapidly and was accompanied by a significant and permanent reduction in size (and also in ornamental features). The permanent and irreversible reduction in volume of the carapace resulting from this reduction rules out any ephemeral type of transition such as could have been engineered by the environment. The passage from *E. aragonensis* to *E. posterior* is characterized by a stochastic replacement of one set of ornamental characters for another. We shall now see whether the ornamental transition was accompanied by a change in shape (i.e. here outline).

The plot formed by the second eigenshape in relation to the third eigenshape is shown in Fig. 6.1. There is a clear separation into discrete clusters corresponding to each of the species. Moreover, points for the oldest representatives of *Echinocythereis aragonensis* are segregated from the youngest members. The constellation formed by the points for *E. isabenana* covers a large part of the graphs. The conclusion promoted by the analysis is

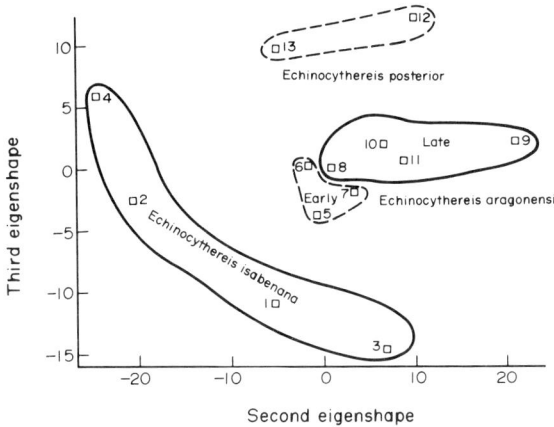

FIG. 6.1 . Plot of the second and third eigenshapes for the *Echinocythereis* data. The breakdown into taxonomically reasonable subdivisions on differences in outline alone is clear and the evolutionary model proposed for this material is supported. Redrawn from Reyment (1988).

that there is differentiation in shape, not only between species, but also within them, thus pointing to some kind of polymorphism in outline.

6.7 Ornamental Polymorphism

The most commonly occurring polymorphism I have found in ostracods concerns the presence or absence of a posterior spine. This variety is of wide occurrence, for example, in species of *Buntonia*. Another kind of polymorphism consists of regular versus breached reticulations. This type occurs in species of *Brachycythere* and *Leguminocythereis*. Although polymorphism is only easily recognizable in adults and instar A-1, a pleiotropic condition in *Leguminocythereis* could be followed back as far as instar A-3. Pitting can also be polymorphic. For example, in the Paleocene species *Buntonia pulvinata* APOSTOLESCU, the opposing states are coarse pits versus fine pitting. In many species of *Cytherella*, a smooth shell surface alternates with a pitted one.

6.7.1 *Variation in Spinosity*

Marginal spinosity seems to display a different variational pattern from the kinds mentioned above. Reyment and Van Valen (1969), reporting on ornamental variation in quasicontinuous characters in a Recent species of *Buntonia* from the offshore zone of Nigeria, found that several features interact in developing the spines, namely, the size of the carapace in relation to the presence or absence of a ventrolateral spine, the number of spines along the anterior margin, and the number of spines along the posterior margin. Both anterior and posterior spines increase in number with successive instars and there is also a sexual dimorphic relationship linked to spinosity to consider as well. Anterior spines are more numerous, on average, in males than in females (presumably concordant with the greater size attained by males), but posterior spines are few in number, on average, in males than in females. This is obviously a highly multivariate situation.

The frequencies of anterior and posterior spines are but weakly associated with each other. However, individuals bearing a ventrolateral spine have, on the average, more anterior and posterior spines than those lacking this spine. The same kind of result was obtained for *Brachycythere oguni* REYMENT, a Maastrichto-Paleocene species of wide occurrence in West and North Africa. The development of spinosities reported in the study by Reyment and Van Valen (1969) may, in part, have the same origin as that found by Abe *et al.* (1988) for Santonian (Cretaceous) ostracods from Israel and in *Cytheridea* (Figs. 6.2 and 6.3).

Hartmann (1982) gave a qualitative appraisal of ornamental variability in Australian marine ostracods noting that there may be very strong differences between juveniles and adults of the same species. It is well documented for

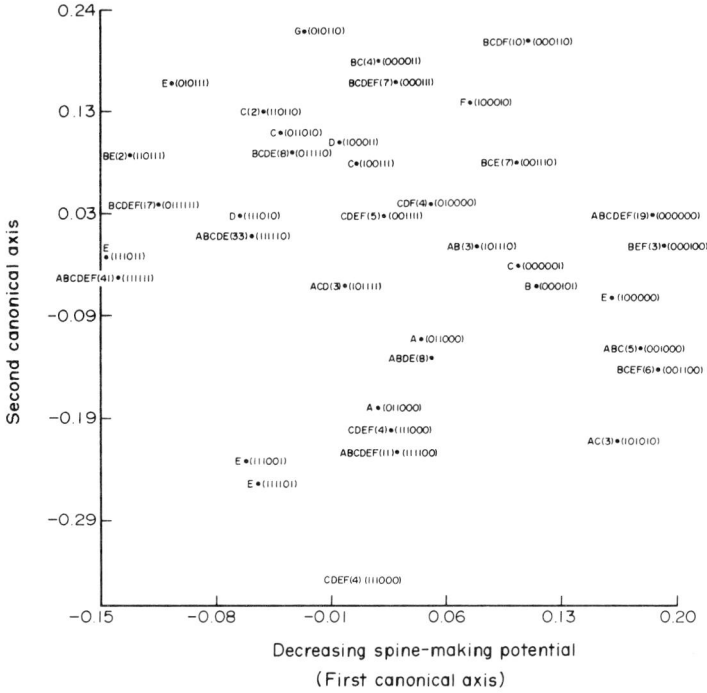

Fig. 6.2. Plot of scores for the first and second canonical variates for the spine data on *Cytheridea* from the Miocene of France. The phenotype for presence/absence of anterior spines is denoted as (*i i i i i i*) with *i* taking the values 1 or 0. The letters *A*, *B*, . . ., *F* denote the stratigraphical levels from which the samples were obtained. The first canonical variate axis could be identified as measuring decreasing spine-making potential. Redrawn from Reyment *et al.* (1977).

many species, including those studied by Hartmann, that larval ornament can, within part of the geographical range of a species, be retained into the adult. This paedomorphic development in ostracods is a subject requiring more attention. I (Reyment, 1966b) have recorded adults with larval ornament to occur in the West African Paleocene species *Actinocythereis bopaensis* APOSTOLESCU. Hartmann (1982) also described size polymorphism in *Xesteroleberis chilensis austrocontinentalis* HARTMANN from Australia. Here, there are two discrete size classes, as well as discrete ornamental patterns. The morphs are intrapopulational and Hartmann concluded that they cannot be of ecological origin.

We shall now look at some examples of the multivariate analysis of polymorphism, expressed in the form of frequencies.

6.8 Determining the Effects of Closure

Fundamental to the multivariate analysis of observations on polymorphic characters is the concept of closed arrays (Aitchison, 1986). Let us just briefly

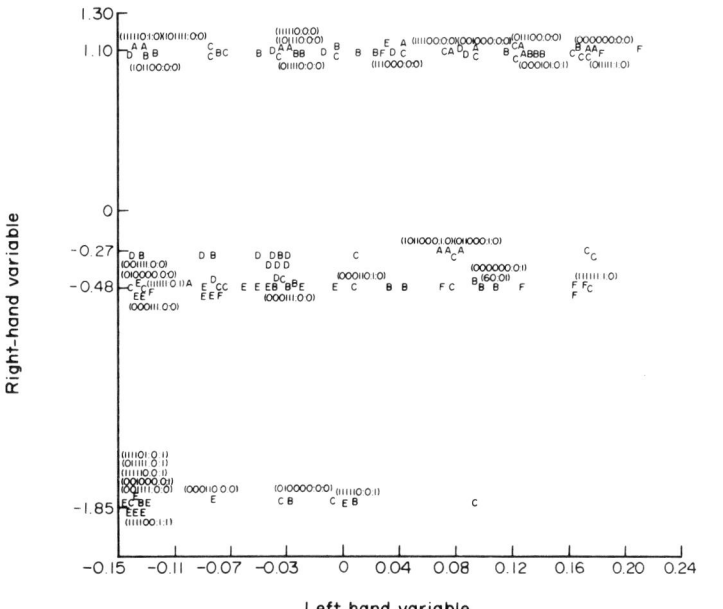

FIG. 6.3. Plot of the scores for the first canonical correlation for the data on *Cytheridea*. The anterior and posterior spines are coded as $(i\,i\,i\,i\,i\,i, j\,j)$ where i and j take the values 0 or 1. The scores for vector of anterior spines are plotted against those for posterior spines. Redrawn from Reyment *et al.* (1977).

consider what the effects of the closure constraint can be on an analysis of polymorphism. I have referred to the human blood groups A, B and O on several occasions in this book. We shall now briefly consider 25 populational averages for Spain and North Africa and computed the correlations in usual space and in simplex space. The simplex correlation matrix (i.e. the correlations computed taking heed of the fact that the rows of the data matrix have a constant sum) is

$$
\begin{array}{c}
\\
A\\
B\\
C
\end{array}
\begin{array}{ccc}
A & B & O
\end{array}
\left[
\begin{array}{ccc}
1.0000 & -0.9006 & 0.5732 \\
-0.9006 & 1.0000 & -0.8724 \\
0.5732 & -0.8724 & 1.0000
\end{array}
\right]
$$

The usual correlation matrix computed for the same data is

$$
\begin{array}{c}
\\
A\\
B\\
C
\end{array}
\begin{array}{ccc}
A & B & O
\end{array}
\left[
\begin{array}{ccc}
1.0000 & -0.4527 & -0.4943 \\
-0.4527 & 1.0000 & -0.5502 \\
-0.4943 & -0.5502 & 1.0000
\end{array}
\right]
$$

The extreme differences in corresponding entries in these arrays should be obvious to anybody. Just compare r_{12} for both. The one is twice that of the other. Hence, any multivariate study involving serological polymorphisms will be doomed to failure if adequate account is not taken of the effects of constraining the data. This is admittedly a very simple illustration but it serves its point.

Example 6.1. Analysis of Ornamental Polymorphism in *Echinocythereis*

Presentation of Problem: We have already met the *Echinocythereis*-lineage on several occasions in the preceding pages. You will recall that the three species are polymorphic for several categories: size, shape and ornament. Some of the more commonly occurring ornamental morphs are listed in Reyment (1985b).

For the purposes of illustrating the correct technique for analysing polymorphism by multivariate methods, I have chosen the three states:

1. Smooth lateral surface.
2. Papillate lateral surface.
3. Papillo-reticulate lateral surface.

The data used in the analysis are listed in Table 6.1.

Some of the changes registered in the lineage are *permanent*, in that, once installed, they were not repeated during the history of the lineage. For example, the abrupt shift from coarse and regularly located papillae (*Echinocythereis isabenana*) to irregularly dispersed, small papillae (*Echinocythereis aragonensis*). After this shift in ornamental properties had been introduced, regularly distributed papillae disappeared for good. Owing to the intrapopulational occurrence of these morphs, the ornamental variability

TABLE 6.1. *The input frequencies and basic statistical results for three morphs of* Echinocythereis

THE INPUT DATA MATRIX		
Morph 1	Morph 2	Morph 3
0.73	0.21	0.06
0.33	0.66	0.01
0.16	0.79	0.05
0.26	0.58	0.16
0.13	0.74	0.13
0.42	0.58	0.00
0.21	0.71	0.07
0.08	0.88	0.04

SIMPLEX CORRELATIONS (upper triangle) and
USUAL CORRELATIONS (lower triangle)

	Morph 1	Morph 2	Morph 3
Morph 1	1.0000	0.0671	−0.8106
Morph 2	−0.9649	1.0000	−0.6386
Morph 3	−0.2362	−0.0271	1.0000

cannot be interpreted in terms of the effects of water chemistry, nor even bathymetrically induced variation.

Method of Analysis: The technique required for analysing multivariate data composed of frequencies of morphs demands the procedures appropriate to **compositional data**. A full account of the theory and practice of compositional data analysis is given by Aitchison (1986), who has also made available a PC statistical package on diskette for doing all the necessary calculations.

The problem facing us is to analyse three morphs occurring in the *aragonensis-posterior* section of the lineage. There are eight stratigraphical levels; the percentages of morphs at each level are listed in Table 6.1. The statistical treatment of the data shows that the effects of closure are about as extreme as you can get. Black becomes white and white becomes black, as it were. Just look at the correlations listed in Table 6.1. The simplex correlation coefficient for r_{12} is almost nought, but the usual counterpart is a massive -0.96. The simplex correlation coefficient for r_{13} is high and negative, but only -0.23 for the inappropriate linear counterpart. The values for r_{23} are also quite different.

You may remember that I pointed out in Chapter 2 that the effects of the constant-sum constraint tend to be most severe for low dimensional situations and to become slighter as the number of dimensions is increased. For ten variables, the effects are usually, but by no means always, trifling; the effects are generally more noticeable in the bivariate correlation coefficients than in, say, the elements of the latent vectors of the correlation matrix.

Proceed now to Table 6.3, which presents the results of the principal component analyses using the log-ratio transformation of Aitchison (1986) for compositional variables, and the inappropriate method of computation, which was designed for Cartesian space. As only to be expected, there is a pronounced difference between the results obtained by the correct procedure and those yielded by standard principal component analysis. In both cases, almost all of the information is concentrated to the first principal component. The log-ratio analysis indicates that Morphs 1 and 3 are locked in a negative association (which is evident from the data matrix). The second principal component represents a small amount of variability deriving from an inverse relationship in Morphs 1 and 2. In no case do we find all morphs linked in a positive association. There is no third principal component, since the rank of the correlation matrix is perforce two (due to the constraint).

Look now at the results obtained by the usual method of principal component analysis. The values in Table 6.2 indicate a bipolar negative association in the first and second morphs, the contribution of the third morph being almost zero. The second principal component is dominated by the effects of the third morph balanced against opposite equal contributions from the first and second morphs. The corresponding latent root is very small.

Note that the foregoing results are for the covariance matrix. If the correlation matrix had been used, there would have been a difference in the

TABLE 6.2. *Principal component analyses for the data of Table* 6.1

ORDINARY PRINCIPAL COMPONENT ANALYSIS		
Latent roots	0.08405	0.00445
Latent vectors	1	2
Morph 1	0.717	−0.392
Morph 2	−0.697	−0.429
Morph 3	−0.021	0.814
LOG-RATIO PRINCIPAL COMPONENT ANALYSIS		
Latent roots	2.1534	0.5537
Latent vectors	1	2
Morph 1	0.537	−0.615
Morph 2	0.264	0.773
Morph 3	−0.801	−0.157

elements of the latent vectors. The difference is, however, not so serious in the present case: the first latent vector of the raw correlation matrix has the elements $(-0.7, -0.1, 0.7)$ and the second vector the elements $(0.4, -0.9, 0.3)$.

So far so good, but there does seem to be a fly in the ointment. By now you will have gathered that the best laid schemes o' mice an' men gang aft agley, and there does seem to be a sensitivity in the log-ratio method to minor fluctuations in the data. I can illustrate this by means of a practical example. Let the sixth vector of morph proportions in Table 6.1 be altered to $0.42:0.57:0.01$. The correlations and principal components of the constrained and unconstrained situations are displayed in Table 6.3. You will see that the simplex correlations have been appreciably influenced for r_{12} and r_{23}, whereas the correlation matrix computed by the usual procedure has hardly been touched. It seems therefore that log-ratio correlations might not be robust to minor sample fluctuations. The effects on the latent vectors are less pronounced and the results obtained do not greatly differ from those yielded for the original data matrix. The latent vectors produced by the usual method of calculation are hardly touched with respect to the first vector and only moderately strongly influenced for the second vector.

The subject of sensitivity of log-ratio correlations to minor deviations in a data matrix seems to require further theoretical work. At the present state of our knowledge I can only recommend caution, particularly if you are adding or deleting a row from the data matrix for some reason or other. Presumably, some kind of a stability study would be a useful way of approaching the problem broached here.

Example 6.2. Shape Polymorphism in *Oertliella* (Ostracoda; Cretaceous)

Presentation of the Problem: Reyment (1982d) applied quantitative genetics to the analysis of ornamental variation in *Oertliella tarfayaensis* REYMENT from the Upper Cretaceous of Morocco. There are two ornamental categories

TABLE 6.3. *Constrained and unconstrained correlations and principal components (correlations and covariances) for the morph data of Table* 6.1 *with one adjustment: demonstration of lack of robustness in the log-ratio method*

SIMPLEX CORRELATIONS

1.0000	−0.3494	−0.7091
−0.3494	1.000	−0.4130
−0.7091	−0.4130	1.0000

USUAL CORRELATIONS

1.0000	−0.9662	−0.2290
−0.9662	1.0000	−0.0291
−0.2290	−0.0291	1.0000

LATENT VECTORS FOR SIMPLEX CORRELATIONS

	1	2
Var. 1	−0.6820	0.4489
Var. 2	−0.0478	−0.8151
Var. 3	0.7298	0.3661

LATENT VECTORS FOR USUAL CORRELATIONS

	1	2
Var. 1	0.7178	−0.3908
Var. 2	−0.6959	−0.4267
Var. 3	−0.0201	0.8156

LATENT VECTORS FOR SIMPLEX COVARIANCES

	1	2
Var. 1	−0.6804	0.4010
Var. 2	−0.0880	−0.8756
Var. 3	0.7275	0.2692

LATENT VECTORS FOR USUAL COVARIANCES

	1	2
Var. 1	0.7092	−0.0262
Var. 2	−0.8901	−0.2302
Var. 3	−0.1442	0.9728

NB: Sixth vector of data matrix in Table 6.2 altered in second decimal place to show sensitivity of the log-ratio method to the composition of the data matrix.

in this species, one of whch is regularly reticulated and the other of which is irregularly ornamented, showing smooth zones with the pattern of reticulations. Although the two classes are distinct, there is some indeterminacy in the smooth-field morph in that the smooth zones vary somewhat in extent from specimen to specimen. Both ornamental types occur in the same sample, so there is no question of subspecific differentiation, ecophenotypic reactions, or the like. In addition to my original study, Levinton (1988) has expressed ideas on the matter of the kind of variability represented in the species.

The question that arises is: are there any features of the shell that vary conjointly with the ornament? Although visual inspection is hardly a reliable guide in a situation such as the present one, I decided that after careful examination under the SEM that *shape* could be such a property.

The tensor-biometric procedure of Bookstein (1986) was selected as being the most sensitive method for the purpose of identifying possible shape morphs. Standard multivariate methods applied to distance traits did not reveal significant differences between the two ornamental morphs.

The Analysis: Seven landmarks and pseudolandmarks were selected for study. These are depicted schematically in Fig. 6.4. They are: *A*, midpoint of rounding of the anterior margin; *B*, posterior tip; *C*, location of the eye-tubercle; *D*, centre of the adductorial tubercle; *E*, postero-dorsal rounding; *F*, intersection of the ventrolateral rib with the ventral margin, G: mid-point of the ventral rounding of the anterior margin. The measurements were made on carefully calibrated SEM photographs using a CAL-COMP digitizing tablet. Only left valves were employed. The study material comprised eight individuals of each of the morphs to be contrasted.

The tensor-biometric analysis succeeded in giving detailed information on shape-variability in these ostracods. With *AB* as baseline, the triangle *ABC*, the location of the eye-tubercle, partly separates the two groups (Fig. 6.5). There is no differentiation in shape with respect to the position of the adductor tubercle, landmark *D*, the postero-dorsal bend, landmark *E*, nor the ventrolateral rib intersection, landmark *F*. Complete segregation occurs, however, for the triangle formed by landmark *G*, the anteroventral marginal rounding. This can be seen in Fig. 6.6. If size is restored to the analysis, the result is to cause total segregation for the morphs on triangle *ABC*, to wit, the triangle formed by the eye-tubercle with the baseline. This is shown in Fig. 6.7. The position for landmarks *D*, *E*, and *F* remains unchanged, i.e. total overlap, whereas the triangle formed with *G* now no longer leads to complete separation.

Discusssion: The tensor-biometric analysis discloses that there is quite marked polymorphism in shape in *Oertliella tarfayaensis* and that the main effects of this are concentrated to the antero-dorsal and antero-ventral marginal zones. Particularly the latter, landmark *G*, characterizes the shape-polymorphism. If size is allowed to enter the analysis, we observe that there is an appreciable size-sensitive component in landmark *C*, the position of the eye-tubercle. This is not so for landmark *G*, which seems to be associated with absolute shape-differentiation in this character. The other characters remain inert throughout the analysis, which shows that the posterior fourth of the carapace, as well as the centrolateral region of the shell, are not affected by polymorphism in shape.

There is, no doubt, always the possibility that the observed polymorphisms in this species are pleiotropically linked. This question cannot be resolved by the present analysis. The current exercise offers a good example, I think, of the sensitivity of tensor-biometrics in quantitative genetics and, especially, for charting polymorphism in morphological traits.

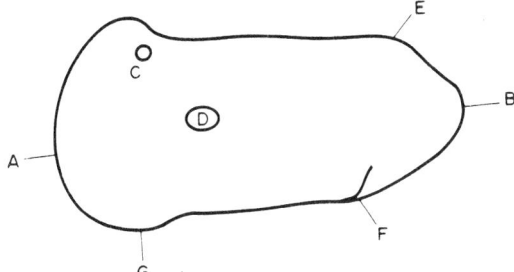

FIG. 6.4. Landmarks and helping points around the carapace of *Oertliella tarfayaensis* from the Cretaceous of Morocco. True landmarks are points *C*, *F*, and *D*, whereas *A*, *B*, *E* and *G* are pseudo-landmarks.

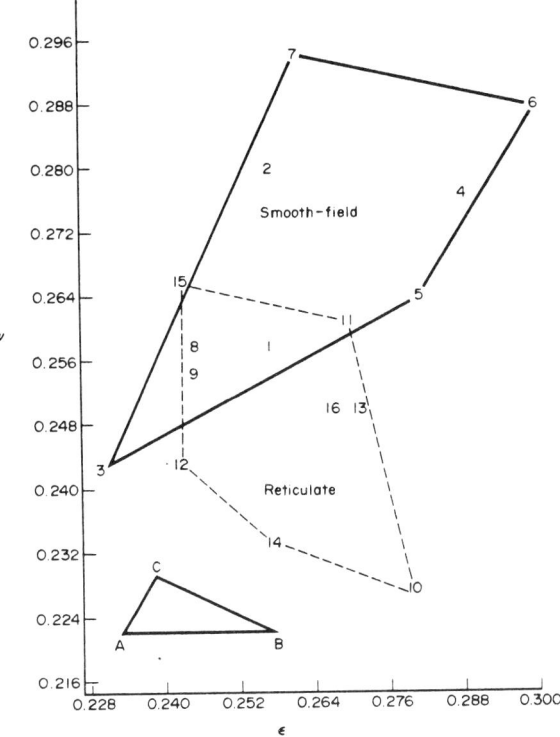

FIG. 6.5. Separation achieved by landmark *C*, the eye-tubercle, on shape differences for *Oertliella tarfayaensis*. There is fairly good separation between smooth field and reticulated individuals.

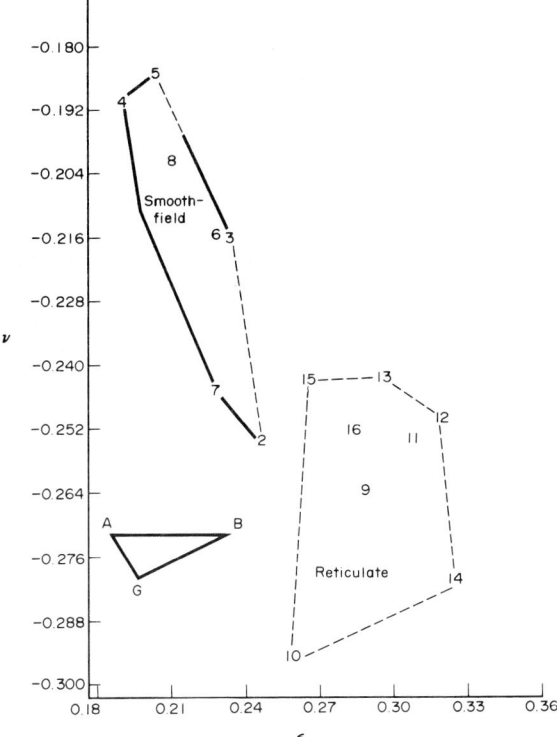

FIG. 6.6. Separation obtained by shape on triangle *ABG* for *Oertliella tarfayaensis*. The differentiation obtained by using helping point *G*, the midpoint of the anteroventral rounding, is complete, thus indicating the possibility of some kind of a pleiotropic relationship between shape and ornament in this species. This does not seem to be an example of the variation recorded by Ikeya and Ueda (1988) and Reyment *et al.* (1988).

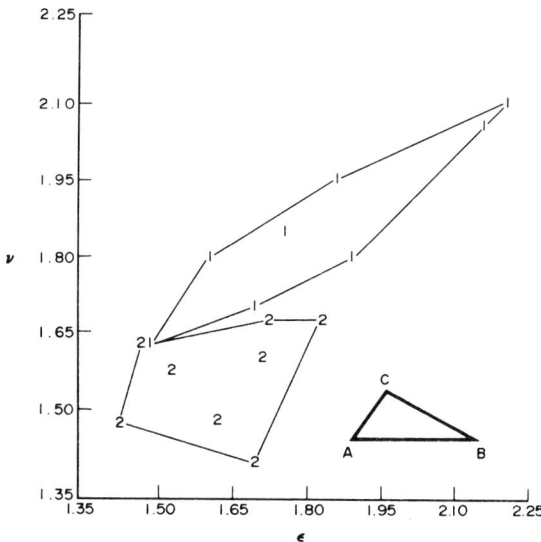

FIG. 6.7. Separation obtained by landmark *C*, location of the eye-tubercle, with size retained in the analysis. Data on *Oertliella tarfayaensis*.

Exercises

1. Using Gower's (1971a) similarity measure and his method of principal coordinate analysis, can you devise a procedure for analysing ornamental polymorphism? Hint: make use of the possibility offered by Gower's method of coding qualitative characters.

2. Jot down all the polymorphisms in our own species you can think of. Before starting, you could increase your knowledge of serological polymorphisms by checking through a few issues of *Nature*, for example, including *Nature* (1990).

3. Brush up your knowledge of environmental polymorphism. An old, though still reliable reference is Huxley (1932). Start with what he wrote on p. 71 and read about wet-season and dry-season forms of certain tropical butterflies: different environmental conditions bring out different expressions of the same gene complex. Another basic reference is Schmalhausen (1949).

4. In *Afrobolivina afra* I found three morphs to exist. A reticulated category, a longitudinally striate one, and a "smooth" one. Given that all three can occur in the same carefully sampled material, how would you relate this to the conditions considered in exercise 3?

5. If you have them available (see References), have a look at the papers by Odette Ducasse and her coworkers. How would you interpret the polymorphic condition described in these papers in terms of what has been said in Chapter 6? Look particularly at the question of ornament in relation to geographical location.

6. Does the work referred to in exercise 5 record coexisting morphs very often? How do you explain the type of ornamental variation recorded most commonly in these publications?

7. See if you can analyse by principal components the following allele frequencies for the ostracod species *Cyprinotus glaucus*. The data were taken from Havel *et al.* (1990). Try both the usual method as well as the appropriate method for compositional data.

Sample size		Alleles	
	1	2	3
Locality A, number of ponds = 8			
48	0.063	0.333	0.604
96	—	0.260	0.740
96	—	0.094	0.906
93	0.011	0.194	0.796
84	—	0.202	0.798
48	—	0.500	0.500
65	0.031	0.108	0.862
68	—	0.059	0.941
Locality B, number of ponds = 8			
72	—	0.153	0.847
82	0.012	—	0.988
72	—	0.028	0.927
61	—	0.049	0.951
72	0.014	—	0.986
60	0.067	—	0.933
48	—	0.542	0.458
95	—	0.074	0.926

8. Here are some populational determinations of Rhesus polymorphisms for the Mediterranean region. See what you can make of them using bivariate correlation coefficients and principal component analysis. Note, that the data are given in the form of percentages (the rows below do not sum to 100 — there are more than four Rhesus alleles, as you will discern from the possible combinations for C-c:D-d:E-e).

Population	CDe	cDE	cDe	cde
Aragon (Spain)	44.07	10.48	3.18	37.11
Barcelona (Spain)	41.85	12.71	3.30	40.52
Basques (Guipuzcoa)	40.32	4.31	7.26	43.62
Galicia (Spain)	43.22	12.72	3.97	38.27
Galicia (Spain)	54.28	10.56	3.72	31.44
Gran Canaria	45.18	7.71	9.73	32.87
Gran Canaria	42.44	11.30	11.14	20.33
Jews (Djerba), Tunisia	45.21	10.60	4.88	35.19
Jews (Sephardim), Israel	38.12	15.39	14.92	20.92
Leon (Spain)	49.13	9.47	2.58	38.82
Leon (Spain)	47.70	10.11	2.17	40.02
Malta	50.45	15.13	3.93	30.79
Portugal (central)	43.45	11.90	4.76	37.87
Portugal (northern)	41.77	10.10	4.67	39.71
Lebanese Shi'ah Moslems	50.36	11.48	7.55	28.43
Sicily	50.51	10.88	2.05	32.12
Sicily	50.92	11.35	2.83	32.76
Tlemcen (Algeria)	56.15	6.97	3.58	33.30
Valencia (Spain)	37.30	14.30	14.98	34.77

Sources of Data:

Mourant, A. E., Kopec, A. C., Domaniewska-Sobczak, K. 1976. *The Distribution of Human Blood-groups and other Polymorphisms*. Oxford, at the Clarendon Press.
Reyment, R. A. 1983. *Annals of Human Biology*, 10:505–522.

9. You are given determinations on five of six possible chemical components of a set of samples. How would you determine the frequencies of the sixth element without doing any chemical analyses?

CHAPTER 7

Multivariate Palaeoecology

Contents

7.1 Introduction

Most organisms live in an environment with at least some seasonal components. Many organisms have evolved the ability to *track* environmental shifts, such as pertain in the rhythmic rotation of the seasons, and some aspects of the importance of this for multivariate morphometrical work were touched upon in Chapter 5. Considerable interest attaches to this category of problems in ecology. Hastings (1984, p. 350) reviewed this subject in the light of the logistic, or r-type, ecological model. The **logistic equation** of Volterra has become one of the classical cornerstones of deterministic ecological models and has given rise to a widely accepted terminology for work in this field: you will doubtless have run into the terms "r-strategies" and "K-strategies", in which parameters from the logistic equation of Volterra have come to be infused with very special meanings. The evolution of what we can conveniently call **tracking ability** can be viewed in this context.

In 1971, I accounted for the application of univariate statistical methods in Palaeoecology. It is, however, often interesting to contemplate the multivariate reaction between organisms and one or more factors of the environment. The kind of statistical model required for treating this situation involves correlations between sets of variables. In the simplest case, we have a multivariate regression where one character is weighed against a set of characters. A generalization of this model is known as canonical correlation: it was introduced in Section 3.7. A recent major modification of the canonical correlation model is the method of Canonical Correspondence Analysis of C. ter Braak (1985, 1986).

I shall not attempt to hide the fact that what we encounter in the field of multivariate palaeoecology is rather subjective, for there is no way known to me of being sure of which factors could possibly have influenced this or that morphological character. Conjecture can sometimes be made firmer by suitably designed laboratory experiments on living relatives of the species of interest. Ostracods lend themselves ideally to such studies, as do also bivalves, gastropods and foraminifers. Pilot investigations on living organisms can also greatly enhance analyses of ecophenotypic variation incurring polymorphism. The occurrence of self-thinning in barnacles is an example of a situation requiring special methods of study and interpretation (Hughes and Griffiths, 1988).

A glance at any recent ecologically oriented palaeontological text, such as Vermeij (1987), will convince you that relatively few palaeoecological problems can be investigated by multivariate statistical methods. The types of subjects amenable to quantitative palaeoecological analysis are:

1. Chemical factors of the environment and their influences on morphological variability.
2. Morphological variation in relation to physical factors of the environment, such as can be measured by electrical logging methods.
3. Analysis of morphological patterns for species from the same suite of samples.
4. Interaction between predator and prey.

5. Environmental reconstructions based on physical and chemical components.
6. Aspects of dispersal and inferred migration.
7. Variations in frequencies of species.

In the first part of the chapter, I take up routine applications of multivariate analysis to the physical and chemical properties of sediments and the relationships between organism and palaeo-environment. The analysis of multivariate assemblages of species comes next: this is what is usually known as Statistical (Palaeo)-ecology. The interaction between predator and prey is the next theme and, finally, the quantification of unlike-neighbour environmental relationships is considered.

7.2 Dispersal and Migration

Most organisms have the capacity to disperse or migrate. Yet, the characteristics of migration and dispersal vary from species to species. Individuals may travel thousands of kilometres every year to breed at established sites (e.g. some whales, some birds, even some dragonflies) returning to non-breeding grounds for the rest of the year. Juveniles of some species actively emigrate from their natal sites to establish a place for living in their adult lives. Our chances of identifying episodes of dispersal and, or, migration in the past are severely restricted. Yet good examples do arise now and then; for example, the distribution of Paleocene ostracods across northern Africa (Section 8.10).

Hence, it should be stressed that it is important to consider time as well as space in a multidimensional palaeoecological study. Variation even within species suggests that phenotypes have undergone major changes—this is the basis for the concept of the *geographical cline*, one of the commonly invoked expressions of dispersion. Ecological migration concerns the physical movement of one or more individuals from one place to another (Bull *et al.*, 1982). Genetic migration is a more specific concept and includes ecological migration and the movement of individuals from one breeding ground to another. This will have a direct influence on gene frequencies, a factor that crops up in human serology (Mourant, 1983). The foregoing concepts are not equivalent: in ecological migration, whole populations may shift as the result of, for example, seasonal changes.

Numerous investigations have been carried out recently on aspects of quantitative genetics in relation to geographical variation and morphometric correlation. Such a study is that of Riska (1985) who analysed regional differences in correlations of species of North American aphids. He concluded that apparently random geographical differences in correlation are caused by chance variations in the mode of response of the organisms to short-term selection. The importance of this study for palaeontological work is that it shows that what seem to be the effects of migration and dispersion really have quite different causes. A classical morphometric study of this problem is that of Jardine (1971) who considered patterns of differentiation between human populations.

Multivariate methods have interested quantitative ecologists for a long time now. Orloci (1975) published a text in multivariate analysis in vegetation research in which various well-tried methods were employed, these being supported by a suite of programs written in BASIC. The two volumes by Legendre and Legendre (1979) are another noteworthy enterprise.

7.3 Correlation of Chemical and Physical Characters

Palaeoecological studies are often difficult to quantify in a realistic manner for variables that are extraneous to the morphological variability of the organisms of interest. There are numerous difficulties to be surmounted before a satisfactory analytical programme can be brought to bear on the problem. The most obvious of these is occasioned by the fact that in many cases, the organisms we find in a sedimentary sample have arrived there post-mortem. There are, of course, a few organisms that are usually found *in situ* where they once lived. It is therefore rather pointless to attempt to relate planktic foraminifers to the physical and chemical properties of the sediment in which they occur. Likewise, ammonites can seldom be usefully connected to the properties of the host sediment in the sense that it is not possible to postulate a relationship between the physical properties of the sediment and the morphological status of the fossils. Ammonites can, however, cast light on aspects of the origin of a sedimentary environment (Reyment, 1958, 1973).

Organisms such as ostracods, bivalves, benthic foraminifers, etc. can, on the other hand, often be realistically linked to the sediment in which they are found. It is in such cases interesting to determine the physical and chemical properties of the host sediment with the end in view of seeking meaningful relationships between morphometry and ecological factors. It may also be relevant to seek connexions between, say, grain size and chemistry. In the following illustration, I consider correlations between various chemical elements and electrical logs measured down a borehole.

Example 7.1. Multivariate Analysis of Chemical and Physical Properties of a Sedimentary Sequence

Presentation of the Problem

The main features of the problem studied here may be summarized briefly as follows (see Reyment and Sturesson, 1987). Multivariate statistical analysis is to be applied to chemical determinations made on sediment cores from a borehole drilled in the Upper Cretaceous of western Nigeria (the Gbekebo borehole we have already met). The elements P, Mn, Ca, K, Fe, Pb and Zn were determined. The electrical logs for the spontaneous potential (Sp), the short-normal resistivity curve and the long-normal resistivity curve were also available. The reason for including potassium among the chemical analyses is that this element provides a means of indicating the possible presence of glauconite in the sediment, a mineral known to be of local importance in the sequence at Gbekebo, and of relevance for interpreting the palaeoenvironment (cf. Degens, 1968). Glauconitic coprolites occur at various levels in the borehole and the interiors of some ostracods and foraminifers may contain grains of this mineral. Manganese and iron can be expected to reflect interstitial chemo-environmental conditions and it is well known that in a nearshore environment, reducing or slightly oxidizing conditions in sediments will be marked by comparatively low concentrations of manganese, whereas the concentration of Mn tends to increase as depth and Eh increase. Moreover, shales rich in organic matter display a tendency to be low in manganese in relation to iron, which is seen by chemists to be an outcome of the higher solubility of manganese over other metals under reducing conditions. Another element of potential interest is phosphorus because in elevated concentrations in the sediment it may be pointing to the activities of photosynthetic organisms, as well as the decomposition of animals and plants.

During the time encompassed by the borehole sequence, the Gbekebo environment was one of sustained regression in the nearshore regime. Hence, the redox-related component of the electrical logs can be expected to react to fluctuations in Mn and Fe (Pirson, 1977).

Statistical methods

The statistical method most useful for studying associations between sets of variables will now be reviewed. The basic technique I employ is *Canonical Correlation Analysis*; this method was briefly introduced in Chapter 3 to which reference is made. The fact that a particular canonical correlation is statistically and formally significant does not necessarily imply that a real association exists between the two sets for a certain pair of linear combinations of the variables. Several confirmatory steps are needed in order to establish the reality of a relationship and to unmask eventual spurious correlations. A desirable property is that the variables should not deviate too markedly from multivariate normality. In the analysis made by Reyment and Sturesson (1987), this requirement is not met by the entire data matrix consisting of 343 observational vectors. It is, however, almost true for the two sub-samples that were found to attract the greatest interpretational possibilities, although skewed univariate distributions were verified for P, Mn, Ca, Zn and, most extremely, for Na.

It is often found that the results of a canonical correlation analysis are difficult to reify. Cooley and Lohnes (1971) and Pimentel (1979) suggest that the possibility of interpretation may be improved by means of a *redundancy analysis*. This is a technique borrowed from the field of classical factor analysis. For relevant details, I refer you to the books mentioned above and to the monograph by Love and Stewart (1968). (This article was reprinted in the compilatory text of Fornell (1982).) The actual cause of indeterminacy in the results of a canonical correlation analysis can often be traced to the same statistical deficiencies as I have already taken up in connexion with the section on Canonical Variate Analysis in Chapter 3. It can therefore be a good idea to try to find out whether the canonical vectors are stable by a preliminary jack-knifing of the data set. If there is pronounced instability in the coefficients, then the logical next step would be to introduce shrinkage estimators (Campbell, 1979). A further matter to be kept under observation concerns the nature of the variables. If you have compositional data, which is usually the case when chemical data occur, you cannot proceed by the usual method of computation. As noted on p. 224 in Chapter 6, you will be obliged to resort to the log-ratio transformation of Aitchison to obtain a solution (1986).

Comparison of Curves

The final proof of how well the chemical variability is tied to the variation exhibited by the physical logs cannot be obtained by visual appraisal alone, nor by the oft-advocated method of cross-correlation from the sphere of econometrics. The most suitable approach for our purposes is the method of stratigraphical slotting (Gordon, 1973). The application of the slotting technique, which was originally devised for pollen-frequency data, to borehole logs on a multivariate basis is given by Gordon and Reyment (1979) and Reyment (1980). The aim of a slotting analysis in the present connexion is to ascertain how close to each other are the two sets of canonical scores for the canonical correlations. In simple terms, the technique can be described as on p. 244.

Findings

A canonical correlation analysis of all 343 samples did not yield a clear picture of eventual relationships. The overlap between sets for the first canonical correlation is so slight that the relationship computed is not statistically significant according to the criteria enunciated in the foregoing section. The reason for this condition seems to lie with the trend exhibited over the entire sampled interval. This trend is interpreted as being due to a directed change in water depth resulting from the regressional phase in the Gulf of Guinea at the close of the Maastrichtian.

The investigation was therefore concentrated to those parts of the sampled sequence that show little or no trending. Two such sections were extracted from the total data set for closer appraisal. The first of these sub-samples,

embracing the part of the borehole ranging from 2140 to 2323 ft, is characterized by subdued oscillations (i.e. fluctuations of modest magnitude) in both the
chemical set and the physical set. Stable ecological conditions are thought to
have prevailed during the laying down of these sediments.

The second sub-sample chosen for analysis is marked by greater agitation in
all curves. It runs from 2853 to 2696 ft in the borehole. The sections of the
sequence displaying agitation in the various measures are presumably indicative of a stressed environment and hence are likely to contain interpretable
chemo-sedimentological information.

The Unstressed Section: The bivariate correlations listed in Table 7.1 for the
11 variables ($N = 60$) are all less than $|0.6|$. Although there are two significant
canonical correlations, the Redundancy Analysis (Love and Stewart, 1968)
indicates only slight overlap for the both sets. Hence, little confidence can be
placed in the canonical correlation coefficients for this sample.

By eliminating the least informative of the variables (i.e. those with near-
zero loadings in the canonical vectors), an improvement in the degree of
overlap between sets could be achieved. With the six variables Mn, Ca, K, Fe,
Sp and short-normal, overlaps of 17% and 20% were obtained. This is hardly
sufficient for a fully satisfactory interpretation of inter-set relationships, but it

TABLE 7.1. *Correlation coefficients for the geochemical and electrical log-variables. The
upper triangle contains values for Gbekebo* 2323–2139 ft *(N = 60). The lower triangles
contains the entries for Gb* 2853–2960 ft *(N = 51)*

	P	Mn	Ca	K	Na	Fe	Pb	Zn
P	1.00	0.25	0.12	−0.10	−0.10	−0.11	−0.03	0.01
Mn	−0.11	1.00	0.48	−0.39	−0.17	−0.27	0.01	−0.22
Ca	−0.17	0.52	1.00	−0.58	0.02	−0.44	−0.12	−0.12
K	0.09	−0.47	−0.29	1.00	0.11	0.67	−0.02	0.33
Na	0.12	−0.71	0.52	0.66	1.00	0.12	0.00	0.16
Fe	−0.11	−0.42	−0.42	0.59	0.61	1.00	−0.01	0.57
Pb	−0.35	0.12	−0.10	0.17	−0.06	0.09	1.00	−0.24
Zn	−0.18	−0.09	0.21	0.14	0.05	0.06	0.05	1.00
Sp	0.09	0.24	0.19	−0.45	−0.41	−0.61	−0.33	0.03
SN	0.01	0.60	0.40	−0.38	−0.42	−0.50	−0.17	−0.13
LN	−0.02	0.73	0.52	−0.41	−0.48	−0.43	−0.11	−0.10

	Sp	SN	LN
P	−0.12	0.23	0.42
Mn	−0.06	0.33	0.16
Ca	0.39	0.38	0.13
K	−0.24	−0.57	−0.17
Na	0.13	−0.03	−0.11
Fe	−0.12	−0.33	−0.18
Pb	−0.25	0.10	0.00
Zn	0.27	−0.03	−0.05
Sp	1.00	0.43	−0.14
SN	0.33	1.00	0.36
LN	0.27	0.88	1.00

Sp = spontaneous potential, SN = short normal, LN = long normal.

TABLE 7.2. *Significant canonical vectors, canonical correlations and morphological structures for the sample Gbekebo* 2853–2696 m

Vector	The geochemical set Mn	Ca	K	Na	Fe	Pb	Zn
1	*The Canonical Vectors*						
	0.746	0.069	−0.073	0.191	−0.448	−0.304	0.004
	Morphological Structure						
	0.833	0.602	−0.634	−0.678	−0.745	−0.284	−0.091
	Overlap = 24.5%						
2	*The Canonical Vectors*						
	0.656	0.595	−0.133	0.493	0.769	0.364	−0.223
	Morphological Structure						
	0.423	0.313	0.194	0.061	0.484	0.387	−0.083
	Overlap = 4.3%						

	The Electrical Logs Sp	SN	LN	Canonical Correlation
1	*The Canonical Vectors*			
	0.412	0.070	0.743	$R^2_{c1} = 0.817$
	Morphological Structure			
	0.633	0.859	0.914	
	Overlap = 44.0%			
2	*The Canonical Vectors*			
	−0.808	−0.747	1.262	$R^2_{c2} = 0.654$
	Morphological Structure			
	−0.718	0.095	0.389	
	Overlap = 9.6%			

does demonstrate an important feature, to wit, that by removing uncorrelated variables from a multivariate study, a more informative analysis can sometimes be produced by the reduction of "fuzziness". Essentially the same point was made by Rao (1952) for redundancy in the generalized statistical distance.

The Stressed Section: In the more agitated part of the sequence, correlations are mostly higher than in the more tranquil stretch (cf. Table 7.1). In particular, more inter-set correlations are statistically significant than in the unstressed data and there are few zero entries in the data-matrix. All but two of the multiple correlation coefficients for this sample have highly significant variance ratios. These are the correlations for P and Zn with the other variables. In Table 7.2 the first canonical correlation coefficients and corresponding canonical structure values (that is, the correlations between the canonical variables themselves and the original variables) are listed. Perusal of Table 7.1 reveals that Zn and Pb are not bound to any of the other variables at the bivariate level. All multivariate correlation coefficients for the electrical logs are significant. The canonical correlation coefficient for the stressed section is accompanied by 24.5% redundancy for the first set and 44% redundancy for the second set. This indicates that there is considerable overlap between sets for the first pair of canonical variates. The canonical correlation of 0.82 is highly significant ($\chi^2_{12} = 78$; $P < 0.005$).

The second pair of canonical vectors is also associated with a significant canonical correlation ($\chi_{12}^2 = 28$; $P = 0.005$). However, the overlaps are too small to permit reification of this canonical correlation.

My first reaction on doing the original analysis was that there seemed to be very little agreement between the canonical variates of the "agitated" stretch and those computed for the presumably calmer environmental sequence. However, the reduced analysis of six variables provides several valuable pointers. A comparison with the corresponding structural elements for the second sample discloses rather close agreement, as transpires in the following (the values for the second sample are enclosed in brackets): Mn = 0.543 (0.833), Ca = 0.632 (0.602), K = -0.975 (-0.634), Fe = -0.568 (-0.745), Sp = 0.387 (0.633), SN = 0.999 (0.859). Hence, a reduction in the level of haphazard variation in the analysis of the unstressed data by eliminating variables providing little information provokes the appearance of inherent inter-set relationships in the results. These relationships are strong enough in the stressed part of the sequence to stand out above the masking level of "background noise". The present case is a good example of the positive effects that can be achieved by applying a planning strategy to a set of data which, if treated routinely, might not have yielded such a wealth of analytical information as just obtained. We shall continue with a more sterotyped approach.

Principal component analysis of the stressed set: The principal components for the complete matrix of correlations is interesting. The first component accounts for 44.4% of Tr **R**, which is not very impressive. The subsequent latent roots fall off slowly: 13.6%, 11.6% and 10.7% of the trace of the correlation matrix. However, after the fifth latent root, the values drop more sharply. The linear relationship defining the first principal component is as follows:

$$0.38\,\text{Mn} + 0.38\,\text{Ca} - 0.34\,\text{K} - 0.38\,\text{Na} - 0.36\,\text{Fe} + 0.27\,\text{Sp} +$$
$$0.37\,\text{SN} + 0.39\,\text{Ln}.$$

The salient point here is that this vector gives almost the same proportional representation as seen for the morphological structure of the canonical analysis of the stressed sample (cf. Table 7.3). Hence, the indications yielded by the first canonical correlation are supported by the analysis of the full set of variables. This result is compatible with the reduced principal component analysis for the "unstressed" sub-set.

Robust principal components: Geochemical data are well known to suffer from deficiencies with respect to deviations from multivariate normality. This can have an unfortunate effect on the results of a principal component analysis. In order to establish the possible influence of non-normality on the data analysed here, robust principal components were computed (Campbell, 1979). Moderately strong deviations were found for the complete sample of 343 observational vectors with respect to lead (usual value = 0.115; robust

TABLE 7.3. *First significant canonical correlation vector, morphological structure and first principal component for the reduced sample from Gbekebo 2323–2319 ft. N = 60. The "stressed" sub-sample*

	Canonical vector	Morphological structure	First principal component
Mn	0.196	0.543	0.313
Ca	0.041	0.632	0.313
K	−0.984	−0.975	−0.510
Fe	0.161	−0.568	−0.418
	Overlap = 17.0%		
Sp	−0.058	0.387	0.271
SN	1.024	0.999	0.426
	$R^2_{c1} = 0.588$		
	$\lambda_1 = 49\%$ TrR		

value = 0.455) in the first component and for manganese (usual value = −0.150; robust value = −0.455) in the second principal component. On the whole, the differences in the values for the two methods of computation are not of great importance.

Concordance in curve patterns

One of the more troublesome aspects of quantitative stratigraphy lies with the difficulty of proving that two curves really do have the same variational pattern. Visual inspection may succeed in deciding the issue for short sequences, but for long series, such as the set of observations in the present example, an automated procedure is absolutely necessary. *Gordon's Slotting Technique* is a method of slotting parallel series of observations (see also Example 7.2). I shall now give a more formal presentation of the method. Borehole logs form oscillatory curves, usually with well defined peaks (often termed spikes in statistical work) and troughs in parts of the sequences. Spikes of wide amplitude are generally associated with geological features that deviate markedly from the main body of strata traversed by the borehole. Sometimes, a palaeoecological event can cause a spike. An oscillatory sequence of variations for one borehole will usually not be directly comparable with a similar sequence in an adjacent borehole. The slotting result can usually be enhanced by the recognition of datum lines, but complications occur, such as differences in rates of sedimentation at the two places pierced by the boreholes. The result is that episodes may be condensed in one sequence and expanded in the other. Gordon's method of slotting was developed for attacking the problem of comparing ordered sequences of objects when the objects in one sequence cannot be identified exactly with objects in the second sequence, but *only resemble them*. This is important to remember as a good deal of the trouble I find people have in using the slotting technique arises just from failure to appreciate what is being done.

Using a dissimilarity measure for the sequences, an object can be shown to fit between a pair of objects in the other sequence if the similarity between both of them is satisfactorily high. A useful measure for this is:

$$DC(j,k) = \sum_{v=1}^{z} |p_{vj} - p_{vk}| \qquad (7.1)$$

where j and k denote two boreholes, v is the variable being operated upon, and z is the total number of variables.

If the two sequences are $S_1 = \{A_1, A_2, \ldots, A_m\}$ and $S_2 = \{B_1, B_2, \ldots, B_n\}$, where there are m observational points A_i in the first borehole and n observational points B_j in the second borehole, they might be slotted together in the following manner:

$$A_1 A_2 \qquad\qquad A_3 \qquad\qquad A_4 A_5 \ldots$$
$$\qquad B_1 \qquad\qquad B_2 B_3 \qquad\qquad B_4 \ldots$$

Thus, A_1 and A_2 fall together, then B_1, followed by A_3, then B_2 and B_3, etc.

The slotting of sequences gives satisfactory local fits, subject to the constraints imposed by the orderings. This indicates that a local bad fit can appear and it will be maintained providing *the overall fit is good*. As with many other graphical techniques in statistics, it requires a fair amount of practice to be able to extract a maximum of information from the slotting technique.

A measure of essentially **multivariate discordance** is given by the following expression:

$$\sigma(S_1, S_2) = \min\left[\sum_{j=1}^{m} \{(d(A_j), B_{s(j)} + d(A_j, B_{s(j)-1})\} \right.$$
$$\left. + \sum_{k=1}^{n} \{(d(A_{t(k)}, B_k) + d(A_{t(k)-1}, B_k)\} \right]. \qquad (7.2)$$

Here, the minimum is formed under the restrictions of consistency and preservation of order; that is, stratigraphical levels may not be shuffled in order to produce a better match. The subscript $s(j)$ is defined for $j = 1, \ldots, m$ by $s(j) = i$ if A_j is placed in the gap between B_{i-1} and B_i; the subscript $t(k)$ is defined for $k = 1, \ldots, n$ by $t(k) = i$ if B_k is placed in the gap between A_{i-1} and A_i. If the dissimilarity d satisfies the triangle inequality, then

$$\sigma(S_1, S_2) \geq \mu(S_1, S_2)$$

where

$$\mu(S_1, S_2) = \sum_{j=1}^{m-1} d(A_j, A_{j+1}) + \sum_{k=1}^{n-1} d(B_k, B_{k+1}). \qquad (7.3)$$

The measure of discordance in fit of the two sequences is defined by Gordon as

$$\psi(S_1, S_2) = [\sigma(S_1, S_2) - \mu(S_1, S_2)]/\mu(S_1, S_2). \qquad (7.4)$$

The procedure operates by sliding the sequences past each other, a step at a time, in order to determine the smallest value of ψ. The sequences being compared need not necessarily have the same number of observational levels, which is a practical advantage. Generally the dissimilarity matrix will be rectangular of order $m \times n$. Gordon's method will always produce a slotting result, although not invariably a good one. The superiority of the method to other attempts at the automated lining-up of borehole sequences, such as some technique borrowed from econometrics, is that it *condenses* similar portions of a particular sequence and *expands* unlike portions. It is **multivariate** in that it can be applied to the comparison of several logs on which several measures are available. Gordon (1979) has published a FORTRAN IV program for doing the calculations. The multidimensional version of the slotting technique

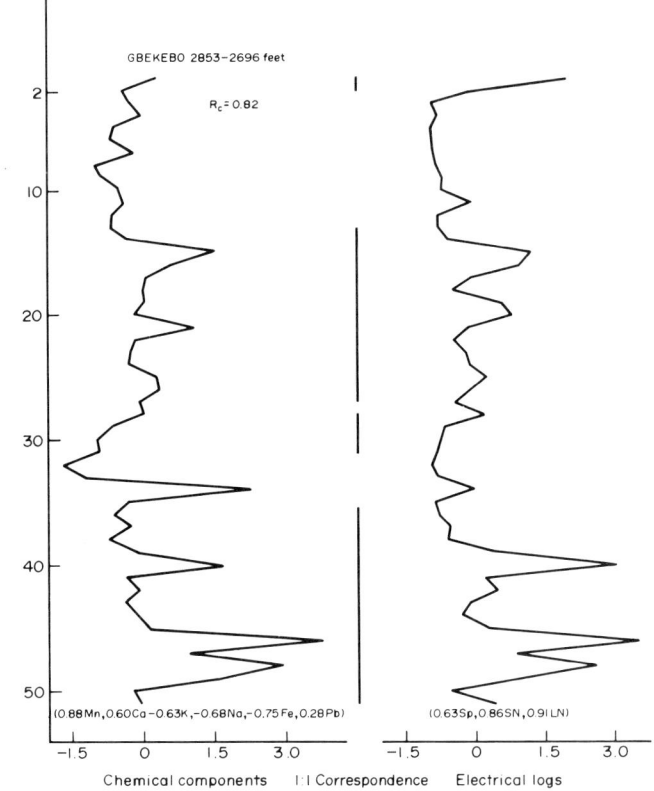

FIG. 7.1. Unconstrained slotting (Gordon, 1973) for the logs obtained from the first canonical correlation scores for the chemical elements and the scores for three electrical borehole logs for 51 observational levels in the ecologically stressed section of the Gbekebo borehole (Nigeria). The value of Gordon's (1973) ψ is 0.315, which is indicative of an excellent fit. The vertical bars denote regions of exact correspondence in the two curves. Sp = spontaneous potential, SN = short normal and LN = long normal. Redrawn from Reyment and Sturesson (1987).

devised by Gordon and Reyment (1979) permits several kinds of logs to be examined simultaneously and contains a provision for the inclusion of marker horizons. I shall now apply the slotting method to the data of the present example.

For the purposes of the present illustration, I examine the analyses for the "stressed" interval ($N = 51$). The two curves produced by the respective first scores of canonical correlations for (a) the seven elements Mn, Ca, K, Na, Fe, Pb, and Zn and, (b) the three electrical logs, were tested for agreement by the modification of Gordon's technique used in Gordon and Reyment (1979) and Gordon (1979). The outcome of the *unconstrained* slotting is shown in Fig. 7.1. The result is remarkably good, as disclosed by the low value of $\psi = 0.315$, which is much less than 1 (cf. Gordon and Reyment, 1979) and the fact that most of the sample points line in a one-to-one correspondence. Apart from some "blocking" in the upper fourth of the comparison (over levels 3 to 12), owing to the slight variation in values from level to level in both curves for this stretch of the sequence, the two series slot well together. Note particularly the bottom third of the slotting result. This analysis leads one to conclude that the variation in the geochemical vector is mirrored by the variability of the vector for the electrical logs, over the interval studied.

Palaeoecological Interpretation

During the late Maastrichtian, the coastline of West Africa was caught up in a large-scale episode of regression, prior to the unleashing of the short-lived, though extensive, early Paleocene regression. Consequently, the sediments at the Gbekebo site can be expected to register the effects of a regressional regime. Pirson (1977) demonstrated in his monograph that the electrical properties of sediments deposited in a nearshore environment under the influence of an ongoing transgression, or regression, are often quite distinctive. The effects of a regressional environment can be expected to carry over to the geochemistry of the sediments. The conclusion pointed to by the present analysis is that bathymetrical factors could well be the driving force of the variation recorded in the data.

Example 7.2. Application of Gordon's Slotting Technique in Palaeoecology: analysis of variation in the ostracod shell in relation to an electrical log

Presentation of the Problem

Many workers have pointed out that it is well nigh impossible to line up two curves, particularly if these curves are

(a) very long, and,
(b) relatively featureless.

TABLE 7.4. *Sequence of means in height of a cara-*
pace for a species of Paleocene ostracods and obser-
vations on the self-potential electric log

Height of carapace in mm	Self-potential electical log (mV)
0.26	1.2
0.30	1.7
0.35	1.9
0.32	2.5
0.30	2.9
0.29	2.7
0.26	1.9
0.24	1.3
0.20	1.5
0.19	1.1
0.17	1.0
0.11	1.2
0.13	1.9
0.20	1.9
0.23	2.0
0.26	2.1
0.31	2.3
0.36	2.7
0.37	2.9
0.41	3.1
0.40	3.2
0.25	1.6

A third complication also merits mention. One curve may be compressed in relation to the second. A question often put in quantitative palaeoecology is whether some character or other is being influenced by an ecological factor. If you ponder over this point, you will no doubt realize that this is not only important for mundane biostratigraphical interests, but also for applications of quantitative genetics to fossils. Bookstein (1987) drew attention to the danger associated with the naive interpretation of sequences of morphological means as expressing evolutionary patterns and he entreated the investigator to try to link such a series of observations to an external variable, such as a reliable indicator of the palaeoenvironment. Pielou (1974) has also contributed to the subject.

The Analysis

I have already demonstrated that the slotting technique of Gordon provides an excellent means of ascertaining just how closely two curves are bound to each other. In the present example, chosen to illustrate an ecophenotypic connexion between the height of the carapace of a species of the ostracod genus *Buntonia* and the self-potential log, which is a measure of the redox potential of the palaeoenvironment, the data are analysed by formulae (7.3) and (7.4). There are 22 levels. The data are listed in Table 7.4.

The results of the slotting for height of the carapace are as follows:

Sequence 1	Sequence 2
1	1
2	
3	
	2
	3
	4
4	
5	
	5
	6
6	
	7
7	
8	
	8
9	
	9
10	
	10
11	
	11
12	
	12
	13
13	
	14
	15
14	
	16
15	
16	
	17
17	
18	
	18
19	
	19
	20
20	
	21
21	
	22
22	

The tests of goodness-of-fit of Gordon (1973) yield the results $\sigma = 193.4$ and $\psi = 2.7$. The latter, which is a standardized measure, is taken as supporting the hypothesis of a "good fit" if it is small (cf. Birks and Gordon, 1985, p. 106).

Discussion

The results indicate that there is quite close agreement in the patterns exhibited by the two sequences. Hence, the variation in morphology of the shell of *Buntonia* sp. and the pattern traced out by the self-potential log are associated. This can be taken as pointing to an ecophenotypic association between the size variation in the ostracod and the fluctuations in the environment, as expressed by the self-potential log.

Example 7.3. Ornamental Polymorphism Under the Influence of Eco-Chemical Factors

Presentation of the Problem

Reyment (1985b) recorded a wide variety of ornamental and conformational polymorphism in species of the ostracod genus *Echinocythereis*. We have met various aspects of this investigation in the foregoing pages. Here, I shall examine possible links between the frequencies of three ornamental categories and the chemical determinations on vanadium, chromium and boron in the sediments in which the ostracods occur. Inasmuch as the three ornamental types considered are not separated from each other in a clear-cut manner—there is some overlap—one might expect a condition driven by ecological causes. More explicitly, I do not wish to postulate that there could be a direct connexion between the frequencies of the morphs and the concentrations of the three chemical elements, but rather that both sets could be reacting in parallel to the same ecological forces.

Method of Analysis

The data are analysed in two ways. Firstly, there is a canonical correlation analysis made on the frequencies and, secondly, these results are contrasted with the analysis made using the log-ratio transformation of Aitchison (1986b). The log-ratio transformation can be easily built into a standard program for canonical correlation but for the present purposes, I carried out this step before starting the multivariate treatment of the data. In addition to the canonical coefficients and variates, I also computed the multiple correlation coefficients for both sets. As already intimated, the data are in the form of percentages. Despite the small size of the sample, a satisfactorily detailed fallout was obtained.

TABLE 7.5. *Basic statistics for the morph-frequency data: log-ratio and usual versions*

		Constrained	Log-ratio values	
Variable	Mean	Standard deviation	Mean	Standard deviation
Morph 1	0.286	0.207	−0.089	0.685
Morph 2	0.642	0.203	0.859	0.493
Morph 3	0.071	0.050	−1.525	0.720
Vanadium	0.347	0.076	0.294	0.180
Chromium	0.428	0.043	0.516	0.163
Boron	0.244	0.042	−0.054	0.262

The values for the chemical elements are expressed as percentages of a trace-element complex.

The Analysis

The correlation matrices for both treatments of the data are displayed in Table 7.6, with the usual values located in the upper triangle of the matrix and the log-ratio correlations in the lower diagonal. The three morph-frequencies constitute the "left-hand" set and the three chemical variables, the "right-hand" set. Basic univariate statistics for the two sets are presented in Table 7.5.

The two sets of correlations show some quite pronounced differences. Some marked examples are the values for r_{56} and r_{15}, just to quote two. The standard deviations differ strongly. The multiple correlation coefficients are listed in Table 7.7 together with the corresponding variance ratios. None of the values of F are significant for the constrained data and only two attain the 5% level

TABLE 7.6. *Correlations for the morph-frequencies and chemical elements: usual and log-ratio versions*

	Morph 1	Morph 2	Morph 3	V	Cr	B
Morph 1	1.0000	−0.9701	−0.1702	−0.4966	−0.5468	−0.0528
Morph 2	−0.7316	1.0000	−0.0730	0.3897	−0.5372	0.0948
Morph 3	−0.3687	−0.2958	1.0000	0.5062	−0.1328	−0.1689
V	−0.5822	0.7323	−0.1764	1.0000	−0.5034	−0.2589
Cr	0.3230	0.0422	−0.6748	−0.2558	1.0000	−0.4477
B	−0.0238	0.3137	−0.6898	0.1008	0.4841	1.0000

Notes: The upper triangle correlations matrix contains the usual values; the lower triangle contains the simplex correlation coefficients of Aitchison (1986).

TABLE 7.7. *Multiple correlation coefficients for the morph-frequency data*

	Constrained multiple R^2	Simplex R^2
Left-set variables		
Morph 1	0.608 (2.07)	0.614 (2.12)
Morph 2	0.563 (1.72)	0.782 (4.79)
Morph 3	0.531 (1.51)	0.834 (6.69)
Right-set variables		
V	0.625 (2.22)	0.543 (1.58)
Cr	0.718 (3.40)	0.535 (1.54)
B	0.035 (0.05)	0.840 (6.97)
Values of $F_{3,4}$ in brackets.		

TABLE 7.8. *Canonical correlation analysis for the morph-frequency data: the first association*

		Constrained data matrix	Simplex data matrix		
Canonical correlation		0.9324	0.9996		
			Correlations with derived canonical		Corelations with derived canonical
Left set	Var.	Loadings	variates	Loadings	variates
	Morph 1	27.26	−0.66	1.76	0.14
	Morph 2	27.55	0.60	1.26	−0.54
	Morph 3	7.02	0.37	1.80	0.79
Right set	V	0.21	0.78	0.42	0.40
	Cr	−0.94	−0.88	0.37	0.59
	B	−0.37	0.00	0.68	0.90
Overlap percentages	Set 2 on set 1	25.79%	31.39%		
	Set 1 on set 2	42.09%	44.00%		

of significance for the log-ratio data, namely, the third and sixth entries, no doubt due to the small size of the sample. The values of the canonical correlations for the first combination are shown in Table 7.8. Both first canonical correlation coefficients are very highly significant. The overlaps indicated by the redundancy analysis (Cooley and Lohnes, 1971; Pimentel, 1979) are large. For the constrained data, the overlap percentages are 25.8% for the left-hand set and 42.1% for the right-hand set. The corresponding values for the log-ratio analysis are 32.4%, for the left-hand set, and 44.0% for the right-hand set. Such values can be taken to be clearly indicative of a genuine correlation between sets. A warning is needed here. The very high first canonical correlation coefficient must incite suspicion in that almost all information collapses into one correlation.

The effects of closure can be most clearly observed if we shift attention to which variables are doing most of the work and hence the reification of the canonical vectors. As an exercise, examine the results presented in Table 7.8 in the light of the foregoing statement. I also calculated the principal components for both data sets. Also here considerable differences in elements occur.

Discussion

The results of the analysis tentatively made on just a few samples certainly point to there being a significant association between the proportions of the morphs by sample and environmental factors, be they directly or indirectly connected to the elements used here. This result strengthens the hypothesis that the ornamental variation in *Echinocythereis aragonensis* is in part under ecophenotypic control, as has indeed been suggested to me by J.-P. Peypouquet. It would be interesting to pursue these results by seeing whether these ornamental features are separated by thresholds.

A statistical outcome of the analysis is that although the general relationships concerning significance of associations appear for both the raw data and the log-ratio data, the canonical vectors yielded by the two data sets are substantially different, so different in fact that they lead to unlike reifications. In view of what has been said about the necessity of choosing the right approach for compositional data, it is difficult to bypass the simplex version of canonical correlation analysis for data composed of frequencies.

7.4 Multivariate Assemblages and Associations Between Species

7.4.1 *Introduction*

The subject matter of this section falls into the category usually considered to be Statistical Ecology for many workers in the field. There are several textbooks available, many of which should prove useful to those who are concerned with data occurring as frequencies of species in relation to ecological factors. The books by Pielou (1974, 1977) provide well written entries to the subject. The volume by Digby and Kempton (1987) is a useful computer-oriented book for the multivariate analysis of ecological communities. The two-volume text by Legendre and Legendre (1979) contains much valuable information of relevance for multivariate applications. I can also recommend relevant sections in Manly (1985). The subject is very broad and I can therefore do little more here than select a few topics of palaeontological relevance, while noting that this selection is no more than a small part of the whole. A few definitions are required now. However, for the balanced treatment of the subject, I refer you to the books mentioned above. Another informative source is Birks and Gordon (1985).

7.4.2 *Association Between Species*

I shall begin this part by considering the way in which associations between species can be formalized. The simplest possible situation is the relationship between a single variate X for species A and a single variate Y for species B. X and Y might be colony proportions for a certain gene or a certain morph, or colony values for a qualitative variable such as length. In most cases, X and Y will be the same measure, observed on different species. The following conditions may pertain:

(a) There can be positive correlation between X and Y in sympatric colonies. This can occur if species A and B are *independently affected* in the *same way* by some selective agency such that conditions which underlie high values of X also promote high values of Y. This could well be the state represented in Example 7.3 above in which polymorphism and chemical factors might be associated.

(b) Negative correlation between X and Y in sympatric colonies can arise if a predator treats species A and B as though they were a single species. The result will be to maintain a constant distribution between A and B such that if the value for X is high, then Y is low and the value of $X + Y$ remains constant. This is not a new idea if you have been following the threads of discourse in this book. This is the natural relationship for compositional data. Note, however, that this situation cannot occur if the level of predation is low, such as is usually the case for drilling gastropods.

(c) A third situation is expressed by the model of *character displacement.* This treats the circumstances in which two species face their severest competition where they coexist. If similar morphs compete for the same resource, the competition is strongest for individuals of A that are most like individuals of B. The outcome may be that individuals of A diverge from B in sympatric colonies and this character displacement will show up in the means of X and Y determined on allopatric and sympatric colonies. A very readable account of character displacement, and the historical development of the concept, can be found in Pielou (1974, p. 332). The mechanism at work is sometimes called *apostatic selection*: predators may identify prey by image recognition for common morphs (Clarke, 1969). Hence, selection acts in the case of one species against any morph that is common in the second species. This is a very general way of describing a complex problem and there are, no doubt, many cases in which apostatic selection cannot be applied as an hypothesis. For example, this theory does not take account of the deterrent effect of effluents which are experienced as distasteful to a predator and which makes prey, otherwise ideal, unacceptable. This is well known for predaceous gastropods. This topic leads us directly to a consideration of variation in population size.

7.4.3 *Fluctuating and Oscillating Populations*

Pielou (1974) has given a well reasoned account of some of the factors liable to be consequential in populations that fluctuate in size. Although we can not assess these in palaeoecological studies, it is necessary to be aware of them and to keep in mind their possible significance on interpreting quantitative results. She listed the following categories:

(a) Fertility and survival rate.
(b) Density dependent response—oscillations can occur in a population if the growth rate is governed by density dependent mechanisms and if there is a delay in the response of the growth rate to density change.
(c) Oscillations in host parasite populations.

Exogenous and Endogenous Oscillations: The three categories just listed are referred to as endogenous oscillations, by which is implied that the temporal variability is caused by the organisms themselves. The opposite state is called exogenous oscillation and is wholly due to environmental effects. Pielou

(1974) discussed Moran's criterion for distinguishing between the two kinds of oscillations. This can be done by the **correlogram**. The curve formed by the correlogram is an undamped wave for endogenous oscillations, whereas it forms a damped wave for exogenous variables. This can be a useful method but you need very long series to be sure of what you have. This subject has already been reviewed in Chapter 5 in the section on time-series analysis.

7.5 On Species Data

7.5.1 *Standardization*

A commonly invoked formula for standardizing species data is (cf. Digby and Kempton, 1987, p. 13) the following measure of association, where the a_{ij} are the number of joint occurrences of species i with species j.

$$2\frac{a_{ij}}{(a_{ii} + a_{jj})}. \tag{7.5}$$

This equation becomes zero for the case when certain species never occur together and 1 for the opposite extreme in which all recorded species always occur together. The observations can be made independent of scale if they are divided by their standard deviations, or by the Yule transformation, that is dividing by the range.

In studies involving presence/absence data, the logarithmic transformation is often useful, since it diminishes the importance of larger values relative to smaller values in the data matrix.

7.5.2 *Similarities and Distances*

A useful thing to know is that similarities and distances can be transformed back and forth. That is, if you have a similarity measure, you can transform directly to a distance measure. Similarities s_{ij} in the range 0 to 1 can be transformed to distances either by calculating

$$d_{ij} = 1 - s_{ij}$$

or

$$d_{ij} = -\log s_{ij}. \tag{7.6}$$

The subject of similarities and dissimilarities is a central one to ecological statistics. A useful reference is the volume on applications of multivariate analysis in statistical ecology in the series edited by G. Patel (1979) for the Parma NATO Conference on Statistical Ecology. This multi-volume work contains a wealth of valuable information for the statistical palaeoecologist. One of the volumes treats multivariate applications and another various kinds of measures of association between species frequencies.

The most rewarding area of application of the analysis of multivariate associations between species occurs in palynology. A valuable reference for those who are concerned with this kind of work is the book by Birks and Gordon (1985), which deals specifically with the quantitative analysis of quaternary pollen data. Mosimann (1968), who seems to be the first person to really identify and isolate the problem connected with sampling pollen, gave an account of the quantitative analysis of frequencies of pollen grains by means of hypergeometric probabilities, which is the appropriate form for sampling without replacement. Mostly, however, binomial probabilities are used for analysing counts of pollen, which is a fully acceptable approximation for data involving such great sample-sizes as are usually encountered. It should be noted, however, that binomial probabilities are the correct statistical form for sampling *with* replacement.

We have already encountered Canonical Correlation Analysis and ascertained that it is a useful procedure for multivariate statistical ecology in a well controlled analytical milieu. We shall now learn about a fairly recent expansion of this technique by the medium of Correspondence Analysis.

7.6 Canonical Correspondence Analysis in Palaeoecology

7.6.1 *Introduction*

A recent generalization, or better, modification, of classical correspondence analysis, termed *Canonical Correspondence Analysis*, has been put forward by ter Braak (1986, 1987). This is based on the results of Hill (1974). The type of problem amenable to analysis by canonical correspondence analysis is seldom realistically encountered in palaeontology, although quaternary palynology and oceanology are rich areas of application. The majority of uses of what is doubtlessly a very versatile graphical technique in multidimensional analysis come from ecological botany and limnology (cf. ter Braak, 1987 and Fängström and Willén, 1987). Pollen-based palaeoecology is a field which is eminently suited for canonical correspondence analysis.

To instill a feeling for what canonical correspondence analysis purports to do, we shall hark back to what we said about the method of canonical correlation. That procedure aims at finding informative linear compounds of two, or more, sets of variables bound in significant correlations to each other. In canonical correspondence analysis, defined as *restricted* correspondence analysis, the ordination diagram displays information on SITES, SPECIES, and ENVIRONMENTAL VARIABLES. The points for sites and species have the same orientation as in simple correspondence analysis: the environmental variables are represented as arrows. The joint plot of species and environmental variables in the ordination diagrams of canonical correspondence analysis can be interpreted similarly to what is done for a *biplot* (Gabriel, 1968). The points for species can be projected on to the "environmental axis" or direction. In ecological botany, this procedure has long gone under the name of *direct gradient analysis*.

Just as in other graphical adaptations of methods based on latent roots and vectors, off-putting geometrical effects can show up. One of these is called the "horseshoe", and was referred to in Chapter 3, another is the sinusoidal "folium", a third, the right lemniscate. You can find these in any table of curves and their pedals. Detrending is often resorted to in order to make the ordination diagram more aesthetically acceptable. Wartenberg *et al.* (1987) criticize this manipulation in no uncertain terms, while Jongman *et al.* (1987) defend the procedure with the aid of algebraic justification. Wartenberg *et al.* point out that detrending is no more than an *ad hoc* adjustment and it can easily lead to a result that is worse than plots that have not been detrended. My opinion is that detrending is a methodology of the category "the proof of the pudding is in the eating" and each situation must be judged on its own particular attributes.

7.6.2 *General Description of Canonical Correspondence Analysis*

A commonly occurring problem in *community ecology* concerns finding how great a number of species respond to external factors, such as ecological factors, both physical and chemical. In applied biology, the same questions arise naturally in ecotoxicology. In the situation covered by canonical correspondence analysis, data are available on

(a) The number of species
(b) The external variables at various points in space and time.

Such data have been collected and analysed for quite a long time. The methods used up to the advent of canonical correspondence analysis have assumed a linear relationship to apply and/or have been restricted to regression analysis if the response of each species separately to the set of external variables is of primary interest. The extension of multivariate regression to canonical correlation is quite obvious, however, and hence to multivariate gradient analysis. (You will remember what we said in Chapter 3 about canonical correlation being a sort of generalization of multi-regression.)

Ordination in ecology has usually been achieved by direct gradient analysis. For example, relative abundances are plotted against values assumed by some ecological factor. Digby and Kempton (1987, p. 49) review direct gradient analysis in a multivariate context and the subject is reviewed in some detail in the book edited by Jongman *et al.* (1987). Canonical correspondence analysis avoids the assumption of linearity; it can also express unimodal relationships between species and external variables. *En passant*, I can note that this method is not necessarily confined to statistical ecology. I can conceive of many situations in morphometry in which the relationship between size and form can be usefuly expressed in relation to external variables. I believe that this will turn out to be a very rewarding area for research.

Multivariate direct gradient analysis (as opposed to indirect gradient analysis by internal estimation: see Chapter 3 for a critique of internal estimation) relates a set of species directly to a set of environmental variables, just as we did in canonical correlation. The treatment is now taken a step further in that an attempt is made to unveil patterns of variation in the composition of a community that can be explained by the environmental variables, hopefully astutely chosen. The technique can be thought of as uniting aspects of ordination, as usually construed, with aspects of direct gradient analysis.

The idea forming the backbone of the canonical correspondence analysis model is derived from a species-packing model in which species are assumed to have Gaussian response surfaces in relation to compound environmental gradients (ter Braak, 1986, p. 1168). These gradients are presumed to be linear combinations of the environmental variables. The term *canonical correspondence analysis* is meant to highlight the fact that it is a technique within the domain of correspondence analysis *sensu* Hill (1974) in which the axes are selected within the framework supplied by the ecological variables. In ter Braak's (1986) own formulation of the technique, canonical correspondence analysis has the following distinguishing properties:

1. It is a multivariate extension of weighted average ordination, which is a simple method for arranging species along environmental variables.
2. It constructs those linear combinations of environmental variables along which the distributions of the species are maximally separated, the degree of separation being expressed by the latent roots of the procedure.
3. Its relationship to the correspondence analysis of Bénzécri (1973) is that it is correspondence analysis with the axes of ordination constrained to be linear combinations of environmental variables.
4. The ordination-graphs display
   a. a pattern of community variations just as in standard ordination;
   b. the main features of the distribution of species along the environmental variables.
5. Its special merit is that it can be used for detecting relationships *between* species as well as the specific responses of species to environmental variables.

7.6.3 *Nature of the Data*

The format-requirements for the data are as follows. There are observations on abundances, or occurrences, of m species at n sites. The third set of observations consists of determinations on q environmental variables ($q < n$). The mathematical details are given by ter Braak (1986, p. 1169). Cajo ter Braak also distributes a computer program called CANOCO which is available from him at cost price.

The algorithmic details are:

1. Begin with arbitrary, but unequal site scores.
2. Calculate species scores by weighted averaging of the site scores.
3. Calculate new site scores by weighted averaging of the species scores.
4. Obtain regression coefficients by weighted multiple regression of the site scores on the environmental variables. The weights are the site totals, which we may denote (y_{i+}).
5. Calculate new site scores—these are, in fact, the fitted values of the regressions of step 4.
6. Centre and standardize the site scores such that

$$\Sigma_i y_{i+} x_i = 0$$

and

$$\Sigma_i y_{i+} x_i^2 = 1$$

where the x_i are the site scores.

7. Stop on convergence. That is, when the new site scores are sufficiently close to the site scores of the previous iteration. Otherwise, return to step 2.

The procedure is akin to the reciprocal averaging algorithm of correspondence analysis (Bénzécri, 1973), with the addition of steps 4 and 5, which introduce restrictions on the site scores. The final regression coefficients are termed *canonical coefficients* and the multiple correlation coefficient of the final regression is called the *species-environment correlation*. This correlation is a measure of how well the extracted variation in community composition can be explained by the environment variables. It equals the correlation between the site scores, which are unweighted mean-species scores, and the site scores, which are a linear combination of the environmental variables. This equality presupposes that sites are weighted proportional to y_{i+} as in steps 4 and 6; this weighting of sites is assumed in the computation of means, variances and correlations.

7.6.4 *Interpretation of the Ordination Diagram*

The computer program CANOCO yields an ordination diagram with sites and species represented by points and environmental variables by arrows (see example in Fig. 7.2). The points for species and sites represent jointly the dominant patterns in the composition of the community in so far as these can be "explained" by environmental variables. The points for species and the arrows for environmental variables jointly indicate the distributions of the species along each of the environmental variables. If an arrow refers to water temperature, for example, the diagram allows one to infer which species occur at, say, the warmest sites, and which species are found at the cooler sites.

The species scores and site scores which are weighted mean species scores represent the approximate community composition at each of the sites. Each site point lies then at the centroid of the species points at that site, which permits inference from the diagram as to which species are likely to occur at a particular site.

The weighted average of the distribution of a species k in relation to an environmental variable j is defined as *the average of the values of that environmental variable at those sites at which the species occurs, the weighting of each site being proportional to species abundance*. This can be written as:

$$\bar{z}_{kj} = \sum_{i=1}^{n} \frac{y_{ik} z_{ij}}{y_{+}k} \tag{7.7}$$

The weighted average indicates the "centre" of the distribution of a species along an environmental variable. Differences in weighted averages between species indicate differences in their distributions along that environmental variable.

The position of the *head of the arrow* for an environmental variable depends on the latent roots of the axes and the intraset correlations of the environmental variable with the axes (ter Braak, 1987). Details of the method of construction are given by ter Braak (appendix in ter Braak, 1986). The ordination diagram is used in the following way.

(a) Perpendiculars are dropped from the species points to the environmental axis of interest.
(b) Note the locations of the endpoints: these mark the *relative positions* of the centres of the distributions of species along that environmental axis.
(c) Observe the relative positions of the pertinent arrow head and the endpoint of a species. The inferred weighted average is greater than average if the endpoint of a species on an environmental axis lies on the same side as the head of the pertinent arrow. It is smaller than average if the origin lies between the endpoint and the head of an arrow (see also Gabriel, 1968).
(d) Important environmental variables tend to be expressed by longer arrows than less important environmental variables.

Example 7.4. Foraminiferal Frequencies and Oceanographical Factors

Presentation of the Problem

Imbrie and Kipp (1971) analysed foraminiferal frequencies for material obtained from deep sea probes in the Atlantic (cf. Jöreskog *et al.*, 1976, p. 161). The data have two sources: firstly, sedimentary samples taken from geographically widespread locations in the Atlantic Ocean, analysed for environmentally significant, modern assemblages of foraminifers. Each sample represents a record of sedimentation over the past 2000–4000 years approximately.

TABLE 7.9. *Weighted correlations, means, standard deviations and variance-inflation factors of environmental variables in the canonical correspondence analysis of the foraminiferal data*

		Weighted correlation matrix (weight = total sample-size)		
	1.0000			
AST	− 0.9594	1.0000		
AWT	− 0.9562	0.1040	1.0000	
ASW	− 0.7895	− 0.3294	− 0.0705	1.0000

Variable	Weighted mean	Standard deviation	Inflation factor
AST	21.392	7.020	18.650
AWT	16.921	8.002	15.336
ASW	35.689	0.988	2.502

Foraminifers from the size-fraction $> 149 \, \mu$m were identified; 27 species and "varieties" were recognized. At each of the core sites, the average winter temperature (AWT), the average summer temperature (AST) and average salinity of the surface water (ASW) were deduced from data available on oceanographical charts. One core taken up in the Caribbean, and the length of which was 10.9 m, was used as a record of Pleistocene depositional history. The core was sampled at intervals of a decimetre and the same foraminiferal species determined as was done for the Recent material.

Findings

The latent roots corresponding to the ordination axes are, in order of magnitude:

$$0.7518, 0.1257, 0.0137.$$

Table 7.9 contains the weighted correlations, means, standard deviations and variance-inflation factors of the three environmental variables AWT, AST, and ASW. The term *variance-inflation factor* is not all that well known and needs explanation. It refers to variables in a *multiple regression equation* and derives from the fact that variances of estimated regression coefficients are proportional to their variance-inflation factors. If this factor is very large, greater than 20, say, the pertinent variable is almost *perfectly correlated* with the other variables and makes, therefore, no unique contribution to the regression. This condition can, in turn, lead to instability in the regression coefficient and hence interpretational inaccuracy. In other words, it can be said that we have encountered a redundant variable if we discover a high variance/inflation factor in the analysis. In the present case, two of the environmental variables have high variance-inflation factors and one is perilously close to the crucial value 20. See Jongman *et al.* (1987) for a detailed explanation of the topic.

TABLE 7.10. *Inter-set correlations of environmental variables with axes*

Environmental factor	First axis	Second axis
AST	−0.956	−0.007
AWT	−0.959	0.104
ASW	−0.789	−0.329
Fraction of variance extracted by axis	0.819	0.040

TABLE 7.11. *Biplot scores of the environmental variables: these coordinates were used for locating the tips of the arrow of the environmental axes (Fig. 7.9). (Note the division by 2.)*

Environmental factor	First axis	Second axis
AST	−0.424	−0.004
AWT	−0.425	0.058
ASW	−0.350	−0.183
R (species, environment)	0.974	0.598

The *species-environment* correlations are 0.974, 0.598 and 0.352. The first two of these are interesting for the analysis.

Environmental variables AST and AWT have equal, large correlations on the first axis. Salinity, ASW, is also highly correlated on this axis. There are no high correlations on the second and third axes, although that for salinity may be significant. These values are listed in Table 7.10. The biplot of the scores and the environmental variables, displayed in Table 7.11, are used for constructing the arrows in the canonical correspondence plot, shown in Fig. 7.2.

The first biplot axis yields negative scores for high temperatures and high salinities, and positive scores for low temperatures and low salinities. The second biplot axis produces negative scores for high salinities and positive scores for high winter temperatures. The plot for species located against the three environmental categories is shown in Fig. 7.2. The plot given in Fig. 7.3 displays the sample scores on the corresponding axes. It will be seen that the samples from the glacial core form a distinct cluster, well isolated from the modern material.

Remarks

There is no doubt in my mind that canonical correspondence analysis provides an informative graphical dissection of complex data and this method, despite its complicated application, may well turn out to be the leading one for ecological studies involving frequencies of species. I am indebted to Professor John Birks, Botany Department, University of Bergen, for performing the analysis for me. Any faults in interpretation are, of course, my own.

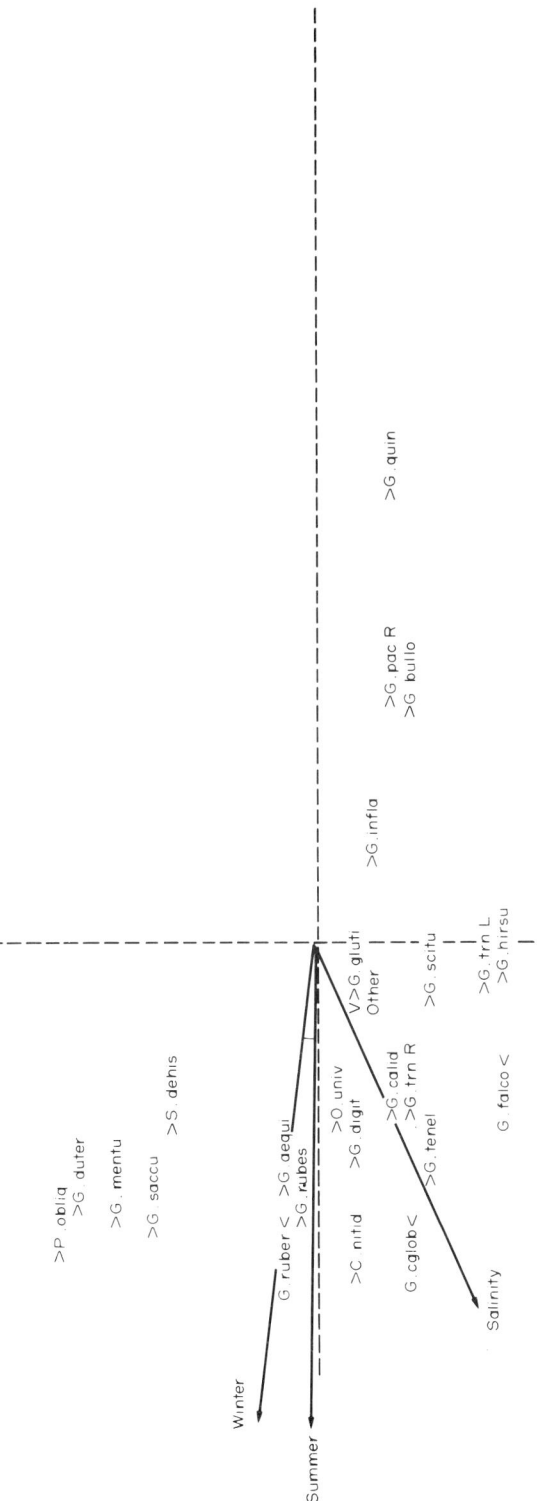

Fig. 7.2. Canonical correspondence analysis (ter Braak, 1986) of planktonic foraminiferal data for species located against three environments. Based on data from Imbrie and Kipp (1971).

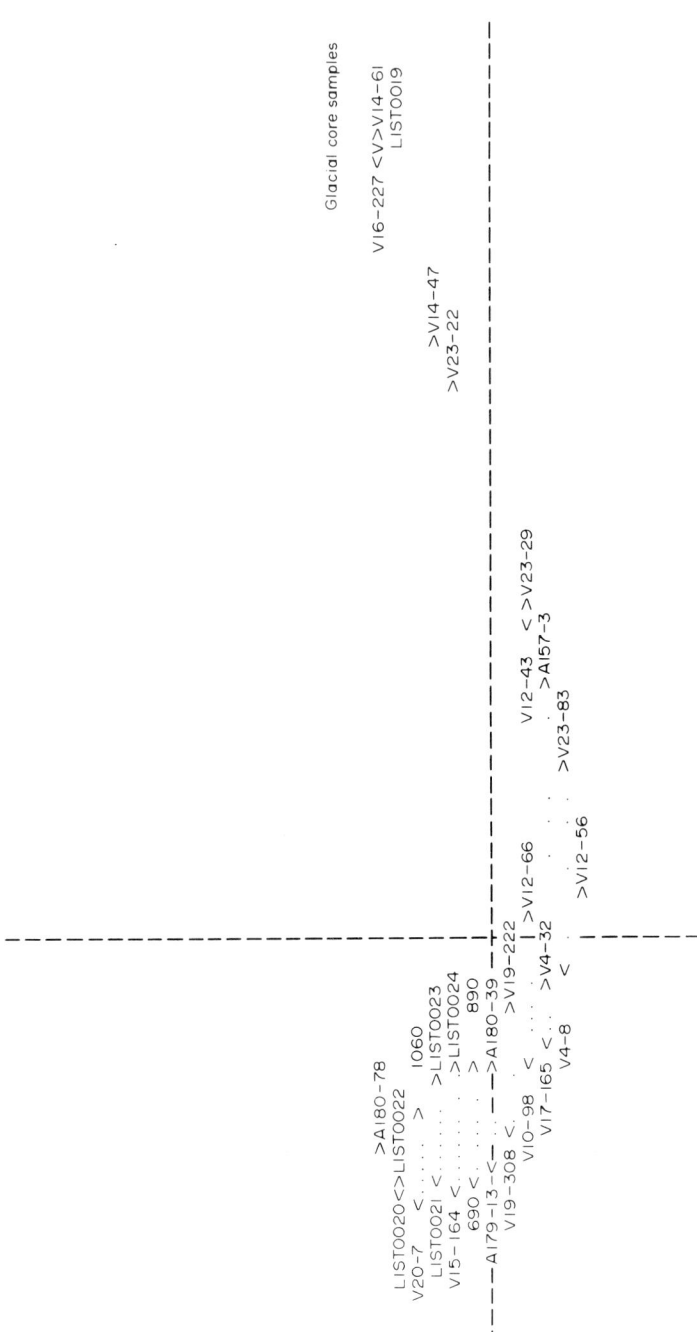

Fig. 7.3. Canonical correspondence analysis. Sample scores on correspondence axes for the foraminiferal data. Glacial core samples lie in a group of their own. Based on data from Imbrie and Kipp (1971).

7.7 Interaction Between Predator and Prey

7.7.1 *Introduction*

In this section I give an account of a multivariate study concerning relationships between predator and prey, one of the classical areas of statistical ecology. It is only occasionally possible to express a quantitative relationship between a fossil predator and its prey and it is only when the predator leaves a permanent imprint of its activities on the remains of its victims that this can be done. Under favourable circumstances, drilling gastropods offer a reasonably good opportunity of doing this.

Naticid and muricid gastropods are efficient predators of some shell-bearing organisms and well-documented case histories of their activities have been published for gastropods (including cannibalism), bivalves, barnacles and ostracods. It surprises many zoologists to learn that even the tiny ostracod can form part of the diet of drills. The juvenile diet of muricids and naticids can include adult ostracods, as is well known to palaeontologists. For economic reasons, most attention has been given to gastropod predation on edible molluscs, mainly oysters. For example, there is much information available on the oyster-whelk *Urosalpinx cinerea*, including detailed observations on the duration of the rasping period (Carriker, 1969, p. 924) and the function of the accessory boring organ (ABO). Analogously well documented information for any naticid species is lacking. Thus the existence of an ABO in naticids remains to receive its final confirmation, although Guerrero and Reyment (1988) have presented indirect evidence for the existence of this organ in that family. The predation of gastropods on ostracods cannot be directly equated with their mode of feeding on pelecypods, other gastropods, and barnacles. In the case of predation on ostracods, the terminal growth stage of the prey corresponds to the lower range of growth of the predator. Hence, the question of the predator having to limit its attack to individuals below a certain maximum size does not arise (the "size refuge" of Colbath, 1985). Thus, in a truncated predational system, the element of choice is eliminated inasmuch as all adult individuals of the prey will be accessible to all drills in a population. Experience shows that drills seldom attack younger larvae of ostracods and there is a size for prey below which the juvenile drills will not be stimulated to attack.

There is also the important factor of *habitat*. Ostracods that live on algae will obviously run little risk of attack, whereas denizens of the water–sediment interface will be in constant danger. Kamiya (1988) reported, albeit anecdotally, for ostracods that inhabit a phytal environment that they tend to have rounded carapaces and a convex ventral surface, whereas benthic forms often possess a flattened ventral surface and an elongated carapace. Phytal species of *Loxoconcha* he studied were, moreover, reported to have a higher mortality for males in that environment than among those inhabiting the benthos. If

Kamiya's observations hold for a wide range of ostracods, it would indicate that there is a form-functional relationship in the carapace of those species which could prove useful in palaeoecological reconstructions.

Palaeoecological aspects

In a palaeo-predational study we ask:

1. Is there a significant correlation between the dimensions of the prey and the size of the hole drilled by the predator and, hence, by extrapolation, the size of the gastropod?
2. Do differences in predational pressure occur in stratigraphically sampled material (cf. Kitchell *et al.*, 1981)?
3. Are there statistically significant differences between naticid and muricid holes?
4. Does the texture of the surface of the shell influence the drill's choice of prey?

All of these questions can be investigated in suitable fossil material as will now be demonstrated.

Example 7.5. Predation on Late Cretaceous and Early Paleocene Ostracods

Presentation of the Problem

The example offered here was taken from Reyment *et al.* (1987). The first question to which we shall seek an answer concerns the eventual association between the size of the prey and the diameter of the borehole drilled in it. The method of canonical correlation was put to service for this study, whereby one set of variables consists of the dimensions of the prey and the complementary set is composed of the dimensions of the drill hole. The second question is to see whether any inferences can be made about the size of the predators. Thirdly, we shall examine whether the surface sculpture of the prey species has had an influence on the preferences of the predators. The data come from the Santonian of Israel and the Paleocene of Nigeria.

Association between size of prey and diameter of borehole

The canonical correlation analysis of the Santonian and Paleocene material yielded the rather surprising result that the correlation between hole diameter and dimensions of the prey for the Santonian ostracods is not statistically significant (for a canonical correlation of only 0.27). On the other hand, the canonical correlation coefficient for the Paleocene individuals was found to be 0.62, which is highly significant for 15 degrees of freedom. The overlap between sets for the Paleocene data is informative. The first set overlaps the

TABLE 7.12. *Correlation matrices, standard deviations and means (in mm) for two samples of drilled ostracod shells (after Reyment et al., 1987). The upper triangle contains the Santonian correlations, the lower diagonal, the Paleocene correlations*

Santonian ($N = 54$)				
	L	H	d	D
Standard deviations	0.098	0.042	0.055	0.067
Means	0.882	0.484	0.167	0.217
L	1.00	0.48	−0.05	0.12
H	0.84	1.00	−0.04	0.09
d	0.38	0.56	1.00	0.90
D	0.40	0.59	0.98	1.00
Paleocene ($N = 33$)				
Standard deviations	0.150	0.082	0.099	0.111
Means	0.719	0.420	0.130	0.169
Ratio d/D	0.823 (Santonian)		0.768 (Paleocene)	

L = length of shell, H = height of shell, d = inner diameter of borehole, D = outer diameter of borehole.

second set to the extent of 25.5% and the second overlaps the first as much as 36%. The corresponding percentages for the Santonian material are only 8, respectively, 5%. The correlation matrices (Table 7.12) show that the two data sets possess substantially different covariance patterns. In both cases, the inner and outer dimensions of the drill holes are highly correlated. However, in the Santonian sample, the dimensions of the hole are not very highly correlated with the size of the ostracods, whereas this is so for the Paleocene ostracods.

The values of the elements of the canonical variates (standardized) are displayed in Table 7.13, together with the corresponding morphological structure and elements of the first principal component. These results support the indications already obtained that the size of the prey and the size of the borehole (hence the size of the predator) are significantly correlated in the Paleocene sample. The poor correlation evinced by the Santonian material can be provided with an explanation. If a sample represents an upward-truncated predational system in which the prey lacked an upper "size-refuge", beyond which freedom from attack would be assured, all adults would be interesting to the predators. Hence, the size-distributional requirements for producing a significant correlation between hole size and size of prey are lacking. This may describe the situation for the Santonian ostracods. The Paleocene material,

TABLE 7.13. *Standardized canonical variate elements, morphological structure and first principal component for the Paleocene ostracod sample* (N = 33)

Var.	Canonical vector elements	Morphological structure	Principal component
L	−0.226	0.636	0.443
H	0.592	0.956	0.515
d	−0.301	0.937	0.515
D	0.712	0.989	0.523

First latent root of principal components = 72% of trace.
Abbreviations as in Table 7.12.

with the stoutly expressed correlation between sets, could have had a different ecological history. A reasonable model is that the drills feeding on ostracods only did this for a short initial period in their lives, after which they migrated to shallower water. This would be manifested as a size-oriented predational pattern by juvenile drills. Such migrational behaviour has indeed been recorded for early juveniles of *Nucella lapillus* (see Moore, 1958, p. 372) and for young *Polinices* preying on *Gemma* (Wiltse, 1980, p. 189). Support for this hypothesis can be drawn from the observation that on the average, the Paleocene holes are significantly smaller than the Santonian ones.

Estimated size of drills

Can we say anything about the size of the gastropods that drilled the ostracods in the two sets of data? Extrapolating from the dimensions of the ABO, I think we can. Carriker and Yochelson (1968) reported that newly hatched drills bore holes less than 0.1 mm in diameter. Wiltse (1980, pp. 187, 198) found the size of the borehole to be directly related to the dimensions of the predator. Guerrero and Reyment (1988) established this connexion for a living naticid species. Reyment *et al.* (1987) were able to produce a table of reconstructed sizes for juveniles on these criteria.

Naticid versus muricid holes

The question we now put is whether undoubted naticid holes differ from undoubted muricid holes? A suitable approach is by means of a discriminant function applied to 21 undoubted naticid holes and 12 undoubted muricid holes in bored ostracod shells from the Nigerian Paleocene. The discriminant function analysis indicates that there is no marked difference in size between holes made by juvenile naticids and juvenile muricids, which is manifested, for example, in the high probability of misidentification of holes. The linear discriminant function wrongly assigns 33.3% of naticid holes, putting them among muricids, and 16.7% of naticid holes, placing them among naticids. There is a summary of the analysis in Table 7.14.

Quite frankly, this looks rather bad. However, it turned out that matters improved vastly once I reacted to the fact that there is curvilinearity in the

TABLE 7.14. *Discriminant function for the Nigerian Paleocene naticid and muricid boreholes*

	L	H	d	D
Difference mean vector	0.011	0.011	0.079	0.094
Linear discriminant coefficients	2.01	−11.35	8.48	19.90

Unbiassed generalized statistical distance = 0.465

Probability of misidentification = 0.37

bivariate relationships. This indication led me to compute the quadratic discriminant function for the samples. This function brought about a betterment in the levels of misidentification such that the error for wrongly allocating naticid holes dropped to a level of 19% and that for the incorrect allocation of holes known to have been drilled by muricids fell to nought.

Ornament and predation intensity

The investigation of the influence of surface texture of the ostracod shell on the interest of the predators was made by a canonical variate analysis of three arbitrary categories of Santonian shells, to wit, smooth, moderately ornamented, and strongly ornamented (reticulated) shells. A preliminary one-way ANOVA disclosed that there are no significant differences between holes drilled in these sculptural types; this result seems to agree with sporadic observations made for living drills.

However, the multivariate analysis gives a more nuanced picture. The Mahalanobis generalized statistical distances indicate that on the basis of the length of the carapace and the inner and outer dimensions of the boreholes, there is a significant difference between the ornamental categories "smooth" and "moderately strongly ornamented", with $P = 0.02$, and between the categories "moderately strongly ornamented" and "strongly ornamented", with $P = 0.05$. This segregation into distinct groups is supported by the canonical variate analysis which groups the data into the categories indicated by the generalized distance (cf. Reyment *et al.*, 1987). The same analysis disclosed that the role of the inner diameter of the borehole is almost insignificant. It seems that the surface texture of the shell is important for the initial emplacement of the hole and the determination of the zone to be scraped by the radula, but it has no relevance for the inner dimensions of the hole.

Discussion

The standard multivariate analysis of the borehole data supplies a wealth of information that was not evident from the univariate study. We find here a further example of an analysis that was greatly enhanced by a hierarchically constructed multivariate analysis. The main interest attaching to a predator–prey study in palaeoecology lies with establishing a pattern for the interactions between species. There is, however, a deeper content, namely, the possibility offered by the gastropods of interpreting the nature of the marine environment. We know that muricids can seek out and attack their prey in the intertidal zone. Naticids are, on the other hand, usually thought to function in relatively deep water (Ziegelmeier, 1954). But, Savazzi and Reyment (1989) report subaerial hunting by naticids in the Pacific and Guerrero and Reyment (1988) described naticid predation in a very shallow coastal environment. In the latter case, the predation took place on the surface of the sediment. These new discoveries make it difficult to generalize about the depth at which extinct naticids functioned and each instance must be interpreted on its own evidence.

7.8 Asymmetric Relationships in Palaeoecology: The Case of Riparian Sedimentary Environments

We now meet a branch of multivariate statistical analysis that you will not find in many standard textbooks on the subject, although it is covered at length in the book by Digby and Kempton (1987). The topic is conceptually one of the more difficult ones taken up in this book and I have therefore striven to be as explicit as possible in presenting it. Although you may not be aware of the fact, asymmetric data occur quite widely in the Geosciences. As soon as you are concerned with *contiguous relationships*, the natural quantification of these is going to involve asymmetries. As examples of asymmetries, consider the following: patterns of species of corals in a reef, rock types distributed over a lithological map, a standard geological map of formations, sediments in a riparian environment, proximities of ornamental fields on an organism, and many more.

Reyment and Banfield (1981) were concerned with analysing sedimentary environmental data that occur in the form of an asymmetric matrix. (The same technique was applied by Reyment (1983) to geological maps used for interpreting the tectonic and hence palaeogeographical history of the Strait of Gibraltar). Let us look at the problem in formal terms.

The relationship between two geological (palaeoecological) entities i and j can be quantified in manifold ways. Such quantities, which we shall designate a_{ij}, are calculated from measurements made on the entities. If the relationship between the entities is **symmetric**, then the quantities a_{ij} and a_{ji} will be equal. In Chapter 3 we considered a multitude of ways for analysing such data in which the property of distance is capitalized upon. When, however, the relationship between the two entities is asymmetric, $a_{ij} \neq a_{ji}$, the familiar techniques are not appropriate. As already intimated, asymmetric data are generated in the quantitative interpretation of **spatial relationships** of all kinds of data that can be represented in the form of a map. Thus, counts on the number of times formation i is the nearest unlike neighbour of formation j, or the number of times sediment i is surrounded by sediment j are typical examples. Here, the sites have a known geographical relationship and the significance therefore lies in the asymmetry of the observations. At the laboratory level, thin sections are a source of asymmetric data. Spatial relationships between mineral species are an example.

A square non-symmetric matrix **A**, the rows and columns of which are classified by the same entities (see, for example, Table 7.15) cannot be treated by the same methods as used in the standard multivariate analysis of square

TABLE 7.15. *Matrix* of the number of nearest unlike neighbours for depositional environ-ments in the Mississippi Delta (Source: McCammon, 1972, p. 424)*

Category	(1) Natural levee	(2) Point bar	(3) Swamp	(4) Marsh	(5) Beach	(6) Lacustrine	(7) Bay sound
(1)	—	117	286	148	0	2	0
(2)	38	—	5	2	0	0	1
(3)	301	10	—	175	1	138	12
(4)	538	3	168	—	29	320	281
(5)	0	0	0	9	—	0	8
(6)	2	0	168	292	0	—	20
(7)	0	1	147	617	161	25	—

*The asymmetric entries reflect the widely different spatial relationships between the seven environmental categories, hence order in the geological pattern. Symmetric entries would have indicated geological disorder to reign.

symmetric matrices. Attempts at analysing non-symmetric data have been made using Multidimensional Scaling for treating the symmetric matrix

$$\tfrac{1}{2}[\mathbf{A} + \mathbf{A}^T].$$

Be aware that this approach does not recognize the asymmetry in the data.

The interpretation of asymmetry can cause difficulties. In many situations, it is dismissed as meaningless "noise". In geological applications, it can, and often does, point to significant relationships. The main credit for the development of asymmetry analysis goes to John Gower and Graham Constantine who together worked out the intricate mathematical statistical details. Gower (1977) gave attention to methods of analysing the asymmetry in square non-symmetric matrices, whereby the differences between a_{ij} and a_{ji} are expressed graphically. There are several ways of going about this, but the one that Colin Banfield and I liked the best is the canonical analysis of asymmetry, since it is the most applicable of the available procedures to the interpretation of data that can be shown in the form of maps. Please note that the technique should not be confused with spatial modelling to which it bears no relation whatsoever. The method is generally available in GENSTAT V, including the PC-system for this language.

7.8.1 *Description of the Method*

Given an $n \times n$ non-symmetric matrix $\mathbf{A}$, the rows and columns of which classify the same entities, then

$$\mathbf{M} = \tfrac{1}{2}[\mathbf{A} + \mathbf{A}^T]$$

is a symmetric matrix and

$$\mathbf{N} = \tfrac{1}{2}[\mathbf{A} - \mathbf{A}^T] \tag{7.8}$$

is a skew-symmetric matrix, $\mathbf{A} = \mathbf{M} + \mathbf{N}$.

Gower (1977) demonstrated that matrices $\mathbf{M}$ and $\mathbf{N}$ can be analysed separately and the results considered together to produce an overall representation of the asymmetry values in $\mathbf{A}$. However, if $\mathbf{M}$ is of little importance, or is unknown, the analysis of $\mathbf{N}$ in its own can still be worthwhile. The canonical form of $\mathbf{N}$ is found by the singular value decomposition of $\mathbf{N}$ (see p. 122 in Chapter 2).

By the singular value decomposition we have that

$$\mathbf{N} = \mathbf{USV}^T \tag{7.9}$$

where $\mathbf{S}$ is a diagonal matrix containing the singular values s_i ($i = 1, 2, \ldots, n$), $\mathbf{U}$ is an orthogonal matrix and $\mathbf{V}$ is the product $\mathbf{UJ}$, where $\mathbf{J}$ is the elementary block-diagonal skew-symmetric matrix made up of 2×2 diagonal blocks

$$\left\{ \begin{matrix} 0 & 1 \\ -1 & 0 \end{matrix} \right\}$$

If n is odd, then the final singular value, s_n, will be nought and the final diagonal element of $\mathbf{J}$ will be 1. Because $\mathbf{N}$ is skew symmetric, the singular values will occur in pairs and can therefore be arranged in descending order of magnitude:

$$(s_1 = s_2) > (s_3 = s_4) > s_5 = s_6) > \ldots$$

Gower showed that the pair of columns of $\mathbf{U}$ (in 7.9) corresponding to each pair of singular values, when scaled by the square root of the corresponding singular value, holds the coordinates of the n specimens in a two-dimensional space—*a plane*—that approximates the skew symmetry. The proportion of the total skew symmetry represented by this plane is given by the size of the corresponding singular value. Hence since the first singular value s_1 ($= s_2$) is the largest, the first two columns of $\mathbf{U}$ give the coordinates of the n specimens in the plane accounting for the largest proportion of skew symmetry. If s_1 is large compared to the other s_i ($i > 2$), the graph of the first plane will give a good approximation to the values in matrix $\mathbf{N}$. If s_1 is not relatively large, then other planes will also be needed to give a good approximation.

Because of the non-metric nature of skew symmetry, the values of $\mathbf{N}$ are represented in the plane by areas of triangles made with the origin. Hence, the area of triangle $(i, j, 0)$ defined by points i, j and the origin, is proportional to the ij-th value in the matrix $\mathbf{N}$ and

$$\text{area } (i, j, 0) = -\text{area } (j, i, 0).$$

The canonical analysis of asymmetry is in a way an analogue of principal component analysis. We have already met other analogues of this fundamental multivariate technique in this book, for example, the method of principal warps in Chapter 5. As we have seen in Chapter 3, principal component analysis seeks to explain distances in terms of differences in transformed values relative to principal axes, with one axis for each latent root. In the analogue for skew symmetry, the explanation is by areas of triangles relative to principal planes, with one plane for *each pair* of singular values.

7.8.2 *Interpreting the Plots*

The plot, or plots, of the *n* entities arising out of the analysis of the symmetric matrix **M** may also be examined with the plane, or planes, obtained as described in the preceding paragraphs. Together they supply a complete graphical picture of the asymmetric values in **A**. The plots for **M** can be found by many methods, including multidimensional scaling. This will yield a set of coordinates for the *n* entities that represent the symmetry of values of **M** with good approximations in two or three dimensions. If two dimensions turn out to be sufficient, then these coordinates can be plotted to produce a graph in which the distances between the entities represent the values of **M**. If two dimensions are not enough, three or more will be required.

The unusual feature of the method reviewed in this section is that the analysis must be made in two compartments, counterparts certainly, but often uneasy of interpretation. I shall now illustrate the procedure by an environmental example.

Example 7.6. Environments in the Mississippi Delta

Presentation of the Problem

The data for illustrating this section were taken from McCammon (1972). They were originally analysed by Reyment and Banfield (1981). The material consists of observations on the nearest neighbour relationships in the following seven environmental categories:

1. natural levee
2. point-bar
3. swamp
4. marsh
5. beach
6. lacustrine
7. bay-sound

The observations on the spatial relationships between these depositional environments were extracted from a map of the Mississippi Delta by McCammon (1972, p. 423). The matrix of nearest unlike neighbours is shown in Table 7.15.

TABLE 7.16. *Asymmetric matrix of Table 7.15 transformed to proportions*

Category	(1) Natural levee	(2) Point bar	(3) Swamp	(4) Marsh	(5) Beach	(6) Lacustrine	(7) Bay sound
(1)	1.0000	0.2116	0.5172	0.2676	0.0000	0.0036	0.0000
(2)	0.8261	1.0000	0.1087	0.0435	0.0000	0.0000	0.0217
(3)	0.4725	0.0157	1.0000	0.2747	0.0016	0.2166	0.0188
(4)	0.4018	0.0022	0.1255	1.0000	0.0217	0.23900	0.2099
(5)	0.0000	0.0000	0.0000	0.5294	1.0000	0.0000	0.4706
(6)	0.0041	0.0000	0.3485	0.6058	0.0000	1.0000	0.0415
(7)	0.0000	0.0011	0.1546	0.6488	0.1693	0.0263	1.0000

TABLE 7.17. *Skew-symmetric matrix formed from the matrix of Table 7.16*

Category	(1)	(2)	(3)	(4)	(5)	(6)	(7)
(1)	0.0000	−0.3073	0.0223	−0.0671	0.0000	−0.0003	0.0000
(2)	0.3073	0.0000	0.0465	0.0206	0.0000	0.0000	0.0103
(3)	−0.0223	−0.0465	0.0000	0.0746	0.0008	−0.0660	−0.0679
(4)	0.0671	−0.0206	−0.0746	0.0000	−0.2539	−0.1834	−0.2195
(5)	0.0000	0.0000	−0.0008	0.2539	0.0000	0.0000	0.1506
(6)	0.0003	0.0000	0.0660	0.1834	0.0000	0.0000	0.0076
(7)	0.0000	−0.0103	0.0679	0.2195	−0.1506	−0.0076	0.0000

The significance of the categories is as given in Table 7.16. Check that $a_{ij} = -a_{ji}$ in the above array.

Asymmetry Analysis

The 7×7 asymmetric matrix is listed in Table 7.15. It was first transformed to a matrix of proportions, **A** (Table 7.16), by dividing each entry by its row total. I should perhaps mention that this transformation is not usually necessary, but McCammon's original matrix has no diagonal elements which means that the above transformation is required in order to ensure constant known diagonal elements for the later analysis of matrix **M**. From matrix **A**, we compute the skew-symmetric matrix **N**, which is listed in Table 7.17.

The pairs of singular values from the singular-value decomposition of **N**, and the corresponding coordinates, are given in Table 7.18. You will see from

TABLE 7.18. *Singular values and coordinates in the corresponding planes obtained from the skew-symmetric matrix of Table 7.17*

Singular values and corresponding planes

	I 0.4241	II 0.4241	III 0.3078	IV 0.3078	V 0.1050	VI 0.1050	VII 0.0000
	First plane		Second plane		Third plane		
(1)	−0.0891	−0.1489	0.4995	0.1779	−0.0276	−0.0116	
(2)	0.1314	−0.0237	0.1993	−0.5010	0.0205	0.0120	
(3)	0.1205	−0.0351	0.0429	0.1057	0.2455	0.0391	
(4)	0.3788	−0.4673	−0.0821	0.0376	0.0053	−0.0859	
(5)	0.2354	0.3610	0.0771	0.0467	0.0194	−0.2361	
(6)	0.2054	0.1893	0.0476	0.0382	0.1161	0.1510	
(7)	0.3785	0.1248	0.0432	0.0947	−0.1722	0.1315	

The categories (1) to (7) as described in previous tables.

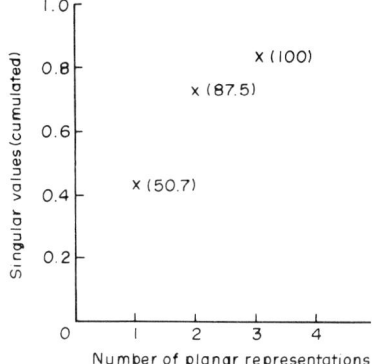

F<small>IG</small>. 7.4. Skew-symmetry analysis. Plot of cumulative singular values against number of corresponding coplanar representations for the Mississippi ecological data. The percentage of the total of the singular values is shown in brackets for the inclusion of successive planes. Redrawn from Reyment and Banfield (1977).

the relative sizes of the pairs of singular values that representation of the skew-symmetry needs are least four dimensions—that is, two planes corresponding to singular values 0.4241 and 0.3078—to represent it with a satisfactory level of approximation.

The distribution properties of the singular values are not known in detail but a reasonable approach is to require values to sum to more than 80% of the total for an acceptable approximation.

In our example, a plot of the first two dimensions (that is the first plane alone) does not show all the skew-symmetry resident in the material, namely only 50%. Therefore, the second plane must also be included. It corresponds to the second pair of singular values and its addition to the analysis improves the fit of **N** to 87.5% (see Fig. 7.4).

The plot of the first plane corresponding to the first pair of singular values is shown in Fig. 7.5. The largest skew-symmetric entry in **N** is $n_{12} = 0.3073$. The corresponding triangle formed by joining environments (1) and (2) to the origin, and here denoted (1, 2, 0) has a relatively small area and, in fact, there are several triangles that greatly exceed it in size. The triangle with the greatest area is (4, 5, 0), corresponding to the second largest skew-symmetric value, which is 0.2539. This indicates that the first planar representation alone is insufficient for a good approximation to the values of **N**. The second plane, shown in Fig. 7.6, discloses, however, that (1, 2, 0) is the triangle with the largest area and that all other triangles have relatively small areas. Hence, this plane accounts for the area not explained by the first planar representation. This was to be expected from what the singular values tell us.

The plot of the first two dimensions, Fig. 7.5, reveals that all points have a large asymmetry with point 4, that is environment (4), Marsh, because all triangles including point 4 have large areas. There seems to be a collinear

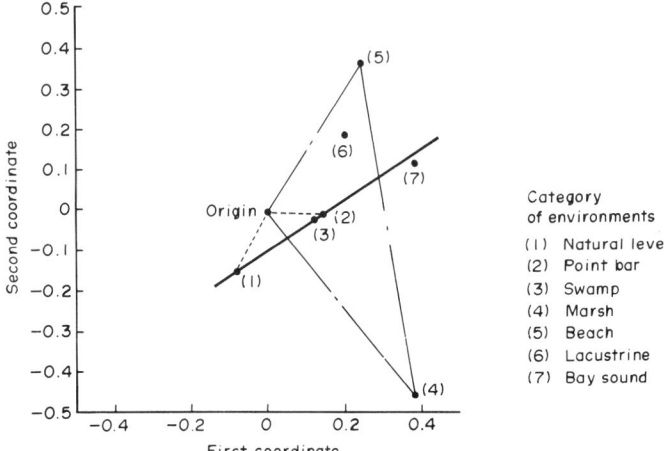

F<small>IG</small>. 7.5. Skew-symmetry analysis. Plot of the coordinates for seven environments in the plane corresponding to the first pair of singular values. Redrawn from Reyment and Banfield (1977).

relationship between the points 1, 3, 2 and 7, in that order (i.e. Natural Levee, Swamp, Point-bar and Bay-sound). Gower (1977) showed that collinear points will possess equal skew-symmetry values with a point on a line through the origin and parallel to the collinearity, because these points will form triangles having the same base and with equal height, hence equal area. Unfortunately, the line through the origin and parallel to the collinearity passes through no other environment, so I cannot exemplify the theory for you on this point. However, I can do this for Fig. 7.6. Points 2 and 3, that is Point-bar and Swamp, lie close together in this plane and thus appear to have similar skew-symmetry values with the other environments because triangles formed by other environments with the origin and these two points have

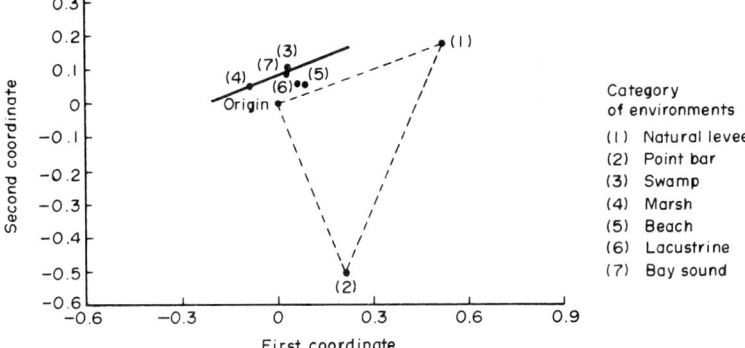

F<small>IG</small>. 7.6. Plot of the coordinates for the seven environments in the plane corresponding to the second pair of singular values. Redrawn from Reyment and Banfield (1977).

similar areas. If, however, we go to the information in the second plane, it will be seen that the foregoing impression is exaggerated, since points 2 and 3 are no longer close.

In Fig. 7.6 the supplementary information of the second plane is dominated by triangle (1, 2, 0). Environments (1) and (2) share the greatest asymmetry, suggesting a possible genetic relationship. This can be seen to be identifiable with the immediate depositional environment of the Mississippi Delta from McCammon's map (McCammon, 1972, figure 1). The points for environments (3), (4) and (7) fall on a line parallel to that joining the origin to environment (1). This indicates that in this planar representation, and this one only, these environments exhibit equal skew-symmetry with the Natural Levee environment. The close proximity of Swamp with Bay-sound and Beach with Lacustrine shows that these pairs have similar skew-symmetries with the other environments.

Principal Coordinate Analysis

In order to reify further the relationships outlined by the graphical portrayals of the skew-symmetry, the symmetrical part $\mathbf{M}$ of the asymmetric matrix $\mathbf{A}$ can be profitably analysed by principal coordinates, noting that this can be regarded as a form of Multidimensional Scaling (see Chapter 3). The results are presented in Table 7.19. The plot of the first pair of coordinates, displayed in Fig. 7.7, represents 54.5% of the total symmetry between the environments. Because these coordinates are plotted in Euclidean space, we can give a metric interpretation to the interpoint distances which denote the similarities between environments. As you might have expected, Bay-sound and Beach are comparatively close together which shows that the degree of organization of the spatial relationships is similar; Natural Levee and Point-bar are also similar. Lacustrine is seen to be very different from Beach and Point-bar and most similar to Swamp and Marsh.

TABLE 7.19. *Principal coordinates obtained for the symmetric matrix derived from the data of McCammon (1972)*

Latent roots

I	II	III	IV	V	VI	VII
1.5372	1.1909	0.8086	0.7107	0.5897	0.1629	0.0000

Corresponding vectors (principal coordinates)

(1)	0.6275	−0.1240	−0.2031	−0.1038	−0.3053	−0.2434
(2)	0.5829	−0.4592	0.4623	−0.0332	0.1939	0.1452
(3)	0.3256	0.3957	−0.5505	0.1937	0.2273	0.1518
(4)	−0.3191	0.1951	0.0526	−0.3383	−0.4770	0.1843
(5)	−0.5247	−0.4387	−0.0402	0.5819	−0.1464	−0.0166
(6)	−0.1451	0.7078	0.4614	0.1355	0.1325	−0.1300
(7)	−0.5472	−0.2767	−0.1825	−0.4357	0.3751	−0.0913

Categories (1) to (7) are as described in previous tables.

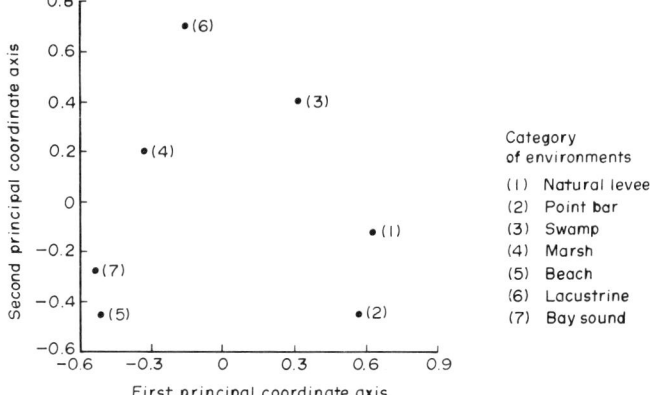

FIG. 7.7. Plot of the principal coordinates for the seven environments accounting for 54.5% of the total symmetry in the data. Redrawn from Reyment and Banfield (1977).

7.9 Further Topics for Research

With this example, we leave the subject of multidimensional palaeoecology, although without having exhausted the supply of suitable topics for study. Much more can be done in the area of secular changes in palaeoecological situations and the interplay between quantitative genetics and palaeoecology is a virtually unexplored field. The application of Leslie matrices to palaeopopulational studies remains to be probed (cf. Pielou, 1974, p. 21) and there are aspects of community ecology that could be invoked with a high probability of success. The multidimensional treatment of species diversity is another subject of potential interest and it is here that pollen-analysts, using modern statistical theory, are making important contributions. In Chapter 8, I have sketched an application of information theory to a multivariate palaeoecological problem. This is an area of research that has been invoked on occasions for the study of particular problems (Miller and Kahn, 1962), and which is open to further development. Asymmetry analysis has a natural field of application in the study of spatial relationships of ornamental features in some species of ostracods, and other crustaceans.

Exercises

1. It is sometimes experienced that a multivariate analysis can profit from having the number of variables reduced (see the relevant sections in Rao, 1952). What did Reyment and Sturesson (1987) find to be at the root of such an improvement in their palaeoecological study of Nigerian boreholes?

2. The "horseshoe-effect" that sometimes occurs in the plotting of scores of an analysis made by means of latent vectors indicates that they align closely along an arch. Can you suggest why some specialists in statistical ecology try to "detrend" such results? (Hint: read the section starting on p. 253). Would you consider detrending to be "fudging" the data?

3. The technique of cross-correlation is often used for correlating between borehole sequences (cf. Davis, 1986, p. 225). Can you give any reason why this method is not statistically and geologically sound? On what grounds is Gordon's slotting technique superior to cross-correlation? Do you think you could embark on a high-dimensional large-sample slotting enterprise on a personal computer?

4. Using the information provided in Table 7.6, write a brief report on the results and make a graph to accompany it.

5. Table 7.14 contains linear discriminant function coefficients for the naticid and muricid borehole dimensions for the ostracods from the Nigerian Paleocene. What can you say about the parts played by the four variables involved (length and height of carapace, inner and outer borehole diameters)? Does consideration of the standard deviations given in Table 7.12 do anything to change your opinion?

6. Take any geological map and try to produce an asymmetric matrix in the manner indicated on p. 270.

7. If you have access to a distributional pattern for the species forming a coral reef, repeat the foregoing exercise for these species. A suitable micrograph depicting microfacies could be substituted for the corals.

8. Would you expect canonical correspondence analysis to do better with data on Pleistocene pollen distributions than, say, data on benthic and planktic foraminifers from deep-sea investigations such as used in the present chapter to exemplify the method?

9. Most often the matrix of associations of species-abundances is symmetric. What kind of a matrix of associations is produced if a_{ij} is the number of times some type of association j follows some other type i in a palaeoecological succession? Refer to Digby and Kempton (1986, p. 10).

CHAPTER 8

Case Histories and Applications

Contents

8.1 Introduction

In this final chapter, some case histories are presented in order to illustrate how a multivariate palaeobiological analysis can be constructed. The choice of examples is by no means exhaustive and I have not attempted to cover the range of all possible applications. Robust estimation, the interpretation of atypical values and the identification of influential observations are examined in turn. The last mentioned point leads to considerations such as determining the correct number of principal components: that is, the smallest number of principal components that can be retained without loss of essential biological information.

The analysis of variation in outline of form is applied to ammonites in one example, whereby the cross-section of whorls is used. The problem here tends to be rather complex inasmuch as ammonite species are notoriously difficult to define, particularly when rich collections are available. That which seems to be a single highly variable species can end up in several taxonomic categories as a result of phenotypic plasticity. An analysis made along the same lines for a species of ostracods discloses multivariate interaction with the environment. Useful reading at this juncture is the text by Schmalhausen (1949), particularly the sections dealing with environmental interaction, phenotypic plasticity and the norm of reaction.

I have repeatedly pointed out that the main mathematical thread uniting the methods and studies in this book is the algebra of latent roots and vectors in all its statistical ramifications. Many procedures turn out to have the necessary prerequisite that the data are multivariate Gaussian and we shall now consider how to test for this condition. A method that is little influenced by even considerable deviations from normality is said to be *robust*. Quite simple procedures can prove valuable in palaeobiogeographical work. An example is the use of the Mahalanobis angle. In Chapter 7, the topic of statistical palaeoecology was delved into. An example appears further on in Section 8.6 in which a different approach is used. Geometric multivariate morphometry is applied to a phylogenetically slanted problem in vertebrate palaeontology.

8.2 Palaeobiological Significance of Multivariate Normality

I have already shown in Chapter 3 that the assessment of normality in samples of fossils can be rather difficult. The sources of this difficulty can be manifold. I have found two of the more common stumbling blocks to be:

1. Purely geological considerations may confuse the issue in that specimens can be damaged or deformed, data can be mixed for various reasons, such as condensed sequences, reworking, water-sorting, etc.
2. The data may be impeccable in the biological sense but owing to polymorphism of some kind or other, seasonal variations in morphology, or some other biological complexity, its statistical properties deviate from multivariate Gaussian form. This situation is usually accompanied by polymodality. For example, in studies on foraminifers, one can almost expect, as a matter of course, to encounter non-normality.

Species may be impossible to differentiate on visual inspection, even at high magnifications (sibling species), but can be held apart on subtle statistical grounds. We shall find that the concept of *influential observations* may be a useful one in this connexion. There are interesting statistical and biological notions to be weighed in relation to multivariate normality. It is still common praxis to test the distributions of each character separately for univariate normality. However, normality at the univariate level does not necessarily carry over to the multivariate case, as I have demonstrated empirically (Reyment, 1971b).

Mardia (1970) introduced the concepts of multivariate skewness and multivariate kurtosis, designated as $B_{1,p}$ and $B_{2,p}$, respectively (see Section 3.8.1), where p denotes the number of variables. Seber (1984, p. 54) gives an illuminating account of these statistics and how to use them and Kres (1977) tables for assessing significance. They are defined as follows:

$$\beta_{1,p} = \mathscr{E}((\mathbf{x} - \boldsymbol{\mu})^T \boldsymbol{\Sigma}^{-1} (\mathbf{y} - \boldsymbol{\mu})^3) \tag{8.1}$$

and

$$\beta_{2,p} = \mathscr{E}((\mathbf{x} - \boldsymbol{\mu})^T \boldsymbol{\Sigma}^{-1} (\mathbf{x} - \boldsymbol{\mu})^2) \tag{8.2}$$

where $\mathbf{y}$ is independent of $\mathbf{x}$ but has the same distribution. When $\mathbf{x}$ is multivariate normally distributed, the value of equation (8.1) is nought and that of the second equation (8.2) is $p(p + 2)$. The above two measures are natural generalizations of the usual univariate measurements

$$\sqrt{\beta_1} = \mu_3/\sigma^3$$

and

$$\sqrt{\beta_2} = \mu_4/\sigma^4.$$

A simplified account of multivariate skewness and kurtosis is given in Reyment (1971b) as well as the main FORTRAN steps for computing the statistics. The computing required for (8.1) is a function of the sample size and clearly becomes a factor to be reckoned with for even moderately large data sets. Mardia and Zemroch (1978) gave an algorithm for computing estimates of (8.1) and (8.2) that increases efficiency.

For gauging the significance of the sample counterparts of (8.1) and (8.2), the following formulae are useful. Firstly,

$$A = 1/6Nb_{1,p}$$

is approximately distributed as χ_f^2, where

$$f = 1/6p((p + 1)(p + 2)). \tag{8.3}$$

The significance of the second formula is found from

$$B = (b_{2,p} - p(p + 2))/(8p(p + 2)/N)^{1/2} \tag{8.4}$$

which is distributed as $N_1(0,1)$.

Mardia's method is not the only way of probing multivariate normality, but it has convenient computational and interpretational properties (see Seber, 1984, p. 150 for an account of this subject). Graphical methods can often be informative, as we shall see in the following example involving Robust Estimational Procedures. A useful preparatory step for this subject is to consult the book by Gnanadesikan (1977) in which detailed accounts of such methods are given. Why do we transform intransigent data you might well ask? After all, everybody knows that taking the logarithms of non-normal data makes things better. Or does it? There are several things you need to keep in mind before transmuting morphometric observations. Consider just the following points:

1. Taking logarithms is not always a beneficial action; experimental work indicates that the logarithmic transformation may actually worsen an analysis (Reyment, 1971b).
2. Logarithmically transformed variables may constitute a theoretical requirement of some multivariate methods. For example, the algebraic procedure for expressing multivariate allometry (Chapter 4, p. 103).
3. In reifying multivariate data, the use of logarithms may make biological interpretations difficult.
4. It is sometimes suggested that only those variables that deviate from multivariate normality should be transformed, thus leaving the well behaved ones alone. This is not a useful procedure and can only be deplored in biological connexions.

We shall now consider two examples.

Example 8.1. Multivariate Normality in Two Species of Foraminifers

First species: *Brizalina mandoroveensis*

Presentation of the Data: *Brizalina mandoroveensis* is a species of benthic foraminifers from the Miocene of Cameroun, West Africa which we have had occasion to consider several times in the preceding chapters. You will no doubt recall that the material comes from a borehole drilled at Ikang in the quest for petroleum carried out by ELF-SPAEF, France. The variables measured on this species number the five traditional ones (1) length of the test, (2) the breadth of the test, (3) the length of the last chamber, (4) the length of the second last chamber, and (5) the diameter of the proloculus. The sample analysed here comprises 36 specimens.

The Analysis: The univariate tests indicate that all five characters conform with the properties of a normal distribution with respect to *skewness*. As regards *kurtosis*, variables 1 to 4 agree with the univariate normal distribution, whereas variable 5, the diameter of the proloculus, displays significant kurtosis ($t = 3.45$). Our interest in checking normality lies with assessing deviations that could indicate polymorphism. Also, deviations of a more serious nature can adversely affect some multivariate statistical procedures (cf. Chapter 3, p. 89). The multivariate values for the foraminifers are:

$$b_{1,5} = 8.37$$

and

$$b_{2,5} = 36.77.$$

The corresponding test-statistics are

$$A = 1/6(36)(8.37) = 50.21$$

for $1/6(5)(6)(7) = 35$ degrees of freedom, which is just statistically significant when compared with $\chi^2_{35} = 49.8$ with $P = 0.05$. The value of the second statistic for significance is

$$B = (36.77 - 5(7))/(40(7)/36)^{1/2}$$
$$= 0.64,$$

which is not significant when compared with $N_1(0,1)$ in a table of the normal curve.

It can therefore be suggested that the data only show a slight deviation from multivariate normality, as was indeed borne out by the results presented in Bookstein and Reyment (1989) and briefly reviewed in Chapter 4.

Second species: *Afrobolivina afra*

Presentation of the Data: *Afrobolivina afra* is also a bolivinid benthic foraminifer. The data come from the Maastrichtian (Upper Cretaceous) of Nigeria, West Africa and have been analysed in several connexions in this book. The same nine characters as were used in previous studies, and illustrated in Fig. 5.1, were employed here. In all, 41 specimens were included in the present study.

Significant values for univariate skewness were obtained for variables 7 and 8. Variable 7 yielded a significant value for kurtosis. The multivariate counterparts are: $b_{1,9} = 42.5$, which corresponds to a chi-square of 290 which for 165 degrees of freedom is significant on the 1% level. The value of $b_{1,9} = 117.74$, which connects to a value of $B = 2.90$, which is significant (entered into a table of the normal curve).

Hence, notwithstanding that the marginal distributions seem to be leaning towards normality, with minor discrepancies, the multivariate estimates of skewness and kurtosis speak out decisively against such a conclusion. I do not intend leaving matters here, however, and the reason for the deviation from normality is taken up in other examples.

8.3 The Interpretation of Atypical Individuals

Introduction

The assessment of multivariate normality for the data on *Afrobolivina*, just considered, shows that there are important deviations in both skewness and kurtosis. The suggestion was ventilated that this could be due to some kind of polymorphic manifestation in the data causing statistical, but not biological, heterogeneity. I have, on several occasions, pointed out that the performance of classical multivariate statistical procedures can be seriously influenced by atypical values, where "atypical" denotes an observation that deviates pronouncedly in some manner or other from its fellows. The feeling of uneasiness that statisticians have felt with such data for quite some time has led to a branch of statistical theory which is concerned with methods of robust estimation procedures. Robust statistical methods that are but slightly affected by atypical observations make an attractive alternative approach in biometrical analyses, particularly where large data-sets are concerned, since even careful scanning can easily fail to detect occasional gross errors, or biologically determined departures from a standard.

It is frequently possible to pick out markedly atypical observations in a single variable by univariate methods (cf. Barnett and Lewis, 1978). We have seen several cases of this in foregoing examples. However, for multivariate data, observations are often found to be atypical only when the value is considered in the same context as all the other variables. Some values may fail to maintain the pattern of relationships evident in the majority of observations.

Interest is twofold in the present case history. In the first instance, we have the non-biological situation of wrong values. Such values can arise in any quantitative work and in palaeontology possible causes always worth considering are wrongly recorded measurements, typing mistakes, and post-mortem deformation (during diagenesis, for example). Secondly, there is the more interesting situation of observations that are biologically unchallengeable, but which fail to conform statistically with the main body of the data. Some morphological difference related to the life-cycle of an organism, or an ecophenotypic reaction may cause considerable differences in characters which are not abnormal in the biological sense, but which can upset the statistical niceties of some sensitive procedure.

Biological Background

Bolivinids, and some other groups of benthic foraminifers, display what is conveniently known as "trimorphism". The idealized life-cycle of a foraminiferal species of this category passes regularly back and forth between gamont and agamont phases. The shells produced during these phases are not seldom characterized by the size of the proloculus, the respective classes being referred to as microspheric (agamont) and megalospheric (gamont).

In some situations, the reasons for which are still obscure, extra megalospheres may develop. These are sometimes termed "pseudomegalospheres". They appear aside from the normal alternation of generations. The pseudomegalospheric individuals have the appearance of true megalospheres to a quite considerable extent, but their sexual status is that of microspheres. Pseudomegalospheres may possess statistical properties differing from those of normal megalospheres and they may therefore be exposed as atypical observations. The discrepancies may not be serious and no doubt many pseudomegalospheres go undetected.

Microspheric tests are more easily identified, as is amply illustrated by the ensuing analysis. One of the morphological characteristics displayed by bolivinids concerns the atypical growth of the last two chambers with the appearance, in many cases, of a loxostomoid termination, that is the final termination occupies the entire breadth-width plane of the test. No loxostomoid individuals were included among the samples analysed. However, the tendency towards a loxostomoid kind of development was picked up by the methods used here.

Example 8.2. Robust Estimation of Multivariate Location and Scatter

Robust multivariate procedures can be thought of as being simple modifications of the standard methods. The contribution of an observation to the statistic or statistics of interest is given *full unit weight* if it is a reasonable observation, otherwise it is downweighted. A robust estimator can be defined by introducing a weight function which depends on the discrepancy between an observation and some robust average value, relative to a robust measure of scatter. The weight function is inversely related to the influence of an observation. We have seen earlier on (p. 64) that this can be done formally by means of the *influence function* of Hampel (1974) and bounding the influence of observations with unduly large discrepancies. Specific details are given below in formulae (8.5) to (8.8). The present discussion was extracted from Campbell and Reyment (1980).

For multivariate data, the fundamental distance or discrepancy is the Mahalanobis generalized statistical distance. If $\bar{\mathbf{x}}$ represents the $p \times 1$ vector of sample means and $\mathbf{S}$ the sample covariance matrix, then the squared distance of the m-th observational vector from the mean vector of the observations is defined by:

$$d_m^2 = (\mathbf{x}_m - \bar{\mathbf{x}})^T \mathbf{S}^{-1} (\mathbf{x}_m - \bar{\mathbf{x}}) \tag{8.5}$$

This is the well-known formula for the Mahalanobis generalized distance for the observations of a sample.

Q–Q probability plots of the d_m^2 may be used to examine the distributional assumptions of an underlying Gaussian distribution (Gnanadesikan, 1977, p. 172; Seber, 1984, pp. 127, 539). Since an atypical observation will tend to deflate the correlations and, possibly, inflate the variances, this will, in general, tend to decrease the Mahalanobis statistical distance and distort the rest of the probability plot. It is, I think, a good idea to discuss briefly at this point what a Q–Q-probability plot is. Not all introductory textbooks by any means tell you about this and it can therefore be troublesome to find out how you should go about making them. I expect you will have gathered by now that they are important in applied biometrical work. The steps required for making a plot in multivariate analysis are as follows:

1. Produce the set of values of d_m^2, the generalized statistical distances of each observation in the sample from the **centroid**.
2. These values are plotted against the quantiles of the gamma distribution, with shape parameter $p/2$. There is a computer program in Roy *et al.* (1971, pp. 286–298) for calculating the gamma quantiles. An alternative procedure is to use the cube-root transformation of a gamma variable to Gaussian form (Healy, 1968; Campbell, 1982).
3. However, I use an approximate method which is very easy to compute and which utilizes the excellent agreement between the percentage points of a Gaussian distribution and those of a lambda distribution with parameter 0.135. The pertinent reference is the article by Ramberg and Schmeiser (1972). Specifically, with

$$u_m = (m - 3/8)/(n + 3/8).$$

the quantiles are given by

$$q_m = \{u_m^{0.1349} - (1 - u_m)^{0.1349}\}/0.1975.$$

4. Plot the ordered values of $d_m^{2/3}$ against the q_m.

Seber (1984, p. 539) devotes his Appendix C to order statistics and probability plotting.

For robust M-estimation of multivariate location and scatter (M stands for Maximum Likelihood), the appropriate measure of discrepancy has been proven to be the Mahalanobis distance. Useful references to this if you want to check the ideas are Maronna (1976), Hampel *et al.* (1986) and Campbell (1980b). The equations used for defining robust estimators of means and covariances are derived by assuming on elliptically symmetrical multivariate density and then associating this density with a contaminated Gaussian density. The theory is nothing for amateurs, so I confine myself to presenting the principal points and holdfasts.

The specific equations needed are:

$$\bar{\mathbf{x}}^c = \frac{\sum_{m=1}^{n} w_m \mathbf{x}_m}{\sum_{m=1}^{n} w_m} \tag{8.6}$$

and

$$\mathbf{S}^c = \frac{\sum_{m=1}^{n} w_m^2 (\mathbf{x}_m - \bar{\mathbf{x}}^c)(\mathbf{x}_m - \bar{\mathbf{x}}^c)^T}{\left(\sum_{m=1}^{n} w_m^2 - 1\right)} \tag{8.7}$$

where

$$w_m = w(d_m^c) = \frac{\omega(d_m^c)}{d_m^c}$$

and

$$d_m^c = \{[\mathbf{x}_m - \bar{\mathbf{x}}^c]^T (\mathbf{S}^c)^{-1} [\mathbf{x}_m - \bar{\mathbf{x}}^c]\}^{1/2}.$$

The solution for the means and for matrix $\mathbf{S}^c$ is iterative.

Here, ω is known as a **Bounded Influence Function**; it is usually linear over the range of values of d_m for most of the data one is likely to encounter in biological applications, but bounded outside this range. Hampel (1974) suggested that the influence, and hence weight, of an extreme atypical observation should be set to nought, so that ω should *redescend* for sufficiently large values of d_m. In the examples for this book, I have employed the following two forms of ω:

(1) The non-descending form is:

$$\omega(d_m) = d_m \text{ if } d_m \leq b_1$$
$$= b_1 \text{ if } d_m > b_1. \tag{8.8}$$

(2) The "redescending" form is:

$$\omega(d_m) = d_m \text{ if } d_m \leq b_1$$
$$= b_1 \exp\{-\tfrac{1}{2}(d_m - b_1)^2/b_2^2\}.$$
$$\text{if } d_m > b_1 \tag{8.9}$$

It has been found by trial and error that b_2 located in the range 2.5 to 1.0 gives a good performance, with a recommended value of $b_2 = 1.25$. The smaller the value of b_2, the quicker is the rate at which ω descends. Seber (1984) has made some useful observations on this subject. The primary references are the papers by Norm Campbell (see Bibliography).

The constant b_1 is yielded by

$$b_1 = \sqrt{v} + b_0/\sqrt{2}$$

where b_0 lies in the range $1.64 - 3.09$. The recommended value is $b_0 = 2.0$ (see Campbell, 1980b).

The most useful approach for most purposes is to base the means and covariance matrix, with associated weights, on the non-descending value of ω for a value of $b_0 = 2.0$. The usual means and covariance matrix provide the initial estimates. Introduce then the redescending ω for up to ten iterations, with $b_2 = 1.25$. Little change in distances and hence weights seem to result after ten iterations.

Robust M-*estimation of Canonical Vectors*: When several groups of data are available, as in multivariate discrimination studies, the robust procedures described here can be applied to each group in turn. This will provide robust estimates of means and covariance matrices for canonical variate analysis: possible atypical observations will also be identified. A more direct approach is to assume that the population means lie on a p-dimensional hyperplane which is specified by the p canonical vectors of interest. Specifically, one assumes that:

$$\boldsymbol{\mu}_k = \boldsymbol{\mu}_0 + \boldsymbol{\Sigma}\boldsymbol{\Psi}\boldsymbol{\zeta}_k$$

Here, $\boldsymbol{\Psi}$ is the unknown $v \times p$ matrix of population canonical vectors, $\boldsymbol{\Sigma}$ is the unknown population covariance matrix and $\boldsymbol{\zeta}_k$ are the group coordinates in the space of the canonical variates. The formal derivation involves the usual form of the canonical variate solution under the assumption of an elliptically symmetric multivariate density and by applying maximum likelihood estimation. A weight function enters the definition of sample group-means and between- and within-groups matrices. Robust M-estimators result by associating the elliptical density with a contaminated Gaussian density. Observations assumed to be from the main body of data are given full unit weight, while others are given reduced weight or influence. The equations used to define robust estimators of means and of covariances for the canonical variate solution are given in Campbell and Reyment (1980, p. 211) to which reference is made for the details. From extensive empirical evidence gained in the application of the procedure, a weight of less than 0.35, with $b_0 = 2.25$ and $b_2 = 1.25$ can be taken to indicate an atypical observation.

(1) Atypical Observations in *Brizalina*

The three stratigraphical levels of *Brizalina mandoroveensis* examined in the previous section were shown to conform well with a multivariate Gaussian form. I was, however, a little uneasy about some of the univariate distributions which, in a seemingly unmotivated manner, differ from the general picture. The sizes of the three samples are 36, 15 and 35.

In the first sample, observation 19 differs from its fellows but slightly, the non-unit weight being 0.98. This deviation does not register on the Q–Q-probability plot. The specimen in question could be slightly deformed, but this is not obvious on direct inspection. The robustly estimated means and variances hardly differ from the normal estimates and at the most only influence the second decimal place. This is a straightforward result which is not difficult to interpret. Fortunately, most cases seem to be of this kind. The following set of data is more of a problem.

(2) Atypical Observations in *Afrobolivina*

Introduction: The data for this Maastrichtian bolivinid foraminifer pose a more complicated problem than the material we have just considered. This will lead us to consider some finer points of multivariate analysis which we have not run into before. I have kept you aware throughout this book of the challenge levelled by atypical values, but we encounter in the next section the concept of *influential observations*.

The Analysis: We have already experienced earlier on in this section that the data considered here differ in several respects from univariate normality and from a multivariate Gaussian form. We shall now ascertain how they behave in relation to robust procedures of estimation. The sizes of three representative samples are 80, 100 and 55 specimens. Using the criteria outlined earlier for determining atypical values, deviating observations were identified as follows:

Sample 1 ($N = 80$)

non-unit weights	0.83	0.91	0.70	0.63	0.73	0.91	0.98
specimens	3	14	16	28	34	38	72

Sample 2

non-unit weights	0.32	0.65	0.57
specimens	85	92	93

There is one quite obviously atypical value that shows up in the Q–Q-probability plot (Fig. 8.1).

Sample 3

No significant deviations were found for this sample and none of the observations qualifies for the role of outlier.

It can be useful to establish how well the means can be reconstructed from the robust estimates. A few examples will suffice. In the case of variable 1, the length of the test, we have the following comparisons for the actual means of the three samples and the robustly reconstructed means.

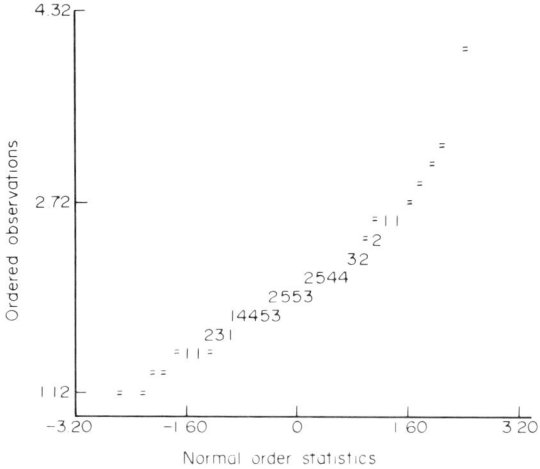

FIG. 8.1. Q–Q probability plot for *Afrobolivina* for sample 1. The numbers indicate coincident points.

Sample	Usual mean	Robust mean
1	67.12	66.48
2	74.57	74.18
3	71.63	71.63

It is clear that not even in the case of the most severely affected sample do the two means differ by much. As only to be expected from the initial indications obtained, sample 3 is the least influenced and it may therefore be suggested that the atypical values separated out by the analysis are not very important. The atypical individuals were identified as belonging to

1. Pseudo-megalospheric tests;
2. Tests with an irregularly developed final chamber;
3. Slightly deformed specimens (presumably due to post-depositional sedimentary compaction);
4. Tests with missing chambers, due to damage.

Obviously microspheric individuals were weeded out prior to analysis. A more comprehensive robust analysis of *Afrobolivina afra* is given by Campbell and Reyment (1980).

8.4 Principal Component Analysis and Cross-Validation

Introduction

One of the aims of multivariate analysis is to bring about a data-analytical treatment of a set of multivariate observations. We have already seen that the most accessible tool for doing this is offered by principal component analysis. It is now proper to ask whether we can obtain further information from a principal component analysis than has already been forthcoming from the may in which I have applied the technique so far? Thus, it is appropriate to enquire:

1. Can one say anything more about atypical values over what has already been alluded to?
2. What can be done about the occurrence of influential values?
3. Can one be more specific about identifying important values and, by the same token, redundant variables?

These, and other questions have been taken up by Krzanowski (1987a, 1987b); a brief review was given in Section 3.6.7). Krzanowski (1987, p. 576) has given a neat differential diagnosis of atypical and influential observations. Thus, an *atypical observation* is one of the values of which on the measured variables show a marked difference from those in the rest of the sample, whereas an *influential observation* is one that causes a great change in the results of an analysis when it is left out, but the values of which on the measured variables show no obvious distinguishing features. As we have seen, atypical observations are likely to show up as outliers in the sample.

The influence function allows a comparison of the results of an analysis with and without a suspected deviating value. A robust estimator can be defined by introducing a weight function, as I have already pointed out.

I shall here introduce the methodology for the biometrically relevant aspects of the topic by summarizing the pertinent steps in the calculations. In essence, the technique is not really new, involving as it does concepts rather well known in psychometry and widely used by chemometricians. It is to a fair extent an *ad hoc* methodology in that some results are judged on arbitrary criteria. The essentially new features concern the isolation of "subtle atypicality", determining the dimensionality of a principal component analysis, and the multivariate recognition of influential values. By subtle atypicality is meant one that does not show up in an obvious manner in graphical displays, but which, when identified, turns out to have an important influence on the outcome of the computations.

Steps in the calculations

1. Compute the latent roots and vectors of the correlation or covariance matrix of the data matrix, $\mathbf{X}$. Thus,

$$\mathbf{S} = \mathbf{VLV}^T \tag{8.10}$$

and

$$\mathbf{V}^T\mathbf{V} = \mathbf{I}.$$

Alternatively, you can use the singular value decomposition, which yields:

$$\mathbf{X} = \mathbf{UDV}^T \tag{8.11}$$

where,

$$d_i^2 = (N - 1)\ell_i,$$

where the ℓ_i are as in equation (8.10).

2. The scores of the principal components are then computed in the usual manner:

$$\mathbf{Z} = \mathbf{XV}$$

as indicated in Chapter 3.

3. For selecting the optimal number of principal components, one may proceed by means of the criterion W_m, which is obtained from the average squared discrepancy between actual and predicted values of the data matrix $\mathbf{X}$, the calculations for which are performed by *cross-validation*. This manipulation is performed by subdividing $\mathbf{X}$ into a number of groups, deleting each group in turn from the data, evaluating the parameters of the predictor from

the remaining data, and predicting the deleted values. In practical terms, the group deleted should be as small as possible. In the present study, I have made each group to be no more than one row of the data matrix **X**. The method put forward by Eastment and Krzanowski (1982) recognizes the need for allowing deletion of variables as well as units. This is a necessary manoeuvre if the estimation of **X** is to succeed.

4. For determining the influence of each of the observations, one may go ahead in descriptive mode by computing the *angles* between subspaces of a common data space. The critical angle put into service here is interpretable as a measure of the influence of each individual on the sample, with

$$a = \cos^{-1} d$$

where d is the smallest element of **D** (diagonal) in the singular value decomposition (8.11). Large values of the angle point to **highly influential observations** in the sample and hence to observations deviating in some special manner from the individuals in the main body of the sample.

5. For scanning the data matrix with respect to those variables that contribute most information, one may advance by means of the two-dimensional representation, whereby the variables are deleted one by one and the resulting sums of squares examined. Small residual sums of squares have small effects on the principal component analysis. A variable with a small residual sum of squares may well be a candidate for omission. Seber (1984, p. 408) is a convenient reference for learning more about the analysis of residuals.

Example 8.3. Analysis of *Veenia fawwarensis*

Presentation of the Problem: *Veenia fawwarensis* HONIGSTEIN is a species of ostracods occurring in the Santonian of Israel which shows interesting polymorphic features that have been studied in detail by various biometrical methods by Abe *et al.* (1988). In the present connexion, some of those data are analysed to see whether the procedures reviewed in the present section can be useful for picking out divergences in the material. You will recall that we have already noted that ecochemical factors seem to play an important part in deciding the morphological manifestations of this species (pp. 134, 139). The characters measured on the specimens are illustrated in Fig. 4.11.

The analysis

A. *The Number of Variables*: The latent roots and vectors of the correlation matrix for 31 specimens of our species are listed in Table 8.1. We note that all variables are equally weighted in the first vector, apart from the third element, which might therefore be a candidate for deletion. The same table also contains the residual sums of squares when the 31 points on the first principal

TABLE 8.1. *Principal components of the correlation matrix for* 31 *specimens of the Santonian (Cretaceous) ostracod species* Veenia fawwarensis *on which seven distance measures of the carapace are available*

Latent vectors	1	2	3	4	5	6	7
Var. 1	0.4348	0.0780	−0.2912	0.1444	−0.2912	0.7786	0.0909
Var. 2	0.3746	−0.1041	0.4618	−0.6744	0.3445	0.2369	−0.0774
Var. 3	0.1543	−0.8941	0.1802	0.2331	−0.2364	−0.0397	−0.1805
Var. 4	0.4356	0.2582	−0.1513	0.0037	−0.1585	−0.2945	−0.7803
Var. 5	0.4443	0.0351	−0.0521	−0.2610	−0.4743	−0.4631	0.5396
Var. 6	0.3912	−0.1948	−0.4915	0.1780	0.6867	−0.1798	0.1785
Var. 7	0.3266	0.2792	0.6343	0.6084	0.1381	−0.0513	0.1458
Latent roots	4.4584	1.0455	0.7206	0.3986	0.1803	0.1080	0.0886

Size of the component spaces being compared: residual sums of squares

Variable deleted	PC1	PC1 + PC2	PC1 + PC2 + PC3
1	2.6072	2.6957	2.9646
2	2.6617	2.8959	6.0464
3	0.7044	39.9630	26.1433
4	2.8077	3.2786	2.7050
5	2.3767	2.3764	2.3894
6	2.8366	3.7992	5.6841
7	2.3889	4.3324	11.0116

Choosing number of components to keep

Number of components	test statistic
0	0.00
1	5.13
2	0.40
3	0.87
4	0.39
5	0.11
6	0.03

In this table, we try to estimate the number of statistically useful principal components. The **rule** is that values much less than unity are not likely to be associated with much information. In the present case, observe that the first component is > 5, hence significant. The second is not significant. The third component falls just short of 0.9 and would usually by taken as acceptable. The conclusion is then that there are three significant principal components.

component are matched successively by Procrustes Analysis, deleting each variable in turn. Thus, column 1 signifies that the successive deletion of each variable gives a relatively small residual sum of squares for variable 3. Hence, this variable has least effect on the first principal component. The second column displays the situation in the first two planes, and the third column, that pertaining for the first three planes. All sums of squares are small for column 2, except for the entry corresponding to variable 3, as before. If you continue to the column for the first three planes of principal components, you will perceive that variable 7 now enters the fray. The conclusion I draw from all of this is that it would not be wise to delete variable 3 from the analysis on the grounds of its mediocre performance in the first principal component.

B. *The Correct Number of Principal Components*: We turn now to the question of ascertaining the correct number of principal components.

TABLE 8.2. *Maximum angles for principal component planes: for sample of* Veenia fawwarensis *from Shiloah, Israel*

Specimen deleted	1	2	3	−3	−2	−1
1	0.56	2.55	5.81	5.97	8.91	16.48
2	0.75	2.20	2.08	2.06	4.42	13.34
3	0.95	6.52	5.81	6.26	7.72	7.75
4	1.39	3.05	2.41	2.43	7.72	9.75
5	1.57	**17.19**	7.55	5.40	6.59	10.99
6	1.63	8.99	6.54	6.70	15.63	0.57
7	0.91	1.27	1.84	2.75	6.26	10.87
8	0.40	1.31	1.92	2.09	3.68	14.67
9	1.80	2.29	2.31	5.29	5.30	5.92
10	0.08	5.88	10.48	9.11	10.93	29.14
11	1.62	1.86	2.56	2.89	15.70	14.94
12	2.62	**17.82**	2.76	1.46	1.25	1.66
13	7.36	**18.77**	9.68	4.70	4.89	12.56
14	0.39	3.13	1.78	1.77	3.08	15.35
15	2.80	6.27	6.32	7.24	7.53	4.31
16	0.66	0.66	4.55	7.05	6.53	13.94
17	0.14	5.62	6.89	6.69	6.24	6.30
18	2.42	4.25	2.06	1.07	1.00	1.11
19	0.75	1.42	1.22	0.67	0.67	1.50
20	0.77	2.39	2.47	4.48	8.73	17.39
21	1.24	1.36	9.42	3.21	2.22	0.49
22	1.68	1.89	2.49	2.77	3.12	10.25
23	1.46	1.64	1.58	0.64	0.63	1.04
24	1.03	3.20	4.81	4.31	6.74	8.94
25	0.14	0.14	0.25	0.42	1.94	4.34
26	0.41	2.52	3.68	3.60	3.78	33.95
27	2.04	4.10	4.04	2.27	2.18	4.03
28	2.59	3.93	2.71	2.54	4.85	2.53
29	0.35	0.41	1.03	0.97	0.57	0.27
30	2.18	7.23	1.09	1.09	1.10	2.52
31	0.66	1.81	1.80	2.79	2.53	2.69

Krzanowski (1987b) advocates the use of a criterion W_m, computed as outlined in Section 3.6.7. This aspires to providing an estimate of the number of statistically informative principal components, the rule being that a value very much less than unity is not liable to be associated with much information. In the present analysis, the test statistic gave 5.129 for one component and 0.403 for two components, 0.864 for three components. One could conclude that three principal components are likely to be significant. The reason for the low value sandwiched between two higher ones lies with the fact that several of the latent roots do not differ greatly from each other.

C. *Use of Critical Angles*: The third topic of interest concerns the identification of atypical values by computing **critical angles**. Table 8.2 lists the largest critical angle between the plane defined by the first two principal components computed from the full sample and the planes defined by the first two principal components on deleting each sample member in turn. The information required for being able to do this is listed in Table 8.2 under the column headed with a "2". For the column headed with a "3", we are concerned with the largest of three critical angles between the three-dimensional spaces defined by the first three principal components computed from the entire sample, deleting each observation in turn, though with subsequent replacement. The highest values in these columns betoken individuals, the omission of which causes the greatest disturbances in the principal component analysis. These individuals are outliers of location or dispersion. The columns headed by negative numerals refer to the **smallest** principal components. Thus, the column bearing the heading "-2" defines the plane of the two smallest principal components and, analogously, for the column headed with a "-3". The smallest principal components usually signify outliers produced by correlations.

In the present case history, you will observe that in the plane of the first two principal components, specimens 5, 12 and 13 deviate from the main body of the sample. These deviations seem to be connected with multidimensional polymorphism (cf. Abe *et al.*, 1988). These ostracod specimens were not exposed by the method of M-estimation (see previous section for what M-estimation is). Directing now attention to the column headed by "-2", you will note that specimens 6, 10 and 11 deviate from the rest of the material. This suggests to me the possibility of a subtle kind of polymorphism occurring in the carapaces of the ostracods, presumably arising from interconnexions between the variables.

Discussion

It can, no doubt, be argued that some outline-registering technique, such as eigenshapes, should be able to disclose the presence of polymorphism in shape in a sample and this may well be for some situations. However, where the variability in shape is continuous, such as in ecophenotypy of shape (e.g. such as results from salinity fluctuations controlling the posterior development of the carapace of some crustaceans), the inspection of computer-drawn images is not likely to lead to useful results for the identification of subtly atypical values.

The procedures united into the methodology employed in the present example require a good deal of programming. The explicit computing details required are given in Krzanowski's papers in which trial data are supplied. Despite the effort required, I think it is well worth the expenditure of a little time to master the method as its application to morphometric data can often yield biologically useful results that do not seem to accrue from competing procedures. The onus of interpretation lies with the palaeontologist, naturally enough, and if influential values can be proven to occur in a data set, the recommended procedure is to go back to the specimens highlighted by the analysis and examine them for properties that could well have been overlooked in the beginning.

TABLE 8.3. *Robust computations for three chemical variables —B, V, Cr in a sedimentary sequence*

Sample No.	Var.	Identity of deviating observations					
1	B	6	22	32			
	V	20	26				
	Cr	2	15	16			
	Usual correlations:	1.000	−0.219	−0.226	1.000	−0.069	1.000
	Robust correlations:	1.000	−0.283	−0.134	1.000	0.054	1.000
	Usual means	35.76	65.50	19.69			
	Robust means	36.05	65.44	15.66			
2	B	2	9	10	19	20	
	V	10	18				
	Cr	—					
	Usual correlations:	1.000	0.605	0.019	1.000	0.072	1.000
	Robust correlations:	1.000	0.582	0.024	1.000	0.081	1.000
	Usual means:	37.94	197.5	11.85			
	Robust means:	37.84	195.7	11.86			
3	B	2					
	V	5					
	Cr	2					

No substantial difference between usual and robust means and correlations.

4	B	—					
	V	4					
	Cr	—					

No substantial differences between usual and robust means and correlations.

Example 8.4. A Palaeoecological Example Using Geochemical Data

To round off this section we shall take a look at another type of data in order to gain a feeling for a more extreme situation from the point of view of a sample displaying many aberrations. The data considered here consist of geochemical determinations on four samples of sediment. We have met these data before: they consist of analyses of vanadium, chromium and boron on samples of Lutetian age from the Rio Isabena in northern Spain.

The sample sizes are 32, 20, 8 and 7 and the analyses were performed in the laboratories of ELF-SPAEF, Pau, France in conjunction with a project aimed at mapping salinity fluctuations in stratigraphical sequences. The effects of robust estimation procedures are more pronounced here than for morphological variables. The results of the analysis are summarized in Table 8.3.

The general impression conveyed by the analysis is that the largest samples are the ones showing the greatest proportion of deviations, doubtlessly the outcome of the comparative procedure underlying the methods of estimation used in robust techniques. The very small samples are too vaguely defined for observations to show up as being atypical. This is an important practical point to bear in mind and implies a caution with respect to sample size. The lesson to be learned here is that robust analysis should be reserved for fairly large samples if any reliance is to be conceded to the results.

The values of the usual and robust correlations do not differ markedly from each other and you might wonder about this. On first sight, the data do not seem to be compositional in nature, since they are not expressed as proportions of some total chemical composition. But danger lurks in the background in the form of the effects of closure which was imposed by the use of equal sample weights for the analyses. I have not tried to adjust for this in the calculations.

8.5 The Statistical Recognition of Moult Stages

Introduction

The moults produced by some arthropods can form a conspicuous constituent in samples of fossils. The quantitative study of instars can often yield useful information on the history of a sediment, its palaeoecological background and also cast light on the palaeobiology of the organisms concerned and their ecological history. The multivariate statistical analysis of crustacean moults can supply indirect evidence of the chemical background of the palaeo-environment (Reyment *et al.*, 1988). I give here briefly two case histories, one for eurypterids and the second for ostracods. These examples illustrate differences in the approach to the analysis of moulted carapaces dictated by the nature of the fossil organism.

1. *Growth stages in a eurypterid*

Presentation of the Problem: Andrews *et al.* (1977) were concerned with identifying moult stages in eurypterids from New York State, U.S.A. Numerous distance traits were measured on the head of *Eurypterus remipes*, which were then subjected to univariate analysis and simple methods of multivariate analysis. Growth stages can often be rather easily recognized in crustaceans by the simple inspection of scatter diagrams. However, in eurypterids, this is not an easy thing to do since there may be considerable overlap between the growth classes. Living *Limulus* is a relevant example of an arthropod in which the earlier stages of growth intergrade but the later moult-stages and adults of which tend to form discontinuous clusters. In such organisms, the identification of moult stages may have to be done on subjective grounds.

Multivariate Recognition of Moults: The characters measured on the head of *Eurypterus* are illustrated in Fig. 8.2. The delineation into moulting classes was originally based on the distribution of the length of the prosoma, denoted

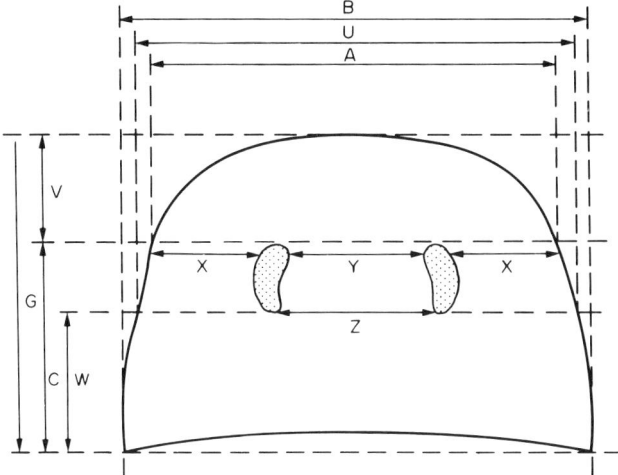

FIG. 8.2. Variables measured on the head of *Eurypterus remipes* from the Silurian of
New York State. After Andrews *et al.* (1977).

G in Fig. 8.2. These moulting classes are mutually exclusive and there is no
overlap between adjacent stages. When examined with respect to the width of
the prosoma for *Eurypterus*, the adjacent moulting classes overlap to the
extent of 11% of the total number of individuals.

In the present connexion, it was thought useful to examine the sequence of
moults in terms of misallocations of specimens, which is in a way a determinant of overlap. A serviceable procedure for doing this is offered by the
technique of stepwise discrimination which can be defined by saying that the
variables are entered into the discriminant analysis in a stepwise manner. The
way in which this technique works is that the first variable maximizes the
between-groups variance with respect to the within-groups variance in the
form of a one-way ANOVA. The second variable entered is the one that
maximizes the between-groups differences, acting in conjunction with the first
variable. The process is continued until all variables that make a significant
contribution to between-group differences have been entered. The variance
ratio is computed for the differences between all groups. An *F*-ratio array with
constant degrees of freedom reflecting the distances between groups is constructed for all possible pairs of groups. Once this has been done, the first
group is tested against all other groups, pooled. Then the second group is
similarly tested, and so on. Finally, each specimen is placed in the group with
which it has the highest probability of membership. Seber (1984, section 6.3),
reviews these procedures.

A further useful multivariate procedure, which was introduced in Chapter
3, is to plot the canonical variate scores for the first two canonical variables.
This plot will often be successful in illustrating graphically the relationships

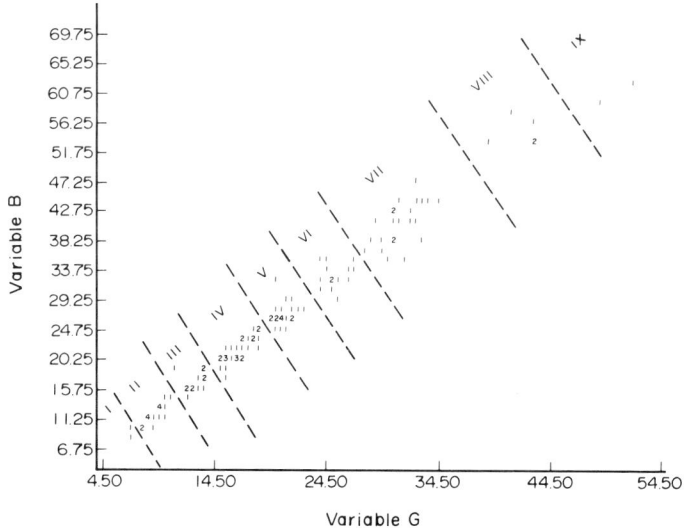

FIG. 8.3. Graph of the prosomal length (*A*) and posterior width of the head (*B*) for *Eurypterus remipes*. The moult stages are indicated, arbitrarily, by dashed lines. After Andrews *et al.* (1977).

between the various groups and individuals. I have already warned you about the wide variability occurring with such scores.

Findings: The simple bivariate scatter between prosomal length (*G*) and prosomal width (*B*) is depicted in Fig. 8.3 in which nine growth stages can be separated out, though not without an element of subjectivity. The graph of the scores on the first canonical variate is shown in Fig. 8.4 for just the two variables *G* and *B*. An increase in the number of variables in the stepwise analysis added no essential information over what was yielded by the bivariate

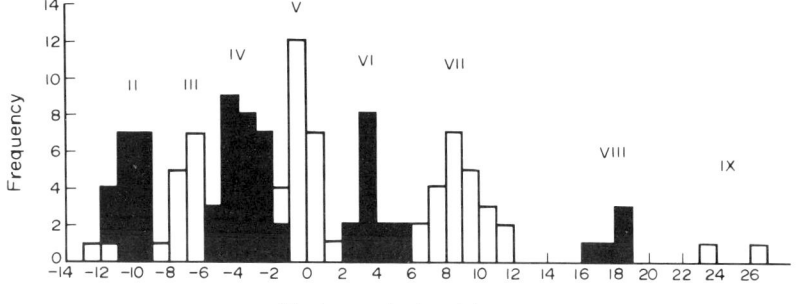

FIG. 8.4. Histogram of the scores on the first canonical vector for *Eurypterus remipes* for the variables anterior and posterior lengths of the prosoma. The moult stages are denoted by Roman numerals. Redrawn from Andrews *et al.* (1977).

TABLE 8.4. *Canonical variables for* Eurypterus remipes

Variable		Canonical variate coefficients		Order of stepwise entry
		I	II	
A:	prosomal width	−0.36	−0.41	5
B:	posterior width of prosoma	0.35	0.59	2
C:	posterior margin to eye	−0.21	0.05	4
G:	prosoma-length	0.96	−0.11	1
U:	width posterior to eyes	0.12	−0.13	9
V:	anterior margin to eye	−0.12	−0.98	7
W:	posterior to eye-posterior	0.14	0.71	8
X:	eyes to lateral margin	−0.13	1.31	6
Y:	anterior distance between eyes	0.24	−0.19	3
Z:	posterior distance between eyes	−0.22	−0.44	10
Percentage of trace		98.0	1.2	
Colgate University collection: $N = 77$.				

study. The principal features of the analysis are displayed in Table 8.4, which presents the canonical variate coefficients and the order in which they were inserted into the stepwise discrimination in relation to discriminatory ability.

Discussion: Andrews *et al.* (1977) concluded that in both *Eurypterus* and *Limulus*, moult stages are not discrete and that their identification is best accomplished by a trial-and-error procedure. Initially, the identification of growth stages was based on the length of the prosoma, which yielded nine classes for both genera, though not unequivocally. When prosomal width was added to the analysis, it was found that the *ad hoc* growth stages began to overlap, thus diminishing the separation already achieved. The difficulty in obtaining a "neat" multivariate analysis in highly coordinated characters such as those considered here has already been remarked upon for the crab data analysed in several places earlier on in this text. There is virtually no variation left over after the extraction of the first latent root in a principal component analysis of such animals. Exercise 8.10 illustrates this variational rigidity for the eurypterid *Euproops* contrasted with *Limulus* (data extracted from Bergström, 1975).

We have also been made aware of the fact that redundant directions of variation in a multivariate analysis can complicate interpretations. Growth in size in relation to moulting class in *Eurypterus* was geometric as is also the case in the king crab. Every time the volume of the body doubled, a moult took place.

The animals studied in the present example tend to be more difficult to deal with than ostracods. But even with ostracods, it can be troublesome at times to define exact boundaries between ontogenetic classes and special statistical techniques may have to be called into play. Such a technique will now be considered.

2. Growth Stages in Ostracods

Presentation of the Problem: In the previous example, I made a point of mentioning the more manageable nature of most ontogenetic material for ostracods. I said there was a tendency for discrete groups to be formed and that it was usually reasonably easy to assign a specimen to its correct moulting class. I have selected now an interesting example in order to cast light on some of these ideas. The data consist of measurements on the length and height of the carapace of the Nigerian Paleocene species *Cytherella sylvesterbradleyi* REYMENT. Abundant material containing shells from stages 2 to 8, inclusive, was available for analysis. The fact that these moulted shells were recovered from the same sample indicates that the material represents occurrences *in situ*.

The Method of Analysis: Granted that the shells could be grouped quite readily on univariate measures, the study of the relationships between the growth classes became an ideal vehicle for canonical variate analysis. I shall begin by accounting for a canonical variate study on seven moulting classes and adults (growth stage 8 probably constitutes adult males and stage 7, adult females). This may seem to be an example of overkill to use a powerful method on just two variables, but my aim is not to represent relationships parsimoniously in fewer dimensions, but to provide a graph of the ontogenetical sequence.

Consideration of the Model: Broadbent (1955) developed a theory of quantum statistics for animals that grow in discrete steps, such as crustaceans. The hypothesis is that the means of the n stages will be equally spaced at intervals of

$$\beta = 2r\delta \quad (r = 0, 1, ..., n).$$

The goal of Broadbent's approach is to estimate δ, the quantum, for the logarithm of the mode of a diagnostic measure of size. I did this (Reyment, 1963a) for *Cytherella sylvesterbradleyi* and found a constant value of $\delta = 0.055$. Now, the present example is concerned with the multivariate distances between successive moulting classes. It is *not* a straight generalization of Broadbent's method in that the distances used are for means and not modes.

The Analysis: The only parts of the canonical analysis of interest in the present instance are the canonical variate means and the generalized statistical distances between successive samples (i.e. moulting classes). The canonical variate means and Mahalanobis generalized distances between successive samples for logarithmically transformed observations are listed in Table 8.5. The means are plotted in Fig. 8.5. It is quite obvious that there is an almost linear relationship between the seven points A (moulting class 2) to G (moulting class 8). There is a reasonably stable distance between successive growth stages, apart from the distance between stages 7 and 8.

TABLE 8.5. *Canonical variate means and pertinent generalized distances for the data on* Cytherella sylvesterbradleyi

Growth stage	N	Mean 1	Mean 2	D_{ij}^2
2	14	5.88	1.02	
3	32	6.13	1.46	0.74
4	15	6.89	1.83	0.72
5	35	7.80	2.23	0.99
6	11	8.61	2.61	0.80
7	18	9.20	2.98	0.49
8	13	12.25	3.52	1.40

Discussion: The graph in Fig. 8.5. yields the direct distances 3.7, 2.9, 3.4, 3.4, 3.2 and 4.5. The Mahalanobis distances computed between successive moulting classes vary more and clearly reflect the differences in variances and covariances that were found for the samples. In both cases, the distance to the final sample deviates quite strongly from preceding samples. This was not apparent in the original univariate quantum-analysis (Reyment, 1963a, Fig. 13). The present study provides useful evidence for the suggestion that stage 8 is actually composed of male shells whereas stage 7 consists of adult female

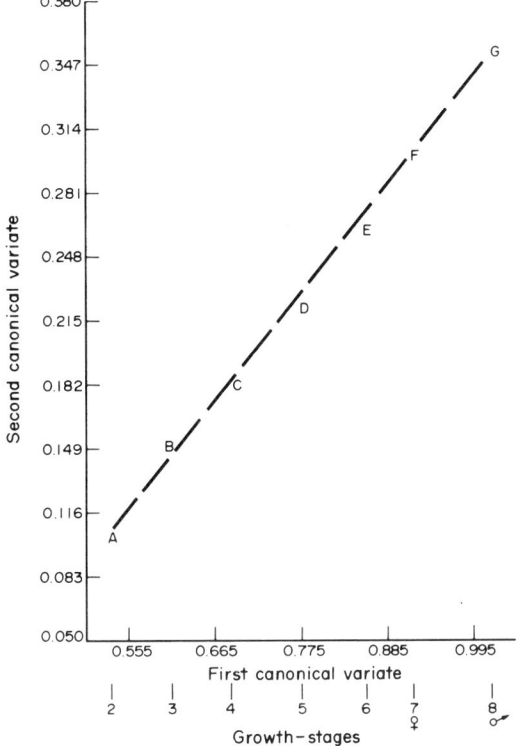

FIG. 8.5. Quantum plot for *Cytherella sylvesterbradleyi* showing means against moulting class.

shells. The usual situation in ostracods is that the female moulting dimensions of the carapace are a linear continuation of the larval proportions, whereas males have a disproportionately exaggerated increase in length; this tends to displace them to the side of the linear relationship. In *Cytherella*, however, males seem to constitute a linear continuation of the sequence formed by larvae and adult females. This interesting situation would no doubt profit from a penetrating investigation of growth in this genus.

The graph of the canonical variate means does not differ in any extraordinary manner from a graph of length and height for the shells and I chose the present rendition in order to highlight the sexual dimorphic size relationship that was hidden in the measurements. It is not really obvious from the plot of the canonical variate means that the eighth growth stage deviates from the linear relationship established for the sequence (the line fitted to the graph was located by "eye-balling"). It was first when the Mahalanobis generalized distances were called into play that the dimorphism was unveiled.

8.6. Discriminant Analysis by Shrunken Estimators: a Cretaceous Foraminifer

Presentation of the Problem: In Chapter 3 dealing with Multivariate Statistical Analysis, we had occasion to scrutinize the question of stability in the latent vectors of canonical variate analysis. There is little doubt in my mind that many morphometrical reifications would profit from careful consideration of the stability of vectorial elements. In the example dealing with stepwise discrimination for *Eurypterus*, in which a standard package program was used, there was no opening for a test of stability and it cannot be ruled out that the order in which the variables were drawn into the analysis could have been influenced by instability in the discriminator.

We shall glance more deeply at the problem that was taken up by Campbell and Reyment (1978). It might be useful if you looked at the relevant section in Chapter 3 again, including the worked example for the Rock Crabs. The species I have selected for study is the by-now-familiar benthic foraminifer *Afrobolivina afra*. This "fruit-fly" of multidimensional palaeobiology is eminently useful for illustrating a wide range of biometrical topics. The data have already been introduced on p. 189.

We learned in Chapter 3 how interpreting canonical variate coefficients (i.e. the elements of the canonical vectors) is a moot point and we should have accepted by now that it is indeed difficult, if not misleading, to attempt to reify the canonical vectors and in particular assign biological significance to them. Rohlf *et al.* (in prep.) have demonstrated that this endeavour cannot hope to succeed owing to the very mathematical nature of the operations that produced the vectors.

Campbell and Reyment (1978) and Campbell (1979) pointed out that one of the properties attendant on high correlations in regression analysis is concerned with the change in sign of coefficients which result from slight changes in the values of ridge constants, where ridge-regressional procedures are employed (Marquardt and Snee, 1975). This is very noticeable in those

TABLE 8.6. *Means for* 46 *samples of* Afrobolivina afra *REYMENT from the borehole at Gbekebo* (N = 997). *To convert to millimetres, divide by* 66

Depth in feet	N	Variables								
		1	2	3	4	5	6	7	8	9
3343	24	70.6	35.5	24.8	14.3	13.2	8.4	19.1	4.5	11.8
3317	12	75.3	32.5	23.3	14.4	13.6	9.1	18.2	4.0	9.9
3313	30	53.3	24.9	18.9	13.3	12.1	5.8	16.2	3.5	8.2
3301	28	86.9	33.2	26.4	15.8	15.1	8.5	20.4	3.9	10.1
3298	62	69.2	31.0	24.8	16.1	14.5	7.9	20.0	3.8	10.2
3290	31	78.7	35.5	27.9	18.7	17.7	11.6	23.1	4.3	11.2
3286	16	61.0	30.9	24.2	15.2	15.2	9.6	20.0	4.1	10.3
3283	16	78.0	30.5	24.4	15.7	14.2	9.5	20.0	3.8	9.9
3274	16	73.5	31.7	25.3	16.0	15.4	11.1	21.6	4.3	9.9
3271	15	77.9	36.5	30.8	20.9	19.1	9.9	28.6	6.1	12.0
3265	16	79.3	34.5	27.9	20.0	19.3	10.1	26.5	4.9	10.1
3253	16	65.2	32.6	25.5	17.3	16.4	9.0	24.0	4.3	10.6
3241	16	62.4	29.1	23.1	18.1	14.7	8.4	20.2	3.5	8.6
3239	12	72.5	29.2	24.1	16.4	15.7	8.3	22.5	4.0	8.5
3227	15	69.9	27.1	23.3	16.9	15.1	8.9	20.8	3.8	7.9
3221	12	68.8	27.7	23.3	15.9	16.0	7.7	21.7	3.9	9.4
3212	11	81.0	34.7	27.9	20.0	17.4	9.7	23.4	4.6	10.3
3209	26	114.3	39.9	33.7	22.7	21.3	11.3	30.8	4.3	11.4
3206	13	75.6	34.6	28.5	21.0	19.7	10.8	24.5	4.0	10.1
3200	26	72.8	31.6	26.0	20.3	18.9	11.0	22.6	3.8	9.5
3197	12	89.2	40.1	31.9	22.1	20.9	12.9	29.7	3.8	12.0
3194	39	74.6	32.1	27.5	20.0	18.5	10.3	24.7	3.9	10.1
3189	13	89.9	34.2	28.8	20.8	18.9	9.4	26.6	4.3	10.0
3183	18	69.1	32.6	26.6	20.5	18.4	9.2	23.3	3.9	9.5
3180	30	68.2	34.1	28.0	22.0	19.4	10.9	23.3	3.9	9.7
3178	17	81.4	34.2	28.6	20.8	19.5	8.2	26.0	3.5	9.5
3174	21	76.9	31.7	27.5	20.5	19.3	9.1	24.7	3.4	9.8
3166	42	77.4	30.5	27.0	21.1	18.2	8.8	24.7	3.5	9.0
3163	31	76.5	30.7	27.1	21.1	19.0	8.9	24.2	3.6	9.4
3160	28	62.1	28.8	24.4	18.8	17.5	9.1	22.4	3.6	9.4
3157	34	85.3	34.7	28.7	20.8	19.9	11.3	27.3	3.7	10.7
3154	13	70.5	30.5	25.8	18.7	18.0	9.0	23.4	3.7	8.8
3149	17	81.1	33.5	28.2	20.8	20.7	10.4	28.2	4.2	11.1
3141	14	76.6	32.8	26.0	19.3	18.6	8.3	24.7	4.0	10.0
3138	14	66.0	30.2	25.4	18.9	17.2	8.0	23.1	4.1	9.3
3132	22	85.3	33.4	28.5	21.1	19.2	7.2	25.5	3.8	9.3
3126	12	78.6	30.5	26.0	20.5	19.0	8.6	23.5	3.3	8.6
2135	14	53.8	28.6	22.9	17.9	16.3	8.7	20.1	3.7	8.4
3112	16	64.7	28.7	22.3	18.5	16.8	7.2	20.8	3.3	8.4
3100	13	67.0	28.8	24.0	19.1	17.3	7.5	23.0	3.9	9.0
3097	14	67.4	30.4	24.5	20.3	17.2	8.0	23.7	3.7	10.7
3084	15	112.9	39.8	34.7	26.1	21.5	11.8	32.2	3.8	13.0
3075	29	113.8	40.0	33.4	26.1	23.5	9.2	33.0	4.0	13.1
3068	31	99.0	36.5	30.7	22.9	21.5	8.1	29.9	4.0	11.3
3065	15	74.5	32.2	26.6	20.7	19.1	10.1	26.9	4.0	10.1
3062	60	113.1	39.8	32.4	25.2	22.8	10.0	31.6	4.0	12.5

The characters 1 to 9 are illustrated in Fig. 5.1.

TABLE 8.7. *Correlation matrix computed from matrix* **W**, *and the pooled standard deviations. Foraminiferal data* (N = 997)

	1	2	3	4	5	6	7	8	9
1	1.00	0.61	0.69	0.52	0.53	−0.02	0.47	0.16	0.19
2	—	1.00	0.82	0.63	0.66	0.33	0.54	0.27	0.33
3	—	—	1.00	0.67	0.64	0.23	0.54	0.25	0.27
4	—	—	—	1.00	0.57	0.22	0.43	0.21	0.23
5	—	—	—	—	1.00	0.19	0.42	0.20	0.22
6	—	—	—	—	—	1.00	0.19	0.12	0.19
7	—	—	—	—	—	—	1.00	0.47	0.46
8	—	—	—	—	—	—	—	1.00	0.31
9	—	—	—	—	—	—	—	—	1.00
σ	19.500	4.991	4.545	3.665	3.288	2.526	3.474	0.636	2.179

directions of the regressor space associated with low variance and low regression sums of squares. Campbell (1979) illustrated that the instability inherent in regression carries over to the similarly constructed discriminant function and thence to canonical variates. The difficulties attaching to interpretation of regression coefficients in such situations are also to be found in many other fields of multivariate analysis, to wit, canonical variates, as already intimated, canonical correlation, and principal components.

The characters analysed on *Afrobolivina* fall naturally into two categories: those describing variation in the dimensions of the chambers (variables 1 through 7) and those registering variation in the location and width of the foramen (variables 8 and 9). Table 8.6 lists the means for each of the 46 borehole samples providing the data. Table 8.7 contains the pooled correlation matrix and the pooled standard deviations for these samples.

Statistical Methods: The method of canonical variates is a suitable multivariate vehicle for attacking the problem posed by the simultaneous analysis of many sampling levels in a stratigraphical succession, such as represented by a borehole, as well as for dissecting their variational patterns, each of which possesses its special interest and which could expose time-correlated differences. In Chapter 3 we have seen how canonical variate analysis can be viewed as a two-stage rotational procedure. Firstly one rotates two orthogonal variables, the principal components of the pooled samples. The second rotation corresponds to a principal component analysis of the group-means in the space of the orthogonal variables. We have already learnt that it can often be convenient to transform the within-groups ellipsoid of concentration into a concentration circle by scaling each latent vector by the square root of the corresponding latent root in the first phase of the computational procedure.

Also in Chapter 3, I pointed out that when there is but little variation between groups along some particular direction, and the corresponding latent root for the orthogonalized variables is also small, marked instability can be expected to show up in some of the coefficients of the canonical variates. A functional solution to this predicament is to add "ridge-type constants" to each latent root before this value is used to standardize the corresponding principal component. As a general rule, it has been found that when the between-groups sum-of-squares for a particular principal component is small, say, less than 5% of the total between-groups variation, and the corresponding latent root is also small, say less than 1–2%, then shrinking of the principal component will probably be beneficial.

Empirical experience shows that although some of the coefficients of the canonical vectors corresponding to canonical variates of interest may change magnitude, and often sign, shrinkage has little effect on the corresponding canonical roots. This indicates that little discriminatory information has been lost. When you have this situation, an obvious conclusion is that one or more of the variables contributing most to the principal component that has been shrunk have slight influence on discriminatory power: these variables are *redundant*. In addition, variables with small standardized canonical variate coefficients can be eliminated. In the ensuing case history, redundancy is one of the problems that confronts us. Although not specifically concerned with the question considered in this section, I think you would find it useful at this juncture to browse through two books I like very much and which I think contain much of general value of applicability in the present connexion. These are Draper and Smith (1966) and Flury (1988). The effort will be rewarding if you are not really at home with all the statistical concepts introduced here. Moreover, they are well written and hence easy to follow.

The Analysis: The latent roots and vectors of the within-groups covariance matrix **W** for the foraminiferal data are listed in Table 8.8. The mean vectors for all levels are listed in Table 8.6. The correlation matrix derived from **W** is given in Table 8.7. We observe that the latent vector corresponding to the *smallest* latent root, which embodies no more than 1.8% of the variation within groups, reflects an association between characters 2 and 4. This makes

TABLE 8.8. *Latent roots and vectors of the within-groups correlation matrix W^* for all nine variables and the between-groups sums of squares for each component*

Latent roots	1	2	3	4	5	6	7	8	9
	4.31	1.25	1.00	0.69	0.48	0.43	0.38	0.29	0.16
Latent vectors	1	2	3	4	5	6	7	8	9
u_1	0.35	0.43	0.42	0.37	0.37	0.15	0.35	0.21	0.23
u_2	0.33	0.09	0.20	0.19	0.21	−0.30	−0.31	−0.55	−0.53
u_3	0.28	−0.15	−0.03	−0.10	−0.07	−0.85	0.23	0.31	0.09
u_4	−0.08	−0.06	0.03	0.09	0.05	0.13	0.01	0.64	−0.74
u_5	−0.44	−0.05	−0.18	0.44	0.51	−0.25	−0.42	0.17	0.24
u_6	−0.01	−0.10	0.03	0.74	−0.66	−0.02	0.01	−0.03	0.03
u_7	0.27	0.26	0.24	−0.20	−0.28	0.03	−0.73	0.33	0.20
u_8	0.64	−0.48	−0.44	0.14	0.19	0.29	−0.11	0.08	0.10
u_9	−0.10	−0.69	0.70	−0.05	0.08	0.06	−0.04	0.00	0.10
diagG_0	1.01	0.25	0.23	0.27	0.26	0.07	1.57	0.18	0.21
trG_0 =	4.05								

NB: The latent vectors are transposed.

TABLE 8.9. *Standardized canonical vectors for nine variables, including shrunken estimates for* Afrobolivina afra

	1	2	3	4	5	6	7	8	9	f
$\mathbf{a}_1^U$	0.54	−0.11	−0.01	0.11	0.06	−0.07	0.79	−0.18	−0.13	—
$\mathbf{a}_2^U$	−0.64	−0.32	0.14	−0.13	−0.44	−0.11	0.39	0.05	−0.32	—
$\mathbf{c}_1^U$	0.00	−0.59	−0.09	0.43	0.44	0.07	0.99	−0.51	−0.10	2.28
$\mathbf{c}_2^U$	0.65	0.81	−0.27	−0.44	−0.40	0.15	0.04	0.28	0.25	0.76
$\mathbf{c}_{1,0}^{GI}$	0.04	−0.37	−0.31	0.43	0.41	0.06	1.01	−0.51	−0.12	2.25
$\mathbf{c}_{2,0}^{GI}$	0.58	0.29	0.31	−0.53	−0.36	0.25	−0.03	0.37	0.32	0.70
$\mathbf{c}_{1,1/2}^{S}$	0.03	−0.42	−0.26	0.43	0.41	0.06	1.00	−0.51	−0.12	2.25
$\mathbf{c}_{2,1/2}^{S}$	0.60	0.41	0.18	−0.51	−0.37	0.23	−0.02	0.35	0.31	0.71
$\mathbf{c}_{1,0}^{GI} = \mathbf{c}_{1\,(0,\,\ldots,\,\infty)}$										
$\mathbf{c}_{1,1/2}^{S} = \mathbf{c}_{1\,(,\,\ldots,\,0.5)}$										

f denotes canonical roots.

it tempting to suspect that there could be instability in these elements of the canonical vectors.

The between-groups sums of squares for all principal components (see Table 8.8) shows that 39% of the variation is associated with the seventh principal component. The variation embraced by this principal component results from a contrast between the seventh character and three other characters. The coefficients for the first canonical variate, denoted $\mathbf{a}_1^U$ in Table 8.9, expose the dominant part played by the seventh principal component. By way of comparison, we note that the first principal component contributes most to the second canonical variate (see $\mathbf{a}_2^U$ in Table 8.9).

The effect of shrinking the contribution of the smallest latent root and corresponding vector, namely the ninth principal component, is summarized in Table 8.9. Here $k_9 = \infty$ implies the elimination of the ninth principal component from the analysis, whereas $k_9 = 0.5$ signifies that the between-groups contribution from this component is diminished. The two sets of coefficients for the original variables, denoted $\mathbf{c}_i^U$ and with shrinkage, denoted $\mathbf{c}_i^{GI}$ and $\mathbf{c}_i^s$ in Table 8.9 are similar. The main differences are for traits 2 and 3 in the first canonical vector, with a more marked shift for variable 2. In the second vector, variable 2 is affected, as is also variable 3, with a change in sign. The canonical roots are little affected by the elimination of the smallest principal component which expresses the fact that little information has been lost by the shrinkage operation.

The eighth principal component was also deleted. This action was found to bring about little more than was already done by the deletion of the ninth component. Referring to Table 8.8, it will be seen that the corresponding latent root is about twice the magnitude of the ninth root and is above the level at which instability usually arises.

The relatively modest standardized coefficients for traits 1, 3, 6 and, to a lesser degree, 9, and the instability observed, suggest that an analysis of a reduced set of the five variables 2, 4, 5, 7 and 8 might merit consideration. As will be apparent from Table 8.10, 34% of the between-groups variation is

TABLE 8.10. *Analysis for five variables for* Afrobolivina

Latent roots	1	2	3	4	5	
	2.82	0.99	0.46	0.43	0.30	
Latent vectors	Breadth var. 2	Last chamber var. 4	2nd last chamber var. 5	Width var. 7	Foramen var. 8	
$\mathbf{u}_1$	0.51	0.47	0.47	0.45	0.30	
$\mathbf{u}_2$	0.21	0.30	0.33	−0.37	−0.79	
$\mathbf{u}_3$	−0.10	0.33	0.18	−0.76	0.52	
$\mathbf{u}_4$	−0.09	0.72	−0.67	0.11	−0.08	
$\mathbf{u}_5$	0.82	−0.25	−0.43	−0.27	0.07	
$\mathrm{diag}\mathbf{G}_{(0,\ldots,0)}$	1.05	0.34	0.67	0.11	0.11	
$\mathrm{tr}\mathbf{G}_{(0,\ldots,0)} = \quad 3.28$						
						f
$\mathbf{c}_1^U$	−0.63	0.42	0.43	0.95	−0.52	2.25
$\mathbf{c}_2^U$	0.99	−0.42	−0.33	−0.21	0.46	0.63
$\mathbf{c}_{1\,(\infty,0,\ldots,0)}^S$	−1.04	0.39	0.39	0.88	−0.76	1.47

f denotes the canonical roots.

linked to the smallest latent root. Hence, little or no instability in the co-efficients is to be expected. From the standpoint of parsimony, it would be sufficient to measure these five characters, at least, for the purposes of a canonical variate study.

Discussion

The present study points to important practical aspects of stability in canonical variate coefficients. The full analysis showed some of the original nine characters to be redundant and it was this redundancy that abetted the instability in the canonical coefficients. Stability could be achieved by deleting one of the within-groups principal components and also by eliminating redundant variables. If the original study had been confined to these five variables, the question of instability would never have originated and the analysis could have been made by standard methods.

How often do you need to worry about instability of the kind discussed here? I have been particularly observant on this point and can honestly claim that serious errors due to instability in coefficients are not something you can expect to run into with every data set you have. The important thing is to be on your watch for the configuration displayed by the elements of the latent vector corresponding to the smallest principal component of the within-groups covariance matrix. If you have some very large elements in this, then you would do well to tread lightly when carrying out a full canonical variate analysis of your data.

The steps involved in programming the calculations are given in Chapter 3 and more explicitly in Campbell and Reyment (1978). I know of no standard computer package containing the procedures required for doing the computations. The methods outlined in the case history presented here apply equally as well to principal component analysis and to canonical correlation.

TABLE 8.11. *The first five latent roots and vectors for the two covariance matrices of* Veenia fawwarensis

A. Reference matrix, S_1

Latent roots	1	2	3	4	5
	0.00874	0.00076	0.00054	0.00031	0.00028

Latent vectors

Variables	u_1	u_2	u_3	u_4	u_5
1	0.77	0.08	0.48	−0.09	−0.38
2	0.17	−0.33	−0.51	−0.35	−0.03
3	0.05	0.16	−0.50	0.00	−0.67
4	0.24	−0.19	0.05	0.00	0.47
5	0.33	−0.21	−0.26	−0.46	0.25
6	0.38	0.59	−0.42	0.42	0.34
7	0.18	−0.65	−0.12	0.69	−0.10

B. Comparison Matrix S_2

Latent roots	1	2	3	4	5
	0.00816	0.00068	0.00039	0.00020	0.00013

Latent vectors

Variables	u_1	u_2	u_3	u_4	u_5
1	0.78	0.00	−0.09	0.54	0.00
2	0.11	−0.53	0.39	−0.21	−0.20
3	0.11	−0.20	0.52	−0.25	−0.09
4	0.34	−0.03	−0.57	−0.59	−0.45
5	0.20	−0.28	0.20	0.21	−0.34
6	0.45	0.46	0.34	−0.43	0.41
7	0.11	−0.62	−0.30	−0.12	0.68

8.7 Multivariate Distributional Patterns

Example 8.5. Patterns in a Species of Cretaceous Ostracods

Presentation of the Problem: The material to be examined here consists of two time-separated samples of the Santonian ostracod species *Veenia fawwarensis* HONIGSTEIN from Israel. An account of variability in shape in this species has been presented by Abe *et al.* (1988). As was proven in this article, the ornamental development of the species and the variability in outline seem to have been under the direct influence of eco-chemical factors. Let us now consider possible environmental interactions with the seven standard measures of morphological distance observed on the carapace. The characters utilized were illustrated in Fig. 4.11.

Method of Analysis: The material studied in this example comprises two samples with $p = 7$ variables and each comprising 31 observational vectors. The standard tests for univariate skewness and kurtosis gave one significant result, to wit, kurtosis in height of carapace for the first sample: this could, of course, be fortuitous. However, at the multivariate level, quite strong divergences could be established. The statistical procedures useful for analysing the type of problem taken up in this section are:

1. Tests for homogeneity of covariance matrices.

2. Tests for determining the orientations of the axes of the dispersion ellipsoids.

The relevant test for Homogeneity of Covariance Matrices is:

$$2I(H_1:H_2(*)) = N_1 \ln\left(\frac{|\mathbf{S}|}{|\mathbf{S}_1|}\right) + N_2 \ln\left(\frac{|\mathbf{S}|}{|\mathbf{S}_2|}\right) \tag{8.12}$$

where $\mathbf{S}$ is the pooled covariance matrix and the $\mathbf{S}_i$ are the respective covariance matrices for the samples being compared. The statistic 8.12 is approximately distributed as χ^2 with $(p(p+1)/2$ degrees of freedom, where p denotes the number of variables.

The large-sample test procedure for investigating the ellipsoidal axes is based on a solution mapped out by Anderson (1963) and applied to biological data by Reyment (1969a). In its simplest form, the test can be written down as:

$$n(d_i \mathbf{b}_1^T \mathbf{S}_1^{-1} \mathbf{b}_i + \frac{1}{d_i} \mathbf{b}_i^T \mathbf{S}_1 \mathbf{b}_i - 2) \tag{8.13}$$

which is distributed approximately as χ^2 with $p-1$ degrees of freedom. Here $\mathbf{S}_1$ is as for equation (8.12) and the d_i are the latent roots of that matrix; the $\mathbf{b}_i$ are the latent vectors of the second matrix, $\mathbf{S}_2$. The procedure is reviewed in more detail in Reyment (1969a). While on the subject, I should mention that Dempster (1969) has studied the question of geometrical homogeneity in covariance matrices by the "shadow-property" of the ellipsoids of scatter. This is a neat way of attacking the issue and one that I have used from time to time.

The Analysis

The main results obtained for the analysis of the axial orientations are presented in Table 8.12. The results yielded by the application of formula (8.12) gave a value of $\chi^2 = 52$, which for 28 degrees of freedom is highly significant. Thus the tentative conclusion may be proposed that there is heterogeneity in the covariance matrices and hence in the covariance-patterns of the two samples. Recalling that we have already established the existence of deviations from the multivariate Gaussian form, we could actually be confronted by the effects of this rather than any geometrical incompatibility. The

TABLE 8.12. *Test for collinearity of principal axes for the two samples of Veenia fawwarensis*

Vector comparison	Chi-squared value	Degrees of freedom
1	75.44	6
2	13.92	6
3	13.67	6
4	12.54	6
5	33.37	6

$\chi_6^2 = 12.6$ at the 95% level.

values of the calculations using formula (8.13) are summarized in Table 8.12 and the latent roots and vectors for the covariance matrices in Table 8.11.

Turning now to Table 8.11, you will see that the elements of the first latent vectors for both covariance matrices are quite close; likewise for the second latent vector. It is first when we come to subsequent latent vectors that important differences begin to forge to the fore. With chi-squared for six degrees of freedom at 12.6 for the 95% level of confidence, we perceive that all differences in axial orientation, except that for the fourth vector (and the discrepancy is minimal) are significant. We are therefore led to conclude that the ellipsoids of scatter must be rotated in relation to each other. This may well be the main cause of heterogeneity in covariance matrices.

A linear discriminant function was constructed for distinguishing between the two populations. Inasmuch as the covariance matrices are not homogeneous, the standard method of Fisher is not appropriate. The multivariate Behrens–Fisher generalization of Anderson and Bahadur (1962) was used for the computations (FORTRAN program in Blackith and Reyment, 1971). The estimated discriminant function is:

$$-31.2x_1 - 109.4x_2 + 12.1x_3 + 21.8x_4 + 4.7x_5 + 32.8x_6 + 56.4x_7$$

This function discloses that most of the discrimination is being done by variables 2 and 7. Character 5 could be deleted from the analysis without loss of efficiency. Two-thirds of the discrimination can be shown to lie with the unequal covariance matrices and only one-third with the differences in means (see Exercise 5). The generalized distance is 4.25, which as Hotelling's T^2 for 7 and 54 degrees of freedom gives a significant value of the variance ratio.

Discussion

The analysis of the empirical distributions of *Veenia* shows that there are fundamental differences in the multivariate patterns for the seven morphometrical characters measured on the carapace. Inasmuch as these are not connected with any recognizable speciation process in the sampled sequence—younger samples of *Veenia* accord well with older samples—the grounds for the divergences must be sought in the influence of the environment. In the present study, an elevated content of magnesium at the crucial level seems to have been responsible for far-reaching effects on the pattern of variability manifested in the carapace.

The moral of our story, if indeed there be one, is that the statistical results of what may seem to be fundamentally important variability from the standpoint of evolutionary biology, must be viewed with caution. We should also ask just how much of the variability observed in material of the kind encountered in this case history is due to chance. The clear difference in variational patterns disclosed by the multivariate analysis, at several levels, could possibly be the outcome of chance fluctuations in the sample. We shall probe this line of thought further by looking at some data for *Bos*.

The Case of the Bisons

Presentation of the Problem: Another case history dealing with the homogeneity of covariance matrices will now receive consideration as a complement to the foregoing example. Professor Leslie F. Marcus, Department of Biology, Queen's College CUNY, New York, kindly provided me with the two sets of measurements on the metatarsals of bison analysed here. One sample derives from Barrington (18 specimens) and the second from Wretton (15 specimens), England. The reason for including this material here is that the standard procedure for testing homogeneity of covariance matrices does not indicate a significant difference to exist between the two samples, whereas the orientations of the principal axes of the ellipsoids of scatter do seem to point to this possibility.

The Analysis: The univariate tests of skewness and kurtosis do not suggest anything other than normally distributed observations to occur. The value of the statistic expressed by formula (8.12) is 47.08 which for 36 degrees of freedom is not a significant value of chi-squared. If we turn to the values yielded by equation (8.13), we find that five out of a possible total of eight axes have non-zero lengths and all of these deviate significantly from collinearity on comparison. Thus, the ellipsoids of scatter are rotated about all non-zero axes. The values of chi-squared for 7 degrees of freedom are, from the first to the fifth axes, 24.3, 16.9, 32.5, 93.2 and 16.0.

The first three angles between latent vectors of the covariance matrices are approximately $16°$, $19°$ and $40°$. The conclusion suggested by these results is that the ellipsoids of scatter do not differ in degree of inflation but do so in their axial orientations.

Discussion: It is perhaps not immediately obvious what the palaeobiological significance of the present analysis might be. Matters liable to affect variability in the bones of mammals would seem to have an ecological background and it is therefore possible that the situation encountered here touches on populations reared under different climatic and/or nutritional conditions. Reyment (1969a) found similar results for microorganisms samples from advantageous as opposed to disadvantageous environments.

The heterogeneous statistical distance squared was computed to be 72.6 which as T^2 is very highly significant. The two samples of bison bones are therefore greatly different in average size, as examined at the multivariate level.

Osteological data are a fertile ground for multivariate problems in palaeontology and archaeology. I have only scratched the surface of this fascinating field with the present example but for further illustrations from the field of multivariate statistics in archaeology, I refer you to the volume by van Vark and Howells (1984).

8.8 Phenotypic Plasticity in Ammonites

Presentation of the Problem

Reyment and Kennedy (1991) studied phenotypic plasticity on Iranian *Knemiceras*. The following account was excerpted from that report. The systematics of the ammonites has no doubt well deserved the reputation of being "difficult". Symptomatic of this impediment is the quite general observation that the complications in getting things right augment with increasing sample size. My years in ammonite taxonomy proved to me that if I had just a few specimens of each purported form, I had little trouble with the taxonomy. If, on the other hand, I had rich collections from the same horizon, the headaches started. Hence, the more material the investigator has at his disposal, the more heavily does the taxonomic onus weigh.

Reyment (1988) considered phenotypic plasticity in Cretaceous vascoceratids and concluded that subtle ecological factors could play an important role in deciding the shape of the shell of a cephalopod in a manner analogous to what occurs for some species of gastropods (Goodfriend, 1986; Oosteroff, 1977). Swan and Saunders (1987) attempted to relate function and form in late Palaeozoic ammonoids using prinicpal component analysis as a tool. The idea is certainly worth hailing, but the statistical analysis reported by those workers suffers from weaknesses, including redundancy (e.g. character UR in Table 1 of Swan and Saunders, 1987) as well as indeterminancy arising from the use of composite variables in the multivariate calculations (cf. Chapter 3).

How do we go about studying variation in ammonite shells? The aim of the present section is to provide an appraisal of suitable quantitative methods of analysis. Several procedures are shown to be useful and both standard traits as well as measurements on points located around the outline of some diagnostic feature of the shell (in the present example, the outline of the whorl section), are utilized. This gives a way of expressing variation in shape. It has the added advantage that the digitization of the outline can be related to a measure of area and to the length of the perimeter. Both of these quantities can prove useful in some analyses.

The traditional taxonomic treatment of *Knemiceras* is available in Maurice Collignon's (1983) posthumously published monograph on the Albian to Cenomanian molluscan associations of the Kazhdumi Marl, Fars, Kuzhekstan, Iran. According to the criteria used by Collignon, the ammonites appearing in the present case history, 17 specimens in all, would qualify for assignation to seven distinct species, two full genera and a subgenus of one of these. All the fossils come from about the same stratigraphical level in the Hamiran area of Kuzhekstan and are presumably late Albian in age.

Method of Analysis

The possible methods relevant for studying the morphological relationships between specimens, and hence for testing the hypothesis of infraspecific variability, are **R**-mode principal component analysis of the standard measures usually made on ammonite shells, to wit, maximum diameter of the conch, maximum whorl height, maximum whorl breadth, and maximum diameter of the umbilicus, four characters all told. Secondly, Q-mode principal coordinate analysis for examining relationships between individuals on the basis of the standard measures, just mentioned. Finally, the Q-R-method of correspondence analysis, which provides a useful graphical technique for illustrating affinities between variables and individuals of a sample, is available.

All of the above-mentioned methods are described in detail in Jöreskog *et al.* (1976) and appraised in Chapter 3 of the present text. Greenacre (1984) is a valuable source for techniques and lore concerning correspondence analysis with its many applications and ramifications.

The second class of methods of interest are directed towards mapping changes in shape, an elusive concept as we have seen in the foregoing pages. If we are concerned with variation in outline, then suitable methods are those based on digitizing the circumference of the object, such as Fourier-methods, and the "eigenshape" procedure of Lohmann (1983). These methods have been discussed in Chapter 4. If the shape-analysis is to be designed to chart changes caused by the displacement of features in terms of landmarks, relative to some arbitrarily selected reference, such as a baseline, then a suitable methodology is that of Bookstein (1986).

The critical changes in variability in ammonites are often concentrated to the whorl section and it is on this feature that the present analysis is centred.

The Analysis

It is now generally conceded that ammonites display sexual dimorphism of one or more kinds. Reyment (1988) reviewed the main features of this and also noted that other varieties of polymorphism have been observed to occur in some species. The material considered in this example seems to be an example of ecophenotypic variation on a grand scale. We shall begin by taking a look at the results of a standard multivariate analysis and then go on to learn how these fit in with those yielded by the method of eigenshapes. The analysis is largely based on graphical results as interest lies primarily with locating each shell in relation to its fellows.

Principal Coordinates: Firstly we consider the results of the Principal Coordinate Analysis. The percentage residual remaining after the two latent roots have been extracted is 43.37%, and after the removal of three roots, 34.29%. Therefore, the information contained in the first two coordinate vectors can be expected to give a reasonably good representation of the distances between individuals. The plot of the first two coordinate vectors is shown in Fig. 8.6. You will see how the shapes spread out over the whole field without any clear tendency to grouping, apart from an inclination for depressed whorl

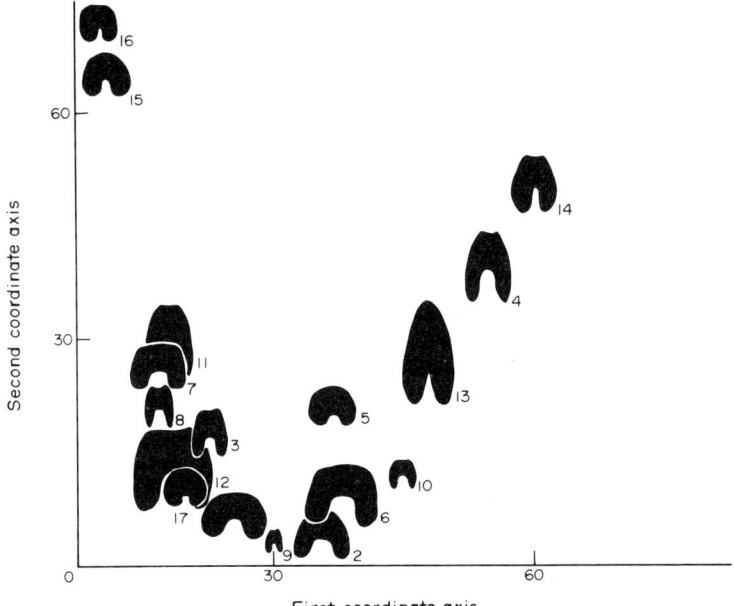

FIG. 8.6. Principal coordinates plot for the silhouettes of whorl sections of *Knemiceras* from Iran. Note that the ordination indicates a largely gradational relationship to pertain. Redrawn from Reyment and Kennedy (1991).

silhouettes to fall in the lower third of the figure. This informs us that the measures of size cannot be well integrated with the outline of the whorl section. That is, there can be no rigid relationship between the parameters determining the degree of evolution (i.e. coiling) of the shell and the shape of the whorls in cross-section.

Constrained Principal Component Analysis: A constrained principal component analysis of the data (see Chapter 3 and Aitchison, 1986) was made. This was done in order to test the possibility of there being a discriminatory effect deriving from variation in the umbilical diameter from specimen to specimen. This was achieved by constraining the four morphological traits to have a constant sum by dividing through each row of the data matrix by the row sums. This standardization showed that there is in fact very little differentiation in growth in the umbilicus. The plot for a *sub-composition* constituted by the diameter of the shell, whorl height, and umbilical diameter is displayed in Fig. 8.7. The appropriate principal axes are fitted to the constellation of points. You will perceive that the values are located near to each other and there are no atypicalities. The plot was produced by means of Aitchison's program CODA (Aitchison, 1986).

Homogeneity of the Data: We now enquire how homogeneous are the data? The foregoing constrained analysis would seem to point to the set of

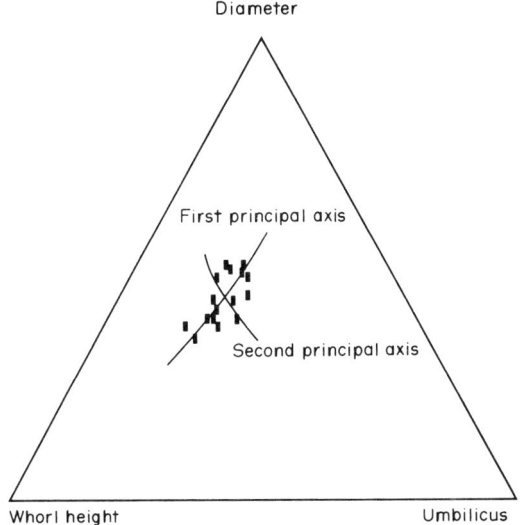

Fig. 8.7. Ternary diagram for the subcomposition of constrained variables maximum diameter, whorl height and umbilical diameter for *Knemiceras*. The first and second principal axes of the points in simplex space are shown.

observations being very homogeneous. We shall now probe this question by a more specifically constructed analysis.

1. The first matter to be taken up is that of Influential Observations (Krzanowski, 1987a, 1987b; Chapter 3). Firstly, the effects of deleting each variable in turn from a principal component analysis were investigated. The least effect was obtained for variable 1, the maximum diameter of the shell, closely followed by variable 3, the whorl breadth. Variable 2 is twice as prominent as variable 3 and the deletion of variable 4 leads to more than twice the effect of eliminating variable 3. Thus, the maximum diameter of the shell is not particularly informative in an analysis based on centred and standardized data.

The successive deletion, with replacement, of specimens gave the result that specimens 7 and 11 have a very strong influence on the outcome of the calculations. Less pronounced, though still appreciable, are the effects caused by removing specimens 5, 8, 11 and 16. A specific test for the number of significant principal components indicates that three should be accepted. According to the criterion used by Krzanowski, the cut-off point for acceptance lies at about unity: here the third root gives a value of 0.9. The reason why three roots are accepted is due to the fact that the second and third latent roots are almost equal. We met this condition in an earlier example.

2. The next step consists of an eigenshape analysis. The results are displayed in Fig. 8.8. The ordination in this figure was produced entirely by the latent-root reduction of a set of coordinates around the peripheries of the silhouettes

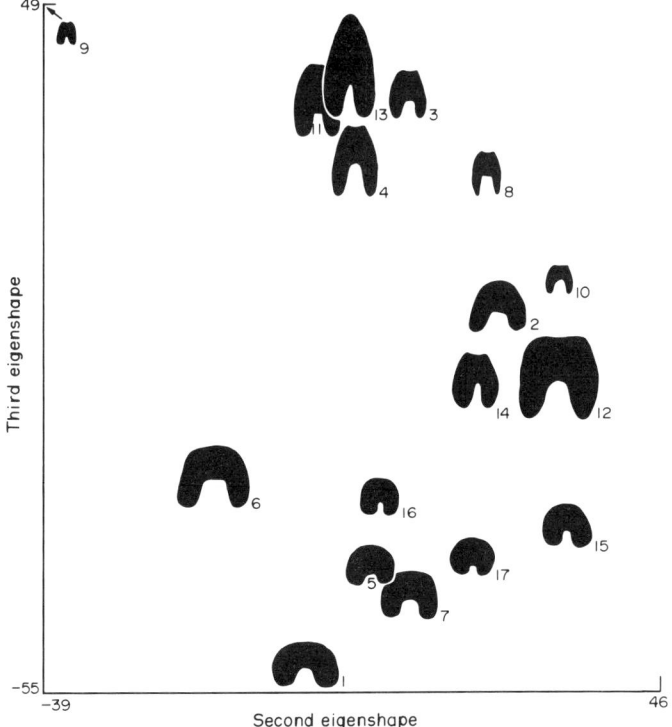

FIG. 8.8. Plot of the second and third eigenshapes for *Knemiceras*. Redrawn from Reyment and Kennedy (1991).

as described in Chapter 4. There is a clear arrangement of the specimens according to the outline of the silhouette of each whorl. The more compressed outlines lie in a bunch in the top of the figure, while squatter shapes form a group located in an intermediate position. The depressed whorl sections are united in the bottom of the graph. This method produced the most informative ordination of the data.

A correspondence analysis was also made. It added little to the picture already obtained and gave a somewhat less satisfactory ordination.

Discussion

It seems reasonable to presume that we are dealing with a single highly variable species and that the Hamiran area was almost certainly close to the place where the animals lived, died and stranded post-mortem. To this attests the abundant presence of encrusting oyster shells on the ammonites. This is analogous to the Tethyan Cretaceous (Turonian) vascoceratids which display an even greater range of morphological variation. The labile environmental conditions that pertained in the vast epicontinental seas of the Cretaceous Period seem to have been responsible for unleashing this variability. The rigidity of the umbilical diameter in the observed variational pattern attests to the likelihood of the hypothesis of infraspecific variation under the control of ecological determinants.

On the methodological plane, the efficiency of ordinating techniques is well demonstrated. The analysis could probably have been expanded so as to encompass more variables, but the results obtained with the present set are sufficient for the purposes at hand.

8.9 Multidimensional Palaeobiogeography

An area of multivariate applications of potential value is that of quantifying palaeobiogeographical analyses. The field is both challenging and fraught with danger. In general, the data must be as good as possible from the statistical point of view if wrong interpretations are to be avoided. I think I have managed to convey by now a feeling for what can be done reliably and what is more conjectural in multivariate analyses in palaeontology. Remember, it is better to desist from a statistical analysis of doubtful material rather than produce, and believe, something wrong or misleading.

In my palaeoecological text (Reyment, 1971a), I accounted for possible areas for quantitative work in palaeobiogeography. Spatial ecology, which is concerned with factors such as climatic influences (largely temperature), was examined by standard methods of univariate analysis. This aspect of ecology is being brought home to us daily as a result of global pollution. The field of palaeobiogeographical analysis by quantitative methods is virtually unexplored. Many statistical treatments of palaeontological data are concerned with following trends through time, usually by means of standard statistical procedures. We have already had ample occasion to regard the joys and pitfalls of such studies and the ever-present risk of drawing false conclusions. Geographically oriented studies are rare largely because of the following reasons:

1. The difficulty of pinpointing in time two or more geographically separated localities.
2. The effects of geological agencies, such as sorting, random dispersal, transport etc. wreak havoc with the reliability of fossil data.
3. Suitable measures for comparing genuine differences between occurrences are often hard to formulate.

Reyment *et al.* (1988) used tensor biometrics for studying temperature-controlled differentiation in a living species of Australian ostracods (already referenced in Chapter 5). In the following I take up a case in which standard multivariate methods were applied to a palaeobiogeographical problem.

TABLE 8.13. *Discriminant function analysis of the palaeobio-geographical data for the Paleocene ostracod species* Anticythereis bopaensis *APOSTOLESCU*

Comparison between Gbekebo (pelitic)/Ilaro (lime-rich)

	Linear discriminant coefficients	Standardized coefficients
Length	2.956	0.108
Height	−80.642	−2.084
Breadth	−3.945	−0.130

This function connects to a D^2 of 4.38, which is highly significant for $F_{3,102} = 26.58$. The probability that the samples are from the same multivariate normal distribution is small on the basis of this result.

The values of the means for these two samples, and for means from Bénin and Libya, substituted into the linear discriminant function are:

Ilaro	29.14
Gbekebo	33.52
Libya	34.00
Bénin	25.27

Gbekebo and Libya are close together in average size, with Ilaro located between them and the vector for the sample from Bénin. Bénin may be even closer to Ilaro than indicated by the discriminant means, as is shown by the following values for the Mahalanobis angle.

Ilaro/Libya	2.29°
Ilaro/Bénin	0.60°

Example 8.6. Palaeobiogeographical Differentiation in a Species of Paleocene Ostracods

Presentation of the Problem: Reyment (1966b) noted that the Paleocene ostracod species *Anticythereis bopaensis* APOSTOLESCU occurs over a wide stretch of North and West Africa, from Nigeria and the Republic of Bénin (formerly Dahomey) to Libya. We ask whether size differences exist at biostratigraphically well-known localities. For the purposes of this brief account, three localities were selected for study. The first of these is located in the Nigerian coastal borehole sequence at Gbekebo, the second comes from the Ilaro limestone around the boundary between Nigeria and Dahomey (Bénin) and the third derives from a borehole in the Sirt Basin of Libya.

Method of Analysis: The method of linear discriminant functions was applied to three measures on the carapace, to wit, length, height and breadth. The results of the calculations are displayed in Table 8.13. The analysis could have been made using the Mahalanobis generalized distance (see Chapter 3, p. 55), but we shall this time use what might be called a mutation of that method. You will recall that this statistical distance differs from the Euclidean distance in that it takes relative dispersions into account as well as correlations. We have also had occasion to use the angle between two vectors as a measure of unlikeness, albeit in an *ad hoc* manner. This carries over to the present line of discussion as the Mahalanobis angle (Seber, 1984, p. 11) which we can conveniently define in the following terms.

The angle between two vectors $\mathbf{x}_r$ and $\mathbf{x}_s$ is given by analogy with formula (2.5) as

$$\cos \theta = \left\{ \frac{\mathbf{x}_r^T \mathbf{S}^{-1} \mathbf{x}_s}{D(\mathbf{x}_r, \mathbf{0}) \cdot D(\mathbf{x}_s, \mathbf{0})} \right\}^{1/2} \tag{8.14}$$

where we define

$$D(\mathbf{x}_r - \mathbf{x}_s)^T = \{(\mathbf{x}_r - \mathbf{x}_s)^T \mathbf{S}^{-1} (\mathbf{x}_r - \mathbf{x}_s)\}^{1/2}. \tag{8.15}$$

The symbolism used in formula (8.14) is that of information theory. The classical reference for information theory applied to multivariate models is the textbook by Kullback (1959). A short but highly informative paper by Mardia (1977) gives the essence of the Mahalanobis angle and its connexion with distance statistics.

The Analysis: The analysis summarized in Table 8.13 tells us that there is no real difference between the samples in the variables selected, to wit, three measures on the size of the carapace and hence no evidence for geographical differentiation in these traits. The strong agreement in the samples from Bénin and Ilaro is doubtless a reflection of the fact that both samples come from the same formation, a limestone. The values of the discriminant mean for the Gbekebo and Libyan samples indicate that there is no cline in size over more than a thousand kilometres, a rather remarkable result. Reyment (1966b) used a different set of samples and found evidence for size differentiation as a function of geographical location for this species. The moral conveyed by the present analysis is that considerable care, and repeated trials, should underlie any interpretations as to palaeobiogeographical differentiation in fossil data. The chance of spurious indications owing to the fickleness inherent in the sampling of fossils must be kept well in mind.

A further point worth recording, and which is hinted at by the analysis of the data on *Anticythereis*, is that environmental pockets, such as the Ilaro lime environment, can have a pronounced effect on geographical clines. Hanson (1966) studied this problem in an appraisal of migrations and fitness of organisms. The effects manifested in the present example could be due to ecological factors (concentration of calcium ions in the original environment) similar to what was found by Reyment and Brännström (1962) for the freshwater species *Cypridopsis vidua*.

8.10 An Example of Statistical Palaeoecology

Presentation of the Problem

I discussed the analysis of species frequencies in Chapter 7. The method of canonical correspondence analysis was applied to foraminiferal data in relation to ecological components. This can be done if we have adequate environmental information on which to base a quantitative study and in that example, such information was forthcoming from oceanological work. What can be done if we have data on species frequencies but little direct evidence for ecological factors? Reyment (1966b) took up this question for Paleocene ostracods by means of an indirect approach. The form of that analysis was as follows:

1. Frequencies for species occurrences were analysed by a variety of factor analysis (cf. Jöreskog *et al.*, 1976, Chapter 3).
2. The factors obtained were treated as biofacies.
3. The biofacies were used as input for an information-theoretic analysis in accordance with Pelto (1954) and Miller and Kahn (1962).

Information Theory and Diversity in Palaeoecology: The basic formula for the application of the concepts of information theory is the Shannon-Wiener Index. This Index can be defined as a function of the relative abundances (i.e. the proportions) of the species in the community. A good reference for getting started in this area is the book by Pielou (1974, p. 290). We can write the Shannon-Wiener Index as follows:

$$H = -\sum_{i=1}^{s} p_i \log p_i. \qquad (8.16)$$

Here, p_i denotes the proportion of the community that belongs to the i-th species. H is quoted in units. The base of logarithms chosen is immaterial as this only alters the unit of diversity. The terminology is:

Nats for logarithms to the base e
Bits for logarithms to the base 2
Decits for logarithms to the base 10.

The Shannon-Wiener Index has the following desirable properties.

1. For a given number of species, H takes its maximum value when $p_i = 1/s$ for all values of i. That is, the maximum value is assumed when all species are present in equal proportions.
2. Given two completely even communities, the one with the greater number of species has the bigger value of H.
3. H can be split into additive components.

Large samples are normally required. Inasmuch as there is no statistical sampling theory available for the Index, H is always determined, never estimated. There is, in fact, a subtle difference involved, if you think about it.

The Analysis

I used (Reyment, 1966b, pp. 542–546) the results of a multivariate palaeoecological study (Reyment, 1963b) to create my "biofacies". I refer you to the cited papers for the full details of the rather lengthy analyses and I shall content myself here with summarizing the main palaeoecological results obtained. I begin by listing the main "biofacies" extracted from the matrix of species associations:

Biofacies 1: This is dominated by two species of *Buntonia*, *B. fortunata* and *B. beninensis*. Other important species in this group are *Brachycythere ogboni* and *Xestoleberis kekere*. This could be a salinity factor.

Biofacies 2: This is dominated by *Nesidea ilaroensis* and to a more modest extent by *Cytherella sylvesterbradleyi*.

Biofacies 3: This group encompasses *Ovocytheridea pulchra*, *Leguminocythereis lagaghiroboensis*, *Buntonia bopaensis*, *Xestoleberis kekere*, and species of *Schizocythere*.

Biofacies 4: This biofacies is dominated by two species, *Cytherella sylvesterbradleyi* and *Trachyleberis teiskotensis*. This could be a bathyal component.

Biofacies 5: This group consists of three species: *Iorubaella ologuni*, *Anticythereis bopaensis*, and *Veenia warriensis*.

The next phase in the analysis was to relate the biofacies to the Shannon-Wiener Index. This was done using a development proposed by Pelto (1954) and Miller and Kahn (1962) which relates this index to the concept of entropy, which can be done by using the fact that $\Sigma p_i = 100\%$, with the addition of a constant, is equivalent in form to the definition of entropy, which can be employed to express uncertainty. The relevant model infers that zones of high entropy ($=$ lack of information) constitute natural barriers between environmental regions where the degree of information available is high. Thus, if, say, i components form a multicomponent system, situations in which all i components are present in equal proportions will contain minimum information. Conversely, the situation where one of the components is represented entirely, the others not at all, will provide maximum information.

Pelto (1954) made use of the concept of **Relative Entropy**, which may be defined as

$$100H_r = 100H/H_m. \tag{8.17}$$

Equation (8.16) expresses H; the maximum entropy H_m is

$$H_m = -N^{-1}\log_e N^{-1}$$
$$= \log_e N \tag{8.18}$$

for an N-component system.

Table 8.14 contains some relative entropy values for the hypothetical environments for the ostracod material. The values in excess of $100H_r \cong 50$

TABLE 8.14. *Relative entropy values for hypothetical environments*
for 17 species of Nigerian Paleocene ostracods

Provenance in borehole	$100H_r$	Dominant "biofacies" association
Gbekebo 700 m	56.1	undecided
Gbekebo 710 m	55.0	undecided
Gbekebo 712 m	54.7	undecided
Gbekebo 713 m	48.7	diagnostic
Gbekebo 715 m	55.0	undecided
Gbekebo 726 m	43.4	non-diagnostic
Gbekebo 733 m	38.7	non-diagnostic
Araromi 305 m	69.9	undecided
Araromi 323 m	35.9	diagnostic
Araromi 340 m	38.5	non-diagnostic
Araromi 422 m	39.5	diagnostic
Ilaro 50 m	44.2	diagnostic
Ilaro 55 m	49.6	diagnostic
Ilaro, outcrop	43.5	diagnostic

usually were not found to be connected with diagnostic types of "biofacies", such as were constructed from the factors of Reyment (1963b).

Discussion

The analysis briefly summarized here was included in order to provide a further example of how latent roots and vectors can be used with advantage to explore relationships between species in palaeoecology. I thought that the analysis showed that strongly calcareous environments were populated mainly by a special association. This point was not strongly brought out in the maximum likelihood factor analysis I used initially in the investigation. The environments in which pelitic sediments were deposited seemed to be favoured by non-specialized ostracod associations. Occasional diagnostic associations in this palaeo-environment were thought to reflect physico-chemical variations.

A recent study by Palmqvist *et al.* (1989) makes use of much the same technique as employed in the present example. These authors apply the Shannon-Wiener Index to Pliocene molluscs in southern Spain in a palaeoecological study.

8.11 Wrightian Factor Analysis and Path Coefficients

Introduction

Wright (1968, and references therein) developed a method of factor-analytical modelling for studying morphological relationships that often permits greater insight into functional affinities, assessed from covariance structures, than routinely applied standard methods of multivariate analysis. Wright's ideas have recently been expanded by Bookstein and associates (Bookstein *et al.*, 1985; Crespi and Bookstein, 1989). Path-analysis has never really attracted much general interest from quantitative biologists, which is surprising once one goes to the trouble of learning what it is about and what a powerful interpretive tool it is. One of the problems lies with the difficulty of understanding how to construct a causal model. Once this *pons* has been crossed, and this may require a good deal of intellectual effort, you will be richly rewarded. Let us look at the main features of the method.

Path-Analytical Factor Model

Consider two variables, x_1 and x_2, each standardized to have unit variance. Their correlation coefficient is r_{12}. A third variable Y is the joint expression of x_1 and x_2. Hence,

$$x_1 = a_1 Y + e_1$$

and

$$x_2 = a_2 Y + e_2$$

where a_1 denotes the direct effect of x_1 on Y and a_2 the same for x_2. The terms e_i are the familiar error terms of the factor model; they are assumed to be uncorrelated.

The variances of the e_1 are $(1 - a_i^2)$ and the correlation between x_1 and x_2 is $a_1 a_2$. If Y is the common cause of these variables, $r_{12} - a_1 a_2$ assesses the strength of this effect. Inference in Wright's factor model is based in the *residuals* of correlations and not on variables.

Contemplate now the situation for three variables, x_1, x_2 and x_3, united by the correlation coefficients r_{12}, r_{13} and r_{23}. There are now three parameters to be estimated by a_1, a_2 and a_3, namely,

$$x_1 = a_1 Y + e_1$$
$$x_2 = a_2 Y + e_2$$
$$x_3 = a_3 Y + e_3 \tag{8.19}$$

The (unique) solution is:

$$a_1 = (r_{12} r_{13} / r_{23})^{1/2}$$
$$a_2 = (r_{12} r_{23} / r_{13})^{1/2}$$
$$a_3 = (r_{13} r_{23} / r_{12})^{1/2}. \tag{8.20}$$

If any of the $a_i > 1$, or there is a negative correlation coefficient, the model collapses, because the error-terms must then be correlated. In the former situation, some error-term(s) would have negative variance and in the latter case, the a_i are imaginary.

The observed correlations between x_i and itself are the ones down the diagonal of the correlation matrix. The dependence of x_i on size is expressed by a_i^2. The residuals $(1 - a_i^2)$ are due to the e_i^2. The a_i^2 are called *communalities* (or self-correlations). The correlation matrix now becomes:

$$
\begin{bmatrix}
a_1^2 & r_{12} & r_{13} \\
r_{21} & a_2^2 & r_{23} \\
r_{31} & r_{32} & a_3^3
\end{bmatrix}
\tag{8.21}
$$

General size for the three variables was defined by Wright as

$$
Y = \frac{a_1 \hat{x}_1 + a_2 \hat{x}_2 + a_3 \hat{x}_3}{a_1^2 + a_2^2 + a_3^3}
\tag{8.22}
$$

where x_i is the explained part of each variable x_i. This quantity cannot be measured. The products of the a_i exactly reproduces the r_{ij} $(i \neq j)$ of the correlation matrix. Wright's criterion of *consistency* requires the diagonal of the adjusted correlation matrix to be a latent vector of that matrix. If you want to learn more about path-analytical factor analysis, I can suggest the following references: Wright (1968), Li (1975) and Jöreskog and Wold (1982).

Example 8.7. Path-Analytical Factor Analysis of Craniometric Traits in the Wild Mongolian Ass

Presentation of the Problem: The data used here were extracted from Miller and Kahn (1962, p. 303) who, in turn, obtained them from Japanese work on *Equus hemionus hemionus*. There were 17 specimens upon each of which 14 traits were measured. I have selected six of these for the present example. The aim of the analysis is to search for evidence of functional

TABLE 8.15. *Size factor and residuals for the Mongolian ass data*

	Primary size factor					
	L_1	L_2	W_1	W_2	D_1	D_2
	0.984	0.948	0.932	0.892	0.887	0.814
	Residual	Correlations				
L_1	0.0323	0.0451	0.0133	0.1021	−0.0728	−0.1004
L_2		0.1006	0.0163	−0.0464	0.0086	−0.0316
W_1			0.1317	0.0584	−0.0467	−0.0482
W_2				0.2035	−0.0918	−0.0361
D_1					0.2129	0.2321
D_2						0.3380

TABLE 8.16. *Primary size vector and residual correlations for sequestered* r_{14} *and* r_{56}: *Mongolian ass data*

y	Primary size factor					
	L_1	L_2	L_3	L_4	L_5	L_6
	0.974	0.972	0.956	0.877	0.829	0.751
L_1	0.0517	0.0311	−0.0008	0.1262	−0.0073	−0.0311
L_2		0.0546	−0.0294	−0.0525	0.0439	0.0100
W_1			0.0863	0.0519	−0.0125	−0.0076
W_2				0.2312	−0.0269	0.0317
D_1					0.3127	0.3316
D_2						0.4364

integration in the characters. The traits selected for analysis are two measures of head length

1. Basilar length
2. Nasal bone length

two measures of head width

3. width of nasal bone (anterior)
4. width of occipital condyle
   and two measures on upper teeth
5. rectangular diameter of incisor I^1
6. rectangular diameter of incisor I^2

The residual correlations for these are displayed in Table 8.15. Note the relatively high correlations between all variables. The loadings on a simple size factor and residual correlations are displayed in Table 8.16. Inspection of the residuals shows that only two of them seem to be important, to wit, r_{14} and r_{56}. The latter relationship for the dentition is not unexpected, but the former points to functional integration that may merit further analysis on more extensive material. The entries in this table were computed as $r_{ij} - a_i a_j$. The diagonal elements are estimates of the squared error-terms.

Wright's method embraces a procedure for *adjusting* the first factor for secondary factors appearing as significant residual correlations (remember, that the residual correlations are differences between correlations near 1 and must therefore be checked by the z-transformation). This is the method of *sequestering* 2×2 blocks from the reduced correlation matrix. In the present example, there are two such blocks, notably,

$$\begin{pmatrix} a_1^2 & r_{14} \\ r_{41} & a_4^2 \end{pmatrix}$$

and

$$\begin{pmatrix} a_5^2 & r_{56} \\ r_{65} & a_6^2 \end{pmatrix}$$

to be sequestered.

TABLE 8.17. *Path coefficients for the Mongolian asses*

	Primary factor	Head	Dentition
Skull-length 1	0.974	0.355	
Skull-length 2	0.972		
Skull-width 1	0.956		
Skull-width 2	0.877	0.355	
Upper tooth L_1	0.829		0.576
Upper tooth L_2	0.751		0.576

Once this has been done, the size factor to be estimated is a latent vector $(a_1, \ldots, a_6)$ of the matrix

$$\begin{bmatrix} a_1^2 & r_{12} & r_{13} & a_1a_4 & r_{15} & r_{16} \\ r_{21} & a_2^2 & r_{23} & r_{24} & r_{25} & r_{26} \\ r_{31} & r_{32} & a_3^2 & r_{34} & r_{35} & r_{36} \\ a_4a_1 & r_{42} & r_{43} & a_4^3 & r_{45} & r_{46} \\ r_{51} & r_{52} & r_{53} & r_{54} & a_5^2 & a_5a_6 \\ r_{61} & r_{62} & r_{63} & r_{64} & a_6a_5 & a_6^2 \end{bmatrix}$$

Bookstein *et al.* (1985) refer to these new diagonal elements a_i as **primary size factors** to underline that they correspond to a slightly different model from the foregoing. The residual correlations and primary size factor derived from this factor-model are displayed in Table 8.16.

The summary in Table 8.17 gives the path-coefficients for the primary factor model. There are two secondary size factors, to wit, the two tooth-traits and the first skull length with the second skull width. The sequestrational factor analysis explains, thus, the observed correlations by three joint-causes: a primary size factor and two secondary factors. All other correlations are very small, being zero or almost zero, showing that the model fitted to the data is performing well. The path diagram for these relationships is shown in Fig. 8.9.

A principal component analysis of the same data yields a first vector which is attached to 85.7% of trace **R**. This is a general "growth vector" (0.973, 0.951, 0.941, 0.914, 0.858) which closely resembles vector **a** in Table 8.16. The subsequent principal components bear no relationship whatsoever to the results of the path-analytic model.

Discussion

The method of analysis illustrated in this example is useful for demonstrating the operational difference between a statistical analysis based on a biological model and multivariate analyses made without a model—automatically, as it were. If your intent is to seek meaningful functional connexions between variables, Wrightian path-analytical factor analysis offers advantages over other ways of attempting to attain this goal.

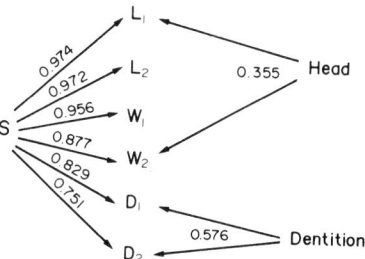

FIG. 8.9. Path-diagram for *Equus hemionus*, the Mongolian ass. There are six elements
in the primary size factor and two secondary size-factors.

8.12 Morphometric Contrasts in Triassic Ichthyosaurs

Presentation of the Problem: The ichthyosaurs were a group of
marine reptile predators that ranged from the Early Triassic to the
Late Cretaceous. Early Triassic ichthyosaurs differ markedly in morphology
of the skull from those of the Jurassic. *Grippia* from the Spathian of
Svalbard is the most primitive in both skull and limb morphology. It can be of
interest to contrast relationships in the skull of this very early genus with a
later representative. With this end in view, *Grippia* is compared morphometrically with *Cymbospondylus* of the Middle Triassic with reference to
sutural intersections, including the parietal frontal suture. The data used in the
following analysis are based on the reconstructions of Massare and Callaway
(1990).

The thin-plate spline offers a useful and informative means of expressing
evolutionary shifts in skull geometry. Active centres of change can be identified and interpreted by this method, as has been well proven in orthodontic
research; the analysis of the effects of the congenital anomaly Apert syndrome
on cranio-facial morphology has become the classical reference (Bookstein,
1986). My choice of genera for this analysis can, of course, be critized. However, my intention is to illustrate how a phylogenetically oriented morphometric analysis can be constructed.

The Analysis

Fourteen landmarks and pseudolandmarks were selected from figures published by Massare and Callaway (1990). I chose these points using the criterion
of what looked to be useful. I have little doubt that an expert on ichthyosaurs
would have made a somewhat different choice. The locations of these features
are shown in Fig. 8.10. These are mostly located at the intersections of sutures
on and around the periphery of the skull in lateral aspect. The analysis can be
conveniently presented in two parts, the **affine** transformation, and the **non-
affine** transformation.

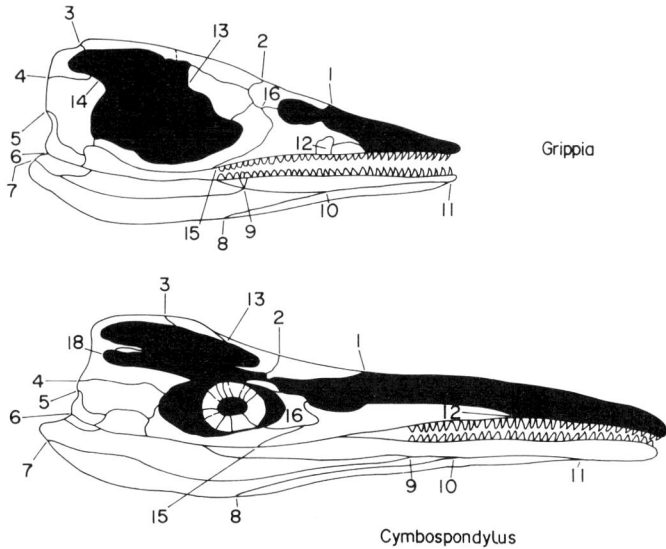

FIG. 8.10. Locations of 14 landmarks and pseudo-landmarks for *Grippia* and *Cymbospondylus*. Redrawn from Massare and Callaway (1990).

The affine transformation

Using the program TPSPLINE, the affine or uniform part of the transformation was calculated to be the matrix:

$$\begin{pmatrix} 1.1742 & 0.5744 \\ 0.1666 & 1.1329 \end{pmatrix}. \tag{8.23}$$

The affine transformation for *Grippia* mapped on to *Cymbospondylus* is illustrated in Fig. 8.11. This is a uniform transformation, there is no bending involved, hence no bending energy is generated (Section 4.6.1). The background grid is still flat after the transformation, but the original squares are now parallelograms: remember—straight lines remain straight lines and parallel lines are still parallel. In addition, proportions along lines are preserved (see Section 2.9(2) and Exercise 14, this chapter).

TABLE 8.18. *First three latent roots and vectors of the bending energy matrix*

Latent roots	Latent vectors
1. 1.2568	$(0.00459, -0.00631, 0.01046, -0.19193, 0.25524, -0.74749, 0.57649,$ $0.01407, 0.00600, -0.01206, 0.00264, 0.00220, 0.00655, 0.07955)^T$
2. 0.63462	$(0.00092, 0.01615, -0.30056, 0.56485, -0.65621, -0.12938, 0.28249,$ $-0.02478, 0.02219, -0.01239, 0.00432, 0.00272, -0.01846, 0.24814)^T$
3. 0.33199	$(-0.25942, 0.06562, 0.02915, -0.10171, -0.01082, 0.05140, 0.02599,$ $-0.46633, 0.70512, -0.26656, -0.01445, 0.30628, -0.15858, 0.09430)^T$

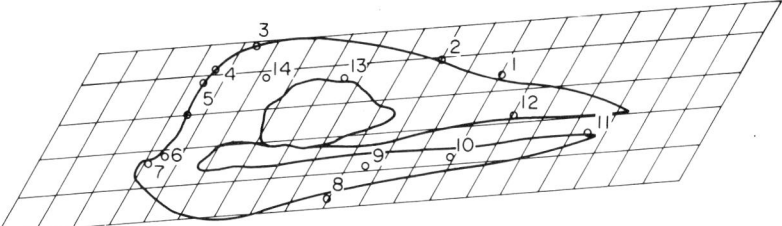

FIG. 8.11. The affine transformation for *Grippia* mapped into *Cymbospondylus*. The little squares of the original grid for the *Grippia*-display have become parallelograms. NB: There is no bending energy involved in this transformation

1. *The Principal Directions of Strain*: We need the above array now. It is used to make a quadratic equation as follows

$$D\alpha^2 + (L_1^2 - L_2^2)\alpha - D = 0 \qquad (8.24)$$

where

$$L_1^2 = 1.1742^2 + 0.5744^2$$
$$L_2^2 = 0.1660^2 + 1.1329^2$$
$$D = (1.1742)(0.1666) + (0.5744)(1.1329)$$

which yields solutions for α of 0.78186 and -0.84637 corresponding to angles 38.39° and $-51.61°$. The angles are obtained from the arc tangent of the values for α.

2. *The Strain Ratios*: The strain ratios are obtained from solving the following quadratic equation:

$$\mu^2 - (L_1^2 + L_2^2)\mu + A^2 = 0$$

where $A^2 = L_1^2 L_1^2 - D$. Inserting the above values for these quantities the strain ratios were found to be $\lambda = \sqrt{\mu} = 1.543$ and 0.800. One of these, the longer, expresses the length of the major axis and the second value, the length at right angles to the first.

The non-affine part

The first three latent roots and vectors of the bending energy matrix $\mathbf{L}_K^{-1}\mathbf{K}\mathbf{L}_K^{-1}$ (i.e. the *principal warps*) are listed in Table 8.18. The bending energy matrix has 11 non-zero latent roots and three zero latent roots. These latter roots correspond to the pattern of landmark displacements that characterize the affine transformation.

The Partial Warps: The energy contribution of each "partial warp" to the spline can be computed as follows: For the warp corresponding to the first

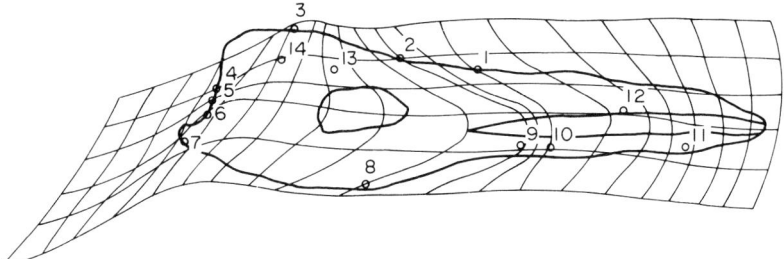

FIG. 8.12. Graph of the deformation represented by the total non-affine transformation associated with the thin-plate spline interpolation.

latent root of the first principal warp we have:

$$1.256832 \times (-0.08046^2 + (-0.23414)^2) = 0.077035.$$

The first number here is the value of the first latent root (see Table 8.18) and the second and third numbers in brackets are the weights for principal warps. The values for the first six energy contributions of each partial warp to the total spline are listed in Table 8.19. The non-affine transformation is shown in Fig. 8.12. This is the total bending associated with the spline and amounts to a bending energy of 2.05719. The whole plate shows evidence of deformation but the most pronounced rucking effect occurs in the region occupied by the posterior of the skull. The geometrical representation of this principal warp given in Fig. 8.13 indicates the significance of this representation. This figure represents the decomposition diagrammed back into the original Cartesian space of the data for the third latent root. It roughly approximates the figure for the total warping effect and is associated with a partial bending energy of 1.64605, with is four-fifths of the total observed bending. The interpretation of the latent vectors is done in a manner analogous to what is usual in interpreting the vectors of principal components, hence the designation Principal Warps.

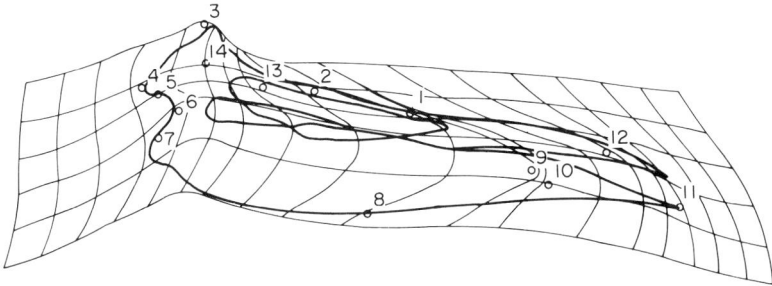

FIG. 8.13. The graphical expression of the third principal warp. This is associated with a partial bending energy of 1.646, which is about 80% of the total bending energy observed.

TABLE 8.19. *Projections of the splines* f_x *and* f_y *(the* X- *and* Y-*components of the computed deformation) upon the first four principal warps*

No.	Weights for principal warps		Energy contribution of each partial warp to the total spline
1	−0.08046	−0.23414	0.077035
2	−0.21632	−0.30475	0.088636
3	2.21687	0.20874	1.645053
4	0.41752	0.08382	0.053132

Examination of the partial bending energies of Table 8.19 discloses that one value, that corresponding to third latent root, overshadows all others in prominence. What does this figure actually stand for? It is not a picture of a deformed "metal plate", as in Fig. 8.12, but a diagram of what the *partial warp* does to the positions of the grid lines when it deforms the initial grid of squares by the appropriate vector multiple.

Discussion

The phylogenetic significance of the spline-analysis is that it illustrates in clear graphical terms just what has taken place over time in the average morphology of the skull. There is no implication of "evolutionary line" here, just an expression of how a Lower Triassic ichthyosaur skull matches up with a Middle Triassic one. Any phylogenetic deductions are the province of the expert. A whole succession of such diagrams can be constructed in which we can identify those changes that are occurring in a regular manner, as evinced by the affine part of the change in shape, and those changes that are taking place through regional accelerations, and embodied in the non-affine part of the transformation. The interesting areas of change can be studied in detail by the method of shape-coordinates and the secular history of the landmarks followed by means of the trajectories mapped out by them.

Exercises

1. It is common practice to assess multivariate normality by a suite of univariate tests for skewness and kurtosis on each of the variables. What objections can you raise to this?

2. We have reviewed the use of multivariate Q–Q-probability plots on several occasions (p. 290 etc.). Do you think such plots can be a useful graphical means of assessing multivariate normality? What does a Q–Q- probability plot of perfectly multivariate Gaussian data look like? (Hint: base your argument around the role of the origin of the graph.)

3. It is often claimed that common multivariate methods (principal components, canonical variates, discriminant functions) are quite robust to deviations from normality. Why then go to all the trouble of doing analyses of the kind outlined on pp. 286–291?

4. Some crustaceans moult all the year round, so that at any given time a population will consist of individuals belonging to different moulting classes, as well as adults. How would this affect a study of the kind made on *Eurypterus* by Andrews *et al.* (1977), and summarized on p. 299?

5. Using the information provided in Table 8.11, compute angles between corresponding latent vectors for the *Veenia* data. Relate your findings to the results listed in Table 8.12.

6. In Chapter 2, we defined the angle between two vectors **a** and **b** in the following manner:

$$\cos \theta = \frac{|\mathbf{a} \cdot \mathbf{b}|}{|\mathbf{a}|^{1/2} \cdot |\mathbf{b}|^{1/2}}$$

Relate this formula to formula (8.14).

7. It is obviously less troublesome to compute the angle as shown in Exercise 8.6 than do the calculations required for formula (8.14). Why then go to all the bother of calculating the Maha-lanobis angle?

8. I think you will hasten to agree that the computations required for achieving stability in canonical variate vectors are quite laborious, at least if you have to write a program in order to get started. What first step can you recommend in order to judge whether data are liable to suffer from instability in vector elements and therefore be able to cut down on unnecessary computing?

9. By the same token, can you propose any course of action that would greatly reduce time spent on having to assess the need for robust methods of estimation? Hint: approach the problem from a graphical point of attack.

10. See what you can do with the following sets of coordinates digitized from *Euproops* and *Limulus* (based on figures published in Bergström, 1975). For example, plot the two sets of coordinates on the same graph and see how much the eurypterid is deformed in relation to the shape of the king crab. In both cases, there are 13 coordinate pairs.

Coordinates for *Euproops*		Coordinates for *Limulus*	
06.44	21.67	15.97	21.96
09.70	19.98	18.80	20.27
10.20	16.97	18.33	17.17
08.44	18.29	17.68	17.97
07.02	15.00	16.45	15.89
06.88	11.22	16.34	11.39
06.52	14.99	15.94	15.88
03.16	16.77	13.97	17.09
03.62	10.12	13.25	19.87
04.92	19.74	14.36	19.87
06.51	20.18	16.05	21.42
08.25	19.86	17.82	19.91
06.61	18.30	16.10	18.95

11. What do you know about the data-analytical techniques of "bootstrapping" and "jack-knifing"? If you are informed on the subject, see where these techniques have been invoked in methods used in the case histories (Krzanowski's procedures, for example). If not, useful references are Hoaglin *et al.* (1985, p. 20) and Efrom and Gong (1983, pp. 36–48). Efrom invented the term "bootstrapping". The latter reference also contains an informative account of "jack-knifing".

12. The latent roots and vectors for the Mongolian asses of Example 8.4 are as follows. What conclusions do you draw about the correlation matrix to which they belong?

			Roots			
Latent	5.1414	0.5799	0.2041	0.1017	0.0215	−0.0486
	I	II	III	IV	V	VI
L_1	0.429	−0.332	−0.063	−0.376	0.085	−0.743
L_2	0.420	−0.103	−0.658	−0.237	−0.384	0.420
W_1	0.415	−0.258	−0.069	0.867	0.022	−0.078
W_2	0.403	−0.384	0.648	−0.216	0.075	0.467
D_1	0.402	0.512	−0.146	−0.060	0.731	0.130
D_2	0.378	0.636	0.341	0.032	−0.552	−0.174

13. The first application of "factor analysis" to zoo-taxonomical problems seems to have been made by G. Teissier in 1938 (Un essai d'analyse factorielle. Les variants sexuels de *Maia squinada*: *Biotypologie* **7**: 73–96). Consult this article and try to form a conclusion about what Teissier actually did, in statistical terms. Useful references are Jöreskog *et al.* (1976), Mulaik (1972) and Thurstone (1947). Can you relate Teissier's study to the first part of Wright's factor model, which predates Teissier by six years (Sewall Wright: General, group and special size factors. *Genetics* 17(5): 603–619)?

14. Write a program to compute the principal directions of strain and the strain ratios using the formulae (8.23) and (8.24). Use the solution for a quadratic equation to find the required two sets of roots. If you want to check your results, show that for the affine transformation matrix

$$\begin{bmatrix} 0.8747 & -0.0289 \\ -0.2956 & 0.9216 \end{bmatrix}$$

the principal axes of transformation are at $53.3°$ and $-36.7°$ to the horizontal and bear strains 1.072 and 0.744 respectively.

Bibliography

Note: Some of the titles listed below are not referenced in the text; they have been included because of their usefulness for further reading and reference.

ABE, K. 1983. Population structure of *Keijella bisanensis* (OKADA), Ostracoda, Crustacea. *Journal of the Faculty of Sciences, University of Tokyo*, **20**: 443–448.

ABE, K., REYMENT, R. A., BOOKSTEIN, F. L., HONIGSTEIN, A., ALMOGI-LABIN, A., ROSENFELD, A. and HERMELIN, O. 1988. Microevolution in two species of ostracods from the Santonian (Cretaceous) of Israel. *Historical Biology*, **1**: 303–322.

AHRENS, H. and LÄUTER, J. 1974. *Mehrdimensionale Varianzanalyse*. Akademie Verlag, Berlin, x + 196 pp.

AITCHISON, J. 1986a. *The Statistical Analysis of Compositional Data*. Chapman and Hall, xv + 416 pp.

AITCHISON, J. 1986b. *CODA: A Microcomputer Package for "The Statistical Analysis of Compositional Data"*. IBM-compatible diskette in BASIC. Chapman and Hall, London, 84 pp.

AITCHISON, J. 1990. Comment on "Measures of variability for geological data" by D. F. WATSON and G. M. PHILIP. *Mathematical Geology*, **22**: 223–226.

ANDERSON, T. W. 1958. *An Introduction to Multivariate Statistical Analysis*. Wiley and Sons, New York, xii + 374 pp.

ANDERSON, T. W. 1963. Asymptotic theory for principal components analysis. *Annals of Mathematical Statistics*, **34**: 122–148.

ANDERSON, T. W. 1984. *An Introduction to Multivariate Statistical Analysis*. Second Edition, Wiley and Sons, New York: xvi + 675 pp.

ANDERSON, T. W. and BAHADUR, R. R. 1962. Two sample comparisons of dispersion matrices for alternatives of immediate specificity. *Annals of Mathematical Statistics*, **33**: 420–431.

ANDREWS, H. E., BROWER, J. C., GOULD, S. J. and REYMENT, R. A. 1977. Growth and variation in *Eurypterus remipes* DE KAY. *Bull. Geol. Instn Univ. Uppsala*, NS 4: 81–114.

BAILEY, D. W. 1956. A comparison of genetic and environmental components of morphogenesis in mice. *Growth*, **20**: 63–74.

BARNETT, V. and LEWIS, T. 1978. *Outliers in Statistical Data*. Wiley and Sons, New York, xi + 365 pp.

BARNETT, V. 1976. The ordering of multivariate data. *Journal of the Royal Statistical Society*, **139**: 318–344.

BARNSLEY, M. F. 1988. *Fractals Everywhere*. Academic Press, London, xii + 395 pp.

BARTLETT, M. S. 1965. Multivariate statistics. In T. H. WATERMAN and H. J. MOROWITZ, Eds. *Theoretical and Mathematical Biology*, pp. 201–224. Blaisdell Publishing Company, New York.

BASSE, E. 1947. Les peuplements malgaches à *Barroisiceras*; révision du genre *Barroisiceras* DE GROSSOUVRE. *Annales de Paléontologie*, **38**: 91–178.

BECKER, W. A. 1987. *Manual of Procedures in Quantitative Genetics*. Academic Enterprises, Pullman, WA, 198 pp.

BELL, M. A. and LEGENDRE, P. 1987. Multicharacter chronological clustering in a sequence of fossil sticklebacks. *Systematic Zoology*, **36**: 52–61.

BELL, M. A., BAUMGARTNER, J. V. and OLSON, E. C. 1985. Patterns of temporal change in single morphological characters of a Miocene stickleback fish. *Paleobiology*, **11**: 258–271.

BELLMAN, R. 1960. *Introduction to Matrix Analysis*. McGraw-Hill, New York, xx + 328 pp.

BENSON, R. H. 1972. The *Bradleya* problem, with descriptions of two new psychospheric ostracode genera, *Agrenocythere* and *Poseidonamicus* (Ostracoda, Crustacea). *Smithsonian Contributions in Paleobiology*, **12:** 138 pp.

BÉNZÉCRI, J. P. 1973. *L'Analyse des Données*: 2, *l'Analyse des Correspondances*. Dunod, Paris. vi + 619 pp.

BERGSTRÖM, J. 1975. Functional morphology and evolution of xiphosurids. *Fossils and Strata*, **4:** 291–305.

BIRKS, H. J. B. and GORDON, A. D. 1985. *Numerical Methods in Quaternary Pollen Analysis*. Academic Press, xiii + 317 pp.

BLACKITH, R. E. 1960. A synthesis of multivariate techniques to distinguish patterns of growth in grasshoppers. *Biometrics*, **16:** 28–40.

BLACKITH R. E. and REYMENT, R. A. 1971. *Multivariate Morphometrics*. Academic Press, London, ix + 412 pp.

BLUM, H. 1973. Biological shape and visual science. *Journal of Theoretical Biology*, **38:** 205–287.

BOAG, P. 1987. Adaptive variation in bill size of African seed-crackers. *Nature*, **392:** 669–670.

BODERGAT, A. M. 1983. Les Ostracodes, temoins de leur environnement: approche chimique et écologique en milieu lagunaire et océanique. *Documents du Laboratoire de Géologie Lyon*, **88,** 246 pp.

BOITARD, MONIQUE 1978. Applications des méthodes d'analyses multidimensionelles à l'étude de populations du complexe *Jaera albifrons* (Isopodes, Asellotes). Thèse, docteur de l'Université, Université de Paris, Pierre et Marie Curie.

BOOKSTEIN, F. L. 1978. *The Measurement of Biological Shape and Shape Change. Lecture Notes in Biomathematics*. Springer Verlag, Berlin-Heidelberg, **24:** vii + 191 pp.

BOOKSTEIN, F. L. 1983. The geometry of craniofacial growth invariants. *American Journal of Orthodontics*, **83:** 221–234.

BOOKSTEIN, F. L. 1984. A statistical method for biological shape comparisons. *Journal of Theoretical Biology*, **107:** 475–520.

BOOKSTEIN, F. L. 1986. Size and shape spaces for landmark data in two dimensions. *Statistical Science*, **1:** 181–242.

BOOKSTEIN, F. L. 1987. Random walk and the existence of evolutionary rates. *Paleobiology*, **13:** 446–464.

BOOKSTEIN, F. L. 1988. Random walk and the biometrics of morphological characters. *Evolutionary Biology*, **23:** 369–398.

BOOKSTEIN, F. L. 1989a. Principal warps: thin-plate splines and the decomposition of deformations. *IEEE Trans. Pattern Anal. Mach. Intell.*, **11:** 567–585.

BOOKSTEIN, F. L. 1989b. Comment to "A survey of the statistical theory of shape" by D. G. KENDALL. *Statistical Science*, **4:** 99–105.

BOOKSTEIN, F. L. 1989c. Four metrics for image variation. *Proc. Intnl Confer. on Information Processing in Medical Imaging*, New York, Alan R. Liss. Inc.

BOOKSTEIN, F. L. 1991. *Morphometric Tools for Landmark Data*. Cambridge University Press, New York.

BOOKSTEIN, F. L. and REYMENT, R. A. 1989. Microevolution in Miocene *Brizalina* (Foraminifera) studied by canonical variate analysis and analysis of landmarks. *Bulletin of Mathematical Biology*, **51:** 657–679.

BOOKSTEIN, F. L. and SAMPSON, P. D. 1987. Statistical models for geometric components of shape change. *Proceedings of the Section on Statistical Graphics*. 1987 *Meeting of the American Statistical Association*, 18–27.

BOOKSTEIN, F. L. and SAMPSON, P. D. 1990. Statistical models for geometric components of shape change. *Communications in Statistics, Theory. Meth.*, **19:** 1939–1972.

BOOKSTEIN, F. L., CHERNOFF, B., ELDER, R., HUMPHRIES, J., SMITH G. and STRAUSS, R. 1985. *Morphometrics in Evolutionary Biology*. Special Publication No. 15, Academy of Natural Sciences, Philadelphia, xvii + 277 pp.

BORESI, A. P. and CHONG, K. P. 1987. *Elasticity in Engineering Mechanics*. Elsevier, xvi + 645 pp.

BOX, G. P. and COX, D. R. 1964. An analysis of transformations. *J. roy. Statist. Soc.*, B, **26:** 211–252.

BROADBENT, S. R. 1955. Quantum hypotheses. *Biometrika*, **42:** 45–57.

BROWN, B. M. and KILDEA, D. G. 1979. Outlier-detection tests and robust estimators based on signs of residuals. *Commun. Statistc. Theor. Meth.*, **A8:** 257–269.

BRUN, L., CHIERICI, M. A. and MEIJER, M. 1982. Evolution and morphological variation of the principal species of Bolivinitidae in the Tertiary of the Gulf of Guinea. *Géologie Méditerranéenne*, **11**: 13–57.

BULL, J. J., VOGT, R. C. and BULMER, M. G. 1982. Heritability of sex ratio in turtles with environmental sex determination. *Evolution*, **36**: 333–341.

BULMER, M. G. 1980. *The Mathematical Theory of Quantitative Genetics*. Oxford University Press, x + 256 pp.

BURNABY, T. P. 1966. Growth invariant discriminant functions and generalized distances. *Biometrics*, **22**: 96–110.

CACOULLOS, T. Ed. 1973. *Discriminant Analysis and Applications*. Academic Press, London, xviii + 434 pp.

CAMPBELL, N. A. 1978. The influence function as an aid in outlier detection in discriminant analysis. *Applied Statistics*, **27**: 251–258.

CAMPBELL, N. A. 1979. *Canonical Variate Analysis: Some Practical Aspects*. Ph.D. Thesis, Imperial College, University of London.

CAMPBELL, N. A. 1980a. Shrunken estimators in discriminant and canonical variate analysis. *Applied Statistics*, **29**: 5–14.

CAMPBELL, N. A. 1980b. Robust procedures in multivariate analysis: I, Robust covariance estimation. *Applied Statistics*, **29**: 231–237.

CAMPBELL, N. A. 1982. Robust procedures in multivariate analysis. II. Robust canonical variate analysis. *Applied Statistics*, **31**: 1–8.

CAMPBELL, N. A. 1984. Canonical variate analysis with unequal covariance matrices: generalizations of the usual solution. *Mathematical Geology*, **16**: 109–124.

CAMPBELL, N. A. and ATCHLEY, W. R. 1981. The geometry of canonical variate analysis. *Systematic Zoology*, **30**: 268–280.

CAMPBELL, N. A. and MAHON, R. J. 1974. A multivariate study of variation in species of rock crab of the genus *Leptograpsus*. *Australian Journal of Zoology*, **22**: 417–425.

CAMPBELL, N. A. and REYMENT, R. A. 1978. Discriminant analysis of a Cretaceous foraminifer using shrunken estimators. *Mathematical Geology*, **10**: 347–359.

CAMPBELL, N. A. and REYMENT, R. A. 1980. Robust multivariate procedures applied to the interpretation of atypical individuals of a Cretaceous foraminifer. *Cretaceous Research*, **1**: 207–221.

CARRIKER, M. R. 1969. Excavation of boreholes by the gastropod *Urosalpinx*: an analysis by light and scanning microscopy. *American Zoologist*, **9**: 917–933.

CARRIKER, M. R. and YOCHELSON, E. L. 1968. Recent gastropod boreholes and Ordovician cylindrical borings. *Professional Paper, U.S. Geological Survey*, 593-B, 1–26.

CARRIKER, M. R., SCHAADT, J. G. and PETERS, V. 1974. Analysis by slow-motion picture photography and scanning electron microscopy of radular function in *Urosalpinx cinerea follyensis* (Muricidae, Gastropoda) during shell penetration. *Marine Biology*, **25**: 63–76.

CARVAJAL, M. C. and LANDERGREN, S. F. C. 1969. Marine sedimentation processes; the interrelationships of manganese, cobalt and nickel. *Stockholm Contributions in Geology*, **18**: 99–122.

CHARLESWORTH, B. 1984. Some quantitative methods for studying evolutionary patterns in single characters. *Paleobiology*, **10**: 308–313.

CHARLESWORTH, B., LANDE, R. and SLATKIN, M. 1982. A Neo-Darwinian commentary on macroevolution. *Evolution*, **36**: 474–498.

CHEETHAM, A. H. 1987. Tempo of evolution in a Neogene bryozoan: are trends in single morphologic characters misleading? *Paleobiology*, 286–296.

CHEVERUD, J. M. 1984. Quantitative genetics and developmental constraints on evolution by selection. *Journal of Theoretical Biology*, **110**: 155–171.

CHEVERUD, J. M. 1988. A comparison of genetic and phenotypic correlations. *Evolution*, **42**: 958–968.

CHEVERUD, J. M. 1989. A comparative analysis of morphological variation patterns in the papionins. *Evolution*, **43**: 1737–1747.

CHEVERUD, J. M. and RICHTSMEIER, J. T. 1986. Finite-element scaling applied to sexual dimorphism in rhesus macaque (*Macaca mulatta*) facial growth. *Systematic Zoology*, **35**: 381–399.

CHEVERUD, J. M., RUTLEDGE, J. J. and ATCHLEY, W. R. 1983. Quantitative genetics of development: genetic correlations among age-specific trait values and the evolution of ontogeny. *Evolution*, **37**: 895–905.

CLARK, W. C. 1976. The environment and the genotype in polymorphism. *Zoological Journal of the Linnean Society*, **58:** 255–262.

CLARKE, B. 1969. The evidence for apostatic selection. *Heredity*, **24:** 347–352.

COLBATH, S. L. 1985. Gastropod predation and depositional environments of two molluscan communities from the Miocene Astoria Formation at Beverly beach state park, Oregon. *J. Paleont.*, **59:** 849–869.

COLLIGNON, M. 1983. Faune albo-cénomanienne de la formation des marnes de Kazhdumi, région du Fars-Khuzestan (Iran). *Documents des Laboratoires de Géologie Lyon*, H. S. **6:** 252–291.

CONDON, E. W. and ODESHAW, H. (Eds.). 1958. *Handbook of Physics*. McGraw-Hill, New York. Irregular pagination.

CONNOR, E. F. 1986. Time series analysis of the fossil record. Patterns and Processes in the History of Life. Eds D. RAUP and D. JABLONSKI. "Dahlem Conference, 1986", 119–147.

COOLEY, W. W. and LOHNES, P. R. 1971. *Multivariate Data Analysis*. Wiley and Son, New York, 364 pp.

CORSTEN, L. C. A. and GABRIEL, K. R. 1976. Graphical exploration in comparing variance matrices. *Biometrics*, **32:** 851–863.

COX, D. R. 1968. Notes on some aspects of regression analysis. *Journal of the Royal Statistical Society*, Series A, 131: 265–279.

COX D. R. and LEWIS, P. A. W. 1966. *The Statistical Analysis of Series of Events*. Methuen, London, 285 pp.

CRAMÉR, H. 1946. *Mathematical Methods of Statistics*. Princeton University Press, xvi + 575 pp.

CRESPI, B. and BOOKSTEIN, F. L. 1989. A path-analytic model for the measurement of selection on morphology. *Evolution*, **43:** 18–28.

CRONIN, T. M. 1985. Speciation and stasis in marine Ostracoda: climatic modulation of evolution. *Science*, **227:** 60–63.

CROW, J. F. 1986. *Basic Concepts in Population, Quantitative, and Evolutionary Genetics*. Freeman and Co., xii + 273 pp.

CUBITT, J. M. and REYMENT, R. A. (Eds.) 1982. *Quantitative Stratigraphic Correlation*. Wiley & Sons, UK, xi + 301 pp.

DARROCH, J. N. and MOSIMANN, J. E. 1985. Canonical and principal components of shape. *Biometrika*, **72:** 241–252.

DAVIS, J. C. 1986. *Statistics and Data Analysis in Geology*. Second Edition, Wiley and Sons, New York, x + 646 pp; with diskette STAT in FORTRAN 77.

DAVIS, P. J. 1965. *The Mathematics of Matrices*. Blaisdell Publishing Company, New York, xiii + 348 pp.

DE ANGELIS, D. L., KITCHELL, J. A., POST, W. M. and TRAVIS, C. C. 1984. Lecture Notes in Biomathematics, **54:** 120–136.

DEFARES, J. G. and SNEDDON, I. N. 1960. *The Mathematics of Medicine and Biology*. North Holland, Amsterdam, xii + 663 pp.

DEGENS, E. T. 1968. *Geochemie der Sedimente*. Ferdinand Enke Verlag, Stuttgart, viii + 282 pp.

DEMPSTER, A. P. 1969. *Elements of Continuous Multivariate Analysis*. Addison Wesley, Reading, Mass., xii + 388 pp.

DIGBY, P. G. N. and KEMPTON, R. A. 1987. *Multivariate Analysis of Ecological Communities*. Chapman and Hall, London. viii + 206 pp.

DO, K. and MCLACHLAN, G. J. 1984. Estimation of mixing proportions. *Applied Statistics*, **33:** 134–140.

DODSON, P. 1975. Functional and ecological significance of relative growth in *Alligator*. *Journal of the Zoological Society of London*, **175:** 315–355.

DRAPER, N. R. and SMITH, H. 1966. *Applied Regression Analysis*. Wiley & Son, New York, ix + 406 pp.

DUCASSE, O. 1981. Etude populationniste du genre *Cytherella* (Ostracodes) dans les facies bathyaux du Paléogène aquitaine. Intérêt dans la reconstitution des paléoenvironnements. *Bulletin de l'Institut de Géologie du Bassin d'Aquitaine, Bordeaux*, **30:** 161–186.

DUCASSE, O. 1983. Etude de populations du genre *Protoargilloecia* (Ostracodes) dans les faciès bathyaux du Paléogène aquitain: deuxième test effectué en domaine profond. Comparaison avec le genre *Cytherella*. *Geobios*, **16:** 273–283.

DUCASSE, O. and CIRAC, P. 1981. La faune de *Mutilus* (Ostracodes: Hemicytheridae) de la région des Zemmours (Maroc nord occidental) à la fin du Miocène et au Miocène. *Géologie Méditerranéenne*, **8**: 87–100.

DUCASSE, O. and ROUSSELLE, L. 1979. Les *Hammatocythere* (Ostracodes) de l'Oligocène aquitain. *Bulletin de l'Institut de Géologie du Bassin d'Aquitaine, Bordeaux*, **25**: 221–255.

EASTMENT, H. T. and KRZANOWSKI, W. J. 1982. Cross-validatory choice of the number of components from a principal component analysis. *Technometrics*, **24**: 73–77.

ECKART, C. and YOUNG, G. 1936. The approximation of one matrix by another of lower rank. *Psychometrika*, **1**: 211–218.

EFROM, B. 1979. Bootstrap methods: another look at the jacknife. *Annals of Statistics*, **7**: 1–26.

EFROM, B. and GAIL GONG, 1983. A leisurely look at the bootstrap, the jacknife and cross-validation. *The American Statistician*, **37**: 36–48.

EHRLICH, R. and FULL, W. E. 1986. Comments on "Relationships among eigenshape analysis, Fourier analysis, and analysis of coordinates" by F. JAMES ROHLF. *Mathematical Geology*, **18**: 855–857.

EISENMAN, R. L. 1963. *Matrix Vector Analysis*. McGraw-Hill, New York, 314 pp.

EISLEY, J. G. 1987. *Mechanics of Elastic Structures*. Prentice-Hall, xiii + 448 pp.

ELDREDGE, N. and GOULD, S. J. 1972. Punctuated equilibria: an alternative to phyletic gradualism. *Models in Paleobiology*, Ed. T. SCHOPF: pp. 82–116. Freeman Cooper and Co., San Francisco.

EMLEN, J. M. 1973. *Ecology, an Evolutionary Approach*. Addison-Wesley, Reading, Mass., 493 pp.

ETTER, R. J. 1988. Physiological stress and color polymorphism in the intertidal snail *Nucella lapillus*. *Evolution*, **42**: 660–680.

EVERITT, B. 1978. *Graphical Techniques for Multivariate Data*. Heinemann, London, 117 pp.

FALCONER, D. S. 1960. *Introduction to Quantitative Genetics*. Oliver and Boyd, Edinburgh, 365 pp.

FALCONER, D. S. 1981. *Introduction to Quantitative Genetics*. Second Edition, Longman Inc., New York, ix + 340 pp.

FÄNGSTRÖM, I. and WILLÉN, E. 1987. Clustering and correspondence analysis of phytoplankton and environmental variables in Swedish lakes. *Vegetatio*, **71**: 87–95.

FERSON, S., ROHLF, F. J. and KOEHN, R. L. 1985. measuring shape-variation of two-dimensional outlines. *Systematic Zoology*, **34**: 59–68.

FISHER, R. A. 1918. The correlation between relatives on the supposition of Mendelian inheritance. *Transactions of the Royal Society of Edinburgh*, **52**: 399–433.

FISHER, R. A. 1936. The use of multiple measurements in taxonomic problems. *Annals of Eugenics, London*, **7**: 179–188.

FISHER, R. A. 1940. The precision of discriminant functions. *Annals of Eugenics London*, **10**: 422–429.

FISHER, R. A. 1960. *The Design of Experiments*. Seventh Edition, Oliver and Boyd, Edinburgh. xv + 248 pp.

FITZGERALD, B. 1983. Observation on surface ornament in fossil ostracods. *Journal of Micropalaeontology*, **2**: 111–117.

FLEISS, J. L. 1973. *Statistical Methods for Rates and Proportions*. Wiley and Sons, New York, xiii + 223.

FLURY, B. 1988. *Common Principal Component Analysis*. Wiley & Son, New York, xiii + 258 pp.

FLURY, B. and RIEDWYL, H. 1988. *Multivariate Statistics*. Chapman and Hall, xi + 296 pp.

FOOTE, M. and COWIE, R. H. 1988. Developmental buffering as a mechanism for stasis evidence from the pulmonate *Theba pisana*. *Evolution*, **42**: 396–399.

FORNELL, C. 1982. *A Second Generation of Multivariate Analysis*; Vol. 1: *Methods*. Praeger Publishers, New York, x + 392 pp.

FOX, L. 1964. *Introduction to Numerical Linear Algebra*. Clarendon Press, Oxford, xi + 295 pp.

FULL, W. F. and EHRLICH, R. 1986. Fundamental problems associated with "eigenshape analysis" and similar "factor" analysis procedures. *Mathematical Geology*, **18**: 451–463.

FUTUYMA, D. J. 1979. *Evolutionary Biology*. Sinauer Associates Inc. Mass., x + 565 pp.

GABRIEL, K. R. 1968. The biplot graphical display of matrices with application to principal components analysis. *Biometrika*, **58**: 453–467.

GALTON, F. 1887. *Natural Inheritance*. Macmillan, London.

GANTMACHER, F. R. 1965. *Matrizenrechnung*. Vol. 1, xi + 324 pp; vol. 2, viii + 244 pp., VEB deutscher Verlag der Wissenschaften, Berlin.

GENSTAT 5 COMMITTEE. 1988. Statistics Department, Rothamsted Experimental Station; GENSTAT 5 Reference Manual. Oxford Science Publications, Oxford, 749 pp.

GIRI, N. C. 1977. *Multivariate Statistical Inference*. Academic Press, New York, xiii + 319 pp.

GNANADESIKAN, R. 1977. *Methods for Statistical Data Analysis of Multivariate Observations*. Wiley and Son, New York. x + 311 pp.

GOLDSTEIN, M. and DILLON, W. R. 1978. *Discrete Discriminant Analysis*. Wiley and Sons, New York, 186 pp.

GOLUB, G. H., HEATH, N. and WAHBA, G. 1979. Generalized cross-validation as a method of choosing a good ridge parameter. *Technometrics*, **21**: 215–222.

GOODALL, C. R. and GREEN, P. B. 1986. Quantitative analysis of surface growth. *Botanical Gazette*, **147**: 1–15.

GOODFRIEND, G. 1986. Variation in land-snail shell form and size and its causes: a review. *Systematic Zoology*, **35**: 204–223.

GORDON, A. D. 1973. A sequence-comparison statistic and algorithm. *Biometrika*, **60**: 197–200.

GORDON, A. D. 1979. SLOTSEQ: A FORTRAN IV program for comparing two sequences of observations. *Computers and Geosciences*, **6**: 7–20.

GORDON, A. D. 1981. *Classification*. Monographs on Applied Probability and Statistics. Chapman and Hall, London, xii + 193 pp.

GORDON, A. D. 1982. On measuring and modelling the relationship between two stratigraphically-recorded variables. In *Quantitative Stratigraphic Correlation*: pp. 241–248. Eds. J. M. CUBITT and R. A. REYMENT, Wiley and Sons, Chiswick.

GORDON, A. D. and BIRKS, H. J. B. 1972. Numerical methods in quaternary paleoecology. I. Zonation of pollen diagrams. *New Phytologist*, **71**: 961–879.

GORDON, A. D. and REYMENT, R. A. 1979. Slotting of borehole sequences. *Mathematical Geology*, **11**: 309–327.

GOSH, A. and KULATILAKE, H. S. W. 1987. A FORTRAN program for generation of multivariate normally distributed variables. *Computers and Geosciences*, **13**: 221–233.

GOULD, S. J. 1971. Geometric similarity in allometric growth: a contribution to the problem of scaling in the evolution of size. *American Naturalist*. **105**: 113–136.

GOULD, S. J. 1977. *Ontogeny and Phylogeny*. Belknap Press, Harvard, 501 pp.

GOWER, J. C. 1966a. Some distance properties of latent roots and vectors used in multivariate analysis. *Biometrika*, **53**: 325–338.

GOWER, J. C. 1966b. A Q-technique for the calculation of canonical variates. *Biometrika*, **53**: 588–589.

GOWER, J. C. 1967. Multivariate analysis and multidimensional geometry. *The Statistician*, **17**: 13–28.

GOWER, J. C. 1968. Adding a point to vector diagrams in multivariate analysis. *Biometrika*, **55**: 582–585.

GOWER, J. C. 1971a. A general coefficient of similarity and some of its properties. *Biometrics*, **27**: 857–874.

GOWER, J. C. 1971b. Statistical methods of comparing different multivariate analyses of the same data. In: *Mathematics in the Archaeological and Historical Sciences* Eds. F. R. HODSON, D. G. KENDALL and P. TOUTA, Edinburgh University Press, pp. 138–149.

GOWER, J. C. 1976. Growth-free canonical variates and generalized inverses. *Bulletin of the Geological Institutions of the University of Uppsala*, **7**: 1–10.

GOWER, J. C. 1977. The analysis of asymmetry and orthogonality. In *Recent Developments in Statistics*, J. BARRA, Ed., North Holland Publishing Company, Amsterdam.

GOWER, J. C. 1987. Introduction to ordination techniques. In *Developments in Numerical Ecology*, Eds. P. and L. LEGENDRE, NATO 151 Series, vol. 614: 2–64. Springer Verlag, Berlin.

GRANLUND, A. 1986. Quantitative analysis of microfossils—a methodological study with applications to Radiolaria. *Meddelanden från Stockholms Universitets Geologiska Institution*, **268**, 99 pp.

GRAYBILL, F. A. 1961. *An Introduction to Linear Statistical Models. Volume I*. McGraw-Hill, New York, xiii + 463 pp.

GREENACRE, M. J. 1984. *Theory and Applications of Correspondence Analysis*. Academic Press, London, xi + 364 pp.

GREENACRE, M. J. and HASTIE, T. 1987. The geometric interpretation of correspondence analysis. *Journal of the American Statistical Association*, **82**: 437–447.

GREIG-SMITH, P. 1964. *Quantitative Plant Ecology*. Butterworths, London, xii + 256 pp.

GRIFFITHS, D. and SANDLAND, R. 1984. Fitting generalized allometric models to multivariate growth data. *Biometrics*, **40**: 139–150.

GRUDZIEN, T. A. and TURNER, B. J. 1984. Direct evidence that the *Ilyodon* morphs are a single biological species. *Evolution*, **38**: 402–407.

GRÜNEBERG, H. 1952. Genetical studies on the skeleton of the mouse. iv. Quasicontinuous variations. *Journal of Genetics*, **51**, 95–114.

GUERRERO, S. and REYMENT, R. A. 1988. Predation and feeding in the naticid gastropod *Naticarius intricatoides* (HIDALGO). *Palaeogeogr., Palaeoclimat., Palaeoecol.*, **68**: 49–52.

HABERMAN, S. J. 1979. *Analysis of Qualitative Data. Vol. 2: New Developments.* Academic Press, London, pp. 369–612.

HALDANE, J. G. S. 1949. Suggestions as to quantitative measurement of rates of evolution. *Evolution*, **3**: 51–56.

HALLAUER, A. R. and MIRANDA, J. B. 1981. *Quantitative Genetics in Maize Breeding.* Iowa State University Press, Ames, Iowa, xii + 468 pp.

HAMPEL, F. R. 1974. The influence curve and its role in robust estimation. *Journal of the American Statistical Association*, **69**: 383–393.

HAMPEL, F. R., RONCHETTI, E. M., ROUSSEEUW, P. J. and STATEL, W. A. 1986. *Robust Statistics.* Wiley and Sons, New York, xxi + 502 pp.

HAND, D. J. 1981. *Discrimination and Classification.* Wiley and Sons, x + 218 pp.

HANSON, W. D. 1966. Effects of partial isolation (distance) migration, and different fitness requirements among environmental pockets upon steady-state gene-frequencies. *Biometrics*, **22**: 453–468.

HARRIS, R. J. 1975. *A Primer of Multivariate Statistics.* Academic Press, New York and London, xiv + 332 pp.

HARTL, F. L. 1980. *Principles of Population Genetics.* Sinauer Associates Inc., Mass. xvi + 488 pp.

HARTL, D. L. and COOK, R. D. 1974. Autocorrelated random environments and their effects on gene frequency. *Evolution*, **28**: 275–280.

HARTMANN, G. 1982. Variation in surface ornament of the valves of three ostracod species from Australia. In *Fossil and Recent Ostracods*, Eds. R. H. BATE, E. ROBINSON and L. M. SHEPPARD. British Micropalaeontological Society Series: pp. 365–380. Ellis Horwood Ltd, Chichester.

HASHIGUCHI, S. and MORISHIMA, H. 1969. Estimation of genetic contribution of principal components to individual variates concerned. *Biometrics*, **25**: 9–15.

HASTINGS, A. 1984. Evolution in a seasonal environment: simplicity lost? *Evolution*, **32**: 350–358.

HAVEL, J. E., HEBERT, P. D. N. and DELORME, L. D. 1990. Genetics of sexual Ostracoda from a low Arctic site. *J. evol. Biol.*, **3**: 65–84.

HAWKINS, D. M. 1980. *Identification of Outliers.* Chapman and Hall, London, x + 188 pp.

HAZEL, L. N. 1943. The genetic basis for constructing selection indices. *Genetics*, **28**: 476–490.

HEALY, M. J. R. 1968. Multivariate normal plotting. *Applied Statistics*, **17**: 157–161.

HEGMAN, N. J. P. and DeFRIES, J. C. 1970. Are genetic and environmental correlations correlated? *Nature*, **226**: 284–285.

HILL, M. O. 1974. Correspondence analysis: a neglected multi-variate method. *Journal of the Royal Statistical Society*, ser. c, **23**: 340–354.

HINTZE, J. L. 1987. *Number Cruncher Statistical System*, Version 5: 01, Utah, USA.

HIRSCHFELD, H. O. 1935. A connection between correlation and contingency. *Proceedings of the Cambridge Philosophical Society* (Math. Proc.), **31**: 520–524.

HOAGLIN, D. C., MOSTELLER, F. and TUKEY, J. W. (Eds.). 1985. *Exploring Data Tables, Trends and Shapes.* Wiley and Son, xxii + 527 pp.

HOENIG, J. M. and RESTREPO, V. R. 1989. Estimating the intermolt periods in asynchronously molting crustacean populations. *Biometrics*, **45**: 71–82.

HOPE, K. 1968. *Methods of Multivariate Analysis.* Unibooks, London, 165 pp.

HOPKINS, J. W. 1966. Some considerations in multivariate allometry. *Biometrics*, **22**: 747–760.

HORST, P. 1963. *Matrix Algebra.* Holt, Rhinehart and Winston Inc., New York, xxi + 517 pp.

HOTELLING, H. 1931. The generalization of Student's ratio. *Annals of Mathematical Statistics*, **2**: 360–378.

HOTELLING, H. 1936. Relations between two sets of variates. *Biometrika*, **28**: 321–377.

HOUSEHOLDER, A. S. 1964. *The Theory of Matrices in Numerical Analysis.* Blaisdell Publ. Co., New York, xi + 257 pp.

HUGHES, R. A. and GRIFFITHS, C. L. 1988. Self-thinning in barnacles and mussels: the geometry of packing. *American Naturalist*, **132**: 484–491.

HUMPHRIES, J. M., BOOKSTEIN, F. J., CHERNOFF, B., SMITH, G. R., ELDER, R. L. and POSS, S. G. 1981. Multivariate discrimination by shape in relation to size. *Systematic Zoology*, **30**: 291–308.

HUXLEY, J. S. 1932. *Problems of Relative Growth*. Methuen and Co. Ltd., London. xix + 276 pp.

IKEYA, N. and UEDA, H. 1988. Morphological variations of *Cytheromorpha acupunctata* (BRADY) in continuous populations at Hamano-ko Bay, Japan. pp. 319–340 in *Evolutionary Biology of Ostracoda*. Eds T. HANAI, N. IKEYA and K. ISHIZAKI. Elsevier-Kodansha, Tokyo, xvi + 1356 pp.

IMBRIE, J. and KIPP, N. J. 1971. A new micropaleontological method for quantitative micro-paleontology: application to a late Pleistocene Caribbean core. In *The Late Cenozoic Glacial Ages*, Ed. K. TUREKIAN: 71–181. Yale University Press, New Haven.

IVERT, H. 1980. Relationship between stratigraphical variation in the morphology of *Gabonella elongata* and geochemical composition of the host sediments. *Cretaceous Research*, **1**: 223–233.

JABLONSKI, D. and RAUP, D. M. (Eds.) 1986. *Patterns and Processes in the History of Life*. Springer Verlag, Berlin, Heidelberg.

JABLONSKI, D., GOULD, S. J. and RAUP, D. M. 1986. The nature of the fossil record; a biological perspective. In JABLONSKI and RAUP (*op. cit.*): 7–25.

JARDINE, N. 1971. Patterns of differentiation between human local populations. *Transactions of the Royal Society (Biology)*, **263**: 1–33.

JOLICOEUR, P. 1963. The degree of robustness in *Martes americana*. *Growth*, **27**: 1–27.

JOLICOEUR, P. and MOSIMANN, J. E. 1960. Size and shape variation in the painted turtle. A principal component analysis. *Growth*, **24**: 339–354.

JOLLIFFE, I. T. 1986. *Principal Component Analysis*. Springer Verlag, New York, xiii + 271 pp.

JOLLIFFE, I. T. 1989. Rotation of ill-defined principal components. *Applied Statistics* **38**: 139–147.

JONGMAN, R. H. G., TER BRAAK, C. J. F. and VAN TONGEREN, O. F. R. (Editors). 1987. *Data Analysis in Community and Landscape Ecology*. Pudoc Waageningen, xix + 299 pp.

JÖRESKOG, K. G. and WOLD, H. (Eds.) 1982. *Systems under Indirect Observation*. Part II. North Holland Publ. Co., Amsterdam, xii + 343 pp.

JÖRESKOG, K. G., KLOVAN, J. E. and REYMENT, R. A. 1976. *Geological Factor Analysis*. Elsevier, Amsterdam, xii + 178 pp.

KAESLER, R. L. and WATERS, J. A. 1972. Fourier analysis of the ostracod margin. *Bull. geol. Soc. Amer.*, **83**: 1169–1178.

KAMIYA, T. 1988. Morphological and ethological adaptations of Ostracoda to microhabitats in *Zostera* beds. In *Evolutionary Biology of Ostracods*, Eds. T. HANAI, N. IKEYA and ISHIZAKI, K. Proc. 9th Intnl Symp on Ostracoda, Shizuoka, Japan (1985): 303–318.

KEEN, M. 1982. Intraspecific variation in Tertiary ostracods. In R. H. BATE, E. ROBINSON and L. M. SHEPPARD (Eds). *Fossil and Recent Ostracoda*, 365–380. Ellis Horwood Ltd, Chichester.

KELLOGG, D. 1975. The role of phyletic change in the evolution of *Pseudocubus vema* (Radiolaria). *Paleobiology*, **1**: 359–370.

KENDALL, D. G. 1989. A survey of the statistical theory of shape. *Statistical Science*, **4**: 87–120.

KENDALL, M. G. 1973. *Time-Series*. Griffin and Co., London, ix + 197 pp.

KIRKPATRICK, M. 1982. Quantum evolution and punctuated equilibria in continuous genetic characters. *American Naturalist*, **199**: 833–848.

KITCHELL, J. A., BOGGS, C. H., KITCHELL, J. F. and RICE, J. A. 1981. Prey selection by naticid gastropods: experimental tests and application to the fossil record. *Paleobiology*, **7**: 533–552.

KITCHELL, J. A., ESTABROOK, G. and MACLEOD, N. J. A. 1987. Testing for equality of rates of evolution. *Paleobiology*, **13**: 272–285.

KOHN, L. A. P. and ATCHLEY, W. R. 1988. How similar are genetic correlation structures? Data from mice and rats. *Evolution*, **42**: 467–481.

KRES, H. 1977. *Statistische Tafeln zur multivariaten Analysis*. Springer Verlag Heidelberg, xvi + 431 pp.

KRISHNAIAH, P. R. (Ed.). 1977. *Multivariate Analysis–IV*. North Holland Publ. Co., Amsterdam, xiii + 549 pp.

KRZANOWSKI, W. J. 1979. Between-groups comparison of principal components. *Journal of American Statistical Association*, **74**: 703–707.

KRZANOWSKI, W. J. 1983. Cross-validatory choice in principal component analysis; some sampling results. *Journal of Statistical Computer Science*, **18**: 294–314.

KRZANOWSKI, W. J. 1984. Principal component analysis in the presence of group structure. *Applied Statistics*, **33**: 164–168.

KRZANOWSKI, W. J. 1987a. Cross-validation in principal component analysis. *Biometrics*, **43**: 575–584.

KRZANOWSKI, W. J. 1987b. Selection of variables to preserve multivariate data structure using principal components. *Journal of the Royal Statistical Society*, Ser. C, **36**: 22–33.

KUHRY, B. and MARCUS, L. F. 1977. Bivariate linear models in biometry. *Systematic Zoology*, **26**: 201–209.

KULLBACK, S. 1959. *Information-theory and Statistics*. Wiley and Sons, New York, xvii + 395 pp.

LANDE, R. 1976. Natural selection and random genetic drift in phenotype evolution. *Evolution*, **30**: 314–334.

LANDE, R. 1977. Statistical tests for natural selection on quantitative characters. *Evolution*, **31**: 442–444.

LANDE, R. 1978. Evolutionary mechanisms of limb loss in tetrapods. *Evolution*, **32**: 73–92.

LANDE, R. 1979a. Quantitative genetic analysis of multivariate evolution, applied to brain: body size allometry. *Evolution*, **33**: 402–416.

LANDE, R. 1979b. Effective deme sizes during long-term evolution estimated from rates of chromosome rearrangement. *Evolution* **33**: 234–251.

LANDE, R. 1980a. The genetic covariances between characters maintained by pleiotropic mutations. *Genetics*, **94**: 203–215.

LANDE, R. 1980b. Microevolution in relation to macroevolution. *Paleobiology*, **6**: 233–238.

LANDE, R. 1986. The dynamics of peak shifts and the pattern of morphological evolution. *Paleobiology*, **12**: 343–354.

LANDE, R. 1988. Quantitative genetics and evolutionary theory. In *Proceedings of the Second International Conference on Quantitative Genetics*, Raleigh, 1987. Edited by B. S. WEIR, E. J. EISEN, M. M. GOODMAN and G. NAMKOONG, Sinauer, pp. 71–84.

LANDE, R. and ARNOLD, S. J. 1983. The measurement of selection on correlated characters. *Evolution*, **37**: 1210–1226.

LAURENT, G. 1987. *Paléontologie et Evolution en France* 1800–1860. Ed. *Comité des Travaux historiques et scientifiques*, 4, x + 533 pp.

LEAMY, L. 1975. Component analysis of osteometric traits in randombred house mice. *Systematic Zoology*, **24**: 176–190.

LEFEBVRE, J. 1976. *Introduction aux Analyses Statistiques Multidimensionelles*. Masson, Paris, xvi + 219 pp.

LEGENDRE, L. and LEGENDRE, P. 1979. *Ecologie Numérique. 1. Le Traitement multiple des données écologiques*, xiv + 197 pp. 2. *La structure des données écologiques*. 247 pp. Masson P. U. Q., Paris.

LEHMANN, U. 1964. *Paläontologisches Wörterbuch*. Ferdinand Enke, Stuttgart, ii + 335 pp.

LEMAN, C. A. and FREEMAN, P. W. 1984. The genus: a macroevolutionary problem. *Evolution*, **38**: 1219–1237.

LEVINTON, S. 1986. Developmental constraints and evolutionary saltations: a discussion and a critique. In *Genetics, Development and Evolution*, ed. J. P. GUSTAFSON, G. L. STEBBINS and F. J. AYALA, pp. 253–288. Plenum Press, New York.

LEVINTON, S. 1988. *Genetics, Paleontology, and Macroevolution*. Cambridge University Press, Cambridge, xiv + 637 pp.

LEWIS, P. A. W. 1964. A branching Poisson process model for the analysis of computer failure patterns. *Journal of the Royal Statistical Society*, Ser. B, **26**: 398–441.

LEWONTIN, P. C. 1974. *The Genetic Basis of Evolutionary Change*. Columbia University Press, New York, 346 pp.

LI, C. C. 1975. *Path Analysis–A Primer*. Boxwood Press, Pacific Grove, California, 347 pp.

LIEBAU, A. 1971. *Homologe Skulpturmuster bei Trachyleberididen und verwandten Ostrakoden*. Dissertation, Berlin, 117 pp.

LINDSEY, J. C., HERZBERG, A. M. and WATTS, D. G. 1987. A method for cluster analysis based on projections and quantile plots. *Biometrics*, **43**: 327–341.

LIVELY, C. M. 1986. Predator-induced shell dimorphism in the Acorn Barnacle *Chthamalus anisopoma*. *Evolution*, **40**: 232–242.

LOFSVOLD, D. 1986. Quantitative genetics of morphological differentiation in *Peromyscus*. I. Tests of the homogeneity of genetic covariance structure among species and subspecies. *Evolution*, **40**: 559–573.

LOFSVOLD, D. 1988. Quantitative genetics of morphological differentiation in *Peromyscus*. II. Analysis of selection and drift. *Evolution*, **42**: 54–67.

LOHMANN, G. P. 1983. Eigenshape analysis of microfossils. A general morphometric procedure for describing changes in shape. *Mathematical Geology*, **15**: 659–672.

LOHMANN, G. P. and SCHWEITZER, P. N. 1989. On eigenshape analysis. Morphometrics in Systematic Biology Workshop, 16–28 May, 1988, Ann Arbor. Eds. F. J. ROHLF and F. L. BOOKSTEIN.

LONG, C. A. 1985. Intricate sutures as fractal curves. *Journal of Morphology*, **185**: 285–295.

LOVE, W. A. and STEWART, D. K. 1968. *Interpreting Canonical Correlations: Theory and Practice*. Palo Alto, American Institutes for Research, 66 pp.

LYNCH, M. and GABRIEL, W. 1983. Phenotypic evolution and parthenogenesis. *American Naturalist*, **122**: 745–764.

MCCAMMON, R. B. 1972. Map pattern reconstruction from sample data. Mississippi Delta region of south-eastern Louisiana. *Journal of Sedimentary Petrology*, **42**: 422–424.

MAHÉ, J. 1974. L'analyse factorielle des correspondances et son usage en paléontologie et dans l'étude de l'évolution. *Bulletin de la Societé géologique de France*, **16**: 336–340.

MALMGREN, B. A. and KENNETT, J. P. 1982. The potential of morphometrically based phylozonation: application of a late Cenozoic planktonic foraminiferal lineage. *Marine Micropalaeontology*, **7**: 285–296.

MALMGREN, B. A., BERGGREN, W. A. and LOHMANN, G. P. 1983. Evidence for punctuated gradualism in the Late Neogene *Globorotalia tumida* lineage of planktonic foraminifera, *Paleobiology*, **9**: 377–389.

MANDELBROT, B. 1983. *The Fractal Geometry of Nature*. Freeman and Co., San Francisco, 468 pp.

MANLY, B. F. J. 1983. Analysis of polymorphic variation in different types of habitats. *Biometrics*, **39**: 13–27.

MANLY, B. F. J. 1985. *The Statistics of Natural Selection*. Chapman and Hall, London, xvi + 484 pp.

MARCUS, L. F. 1969. Measurement of selection using distance statistics in the prehistoric Orangutan *Pongo pygmaeus palaeosumatriensis*. *Evolution*, **23**: 301–307.

MARDIA, K. V. 1970. Measures of multivariate skewness and kurtosis with applications. *Biometrika*, **57**: 519–530.

MARDIA, K. V. 1977. Mahalanobis distances and angles. In P. R. KRISHNAIAH (Ed.), *Multivariate Analysis—IV*. North Holland Publ. Co., Amsterdam, pp. 495–511.

MARDIA, K. V. 1989. Shape analysis of triangles through directional techniques. *J. roy. Statist. Soc.*, B 51: 449–458.

MARDIA, K. V. and ZEMROCH, P. J. 1978. *Tables of the F- and Related Distributions*. Academic Press, London.

MARDIA, K. V., KENT, J. T. and BIBBY, J. M. 1979. *Multivariate Analysis*. Academic Press, xv + 521 pp.

MARONNA, R. A. 1976. Robust *M*-estimators of multivariate location and scatter. *Annals of Statistics*, **4**: 51–67.

MARQUARDT, D. W. and SNEE, R. D. 1975. Ridge regression in practice. *American Statistician*, **29**: 3–19.

MASSARE, J. A. and CALLAWAY, J. M. 1990. The affinities of Early Triassic ichthyosaurs. *Bull. Geol. Soc. Amer.*, **102**: 409–416.

MATSUOKA, H. and OKADA, H. 1990. Time-progressive changes of the genus *Gephyrocapsa* in the Quaternary sequence of the tropical Indian Ocean. *Proc. Ocean Drilling Program*, **115**: 255–270.

MAXWELL, E. A. 1958. *Coordinate Geometry with Vectors and Tensors*. Oxford University Press, ix + 194 pp.

MAYNARD, J. B. 1983. *Geochemistry of Sedimentary Ore Deposits*. Springer Verlag, New York, xi + 305 pp.

MAYNARD SMITH, J. 1983. The genetics of stasis and punctuation. *Annual Review of Genetics*, **17**: 11–25.

MAYNARD SMITH, J. 1989. *Evolutionary Genetics*. Oxford University Press, xii + 325 pp.

MAYNARD SMITH, J. R., BURIAN, S., KAUFFMAN, S., ALBERCH, P., CAMPBELL, J., GOODWIN, B., LANDE, R., RAUP, D. and WOLPERT, L. 1985. Developmental constraints and evolution. *Quarterly Review of Biology*, **60**: 256–287.

MAYR, E. 1963. *Animal Species and Evolution*. Belknap Press, Harvard University, xiv + 797 pp.

MAYR, E. 1969. *Principles of Systematic Zoology*, McGraw-Hill, 428 pp.

MILLER, R. L. and KAHN, J. S. 1962. *Statistical Analysis in the Geological Sciences*. Wiley and Sons, New York, 357 pp.

MILLIGAN, B. G. 1986. Punctuated evolution induced by ecological change. *American Naturalist*, **127:** 522–532.

MONRO, D. M., FORTRAN 77. 1987. Edward Arnold, London, vii + 360 pp.

MOORE, H. B. 1958. *Marine Ecology*. Wiley and Sons, xi + 493 pp.

MORAN, N. A. and WHITHAM, T. G. 1988. Evolutionary reduction of complex life cycles: loss of host-alternation in *Pemphigus* (Homoptera: Aphididae). *Evolution*, **42:** 717–728.

MORRISON, D. F. 1976. *Multivariate Statistical Methods*. Second edition, McGraw-Hill Inc., USA, xv + 415 pp.

MOSIMANN, J. E. 1968. *Elementary Probability for the Biological Sciences*. Appleton-Century-Crofts, New York, xii + 255 pp.

MOSIMANN, J. E. 1970. Size allometry: size and shape variables with characterizations of the lognormal and generalized gamma distributions. *Journal of the American Statistical Association*, **65:** 930–945.

MOSIMANN, J. E. and JAMES, F. C. 1979. New statistical methods for allometry with applications to Florida red-winged blackbirds. *Evolution*, **33:** 444–459.

MOURANT, A. 1983. *Blood Relations*. Oxford University Press, vi + 146 pp.

MULAIK, S. A. 1972. *The Foundations of Factor Analysis*. McGraw-Hill, New York, xvi + 451 pp.

MURDOCH, D. C. 1957. *Linear Algebra for Undergraduates*. Wiley and Sons, New York, vii + 239 pp.

NOORDWIJK, A. J. VAN, BALEN J. H. VAN and SCHARLOO, W. 1980. Heritability of ecologically important traits in the great tit. *Ardea*, **68:** 193–203.

OKADA, Y. 1981. Development of cell arrangement in ostracod carapaces. *Paleobiology*, **7:** 276–280.

OLSON, E. C. and MILLER, R. L. 1958. *Morphological Integration*. University of Chicago Press, Chicago. xv + 317 pp.

OOSTEROFF, L. M. 1977. Variation in growth rate as an ecological factor in the landsnail *Cepaea nemoralis (L.)*. *Netherlands Journal of Zoology*, **27:** 1–132.

ORLOCI, L. 1975. *Multivariate Analysis in Vegetation Research*. Junk B. V., The Hague, ix + 276 pp.

PALMER, A. R. 1985. Quantum changes in gastropod shell morphology need not reflect speciation. *Evolution*, **39:** 699–705.

PALMQVIST, P., GUERRERO, S. and SALVA, M. I. 1989. Estudio paleoecológico de la fauna de Moluscos de un afloramiennto de materiales pliocénicos en Estepona (Málaga). *Revista Española de Palaeontología*, **4:** 29–38.

PEARSON, K. 1901. On lines and planes of closest fit to systems of points in space. *Philosophical Magazine*, **2** (ser. 6): 559–572.

PELTO, C. R. 1954. Mapping of multi-component systems. *Journal of Geology*, **62:** 501–511.

PENROSE, L. S. 1954. Distance, size and shape. *Annals of Eugenics*, **18:** 337–343.

PETRY, D. 1982. The pattern of phyletic speciation. *Paleobiology*, **8:** 56–66.

PEYPOUQUET, J. P. 1977. *Les Ostracodes et la connaissance des paléomilieux profonds. Applications au Cénozoique de l'Atlantique Nord-Oriental*. Thèse, Doctorat d'Etat ès Sciences; Université de Bordeaux I.

PHILLIPS, P. C. and ARNOLD, S. J. 1989. Visualizing multivariate selection. *Evolution*, **43:** 1209–1222.

PIELOU, E. C. 1974. *Population and Community Ecology*. Gordon and Breach, New York, viii + 424 pp.

PIELOU, E. C. 1977. *Mathematical Ecology* (Second Edition). Wiley—Interscience, New York, x + 385 pp.

PIMENTEL, R. A. 1979. *Morphometrics: the Multivariate Analysis of Biological Data*. Kendall-Hunt, Dubuque Iowa, x + 276 pp.

PIRSON, S. J. 1977. *Geologic Well Log Analysis*. Gulf Publishing Co., Houston, xiii + 377 pp.

PREISENDORFER, R. W. 1988. *Principal Component Analysis in Meteorology and Oceanography*. Developments in Atmospheric Science 17, Elsevier, Amsterdam, xviii + 425 pp.

PRESS, W. H., FLANNERY, B. P., TEUKOLSKY, S. A. and VETTERLING, W. T. 1986. *Numerical Recipes.* Cambridge University Press, xx + 818 pp.

PRIM, R. C. 1957. Shortest connection networks and some generalizations. *Bell Systems Technical Journal*, **36**: 1389–1401.

RAMBERG, J. S. and SCHMEISER, B. W. 1972. An approximate method for generating symmetric random variables. *Communications of the ACM*, **15**: 987–990.

RAO, C. R. 1952. *Advanced Statistical Methods in Biometric Research.* Wiley and Sons, New York, 292 pp.

RAO, C. R. 1964. The use and interpretation of principal components analysis in applied research. *Sankhya*, **12**: 229–246.

RAO, C. R. and MITRA, S. K. 1971. *Generalized Inverse of Matrices and its Applications.* Wiley and Sons, New York, xiv + 240 pp.

RAUP, D. M. 1977. Stochastic models in evolutionary palaeontology. In A. HALLAM Ed., *Patterns of Evolution, as illustrated by the Fossil Record.* Developments in Palaeontology and Stratigraphy 5: 59–78.

RAUP, D. M. and CRICK, R. E. 1981. Evolution of a single character in the Jurassic ammonite *Kosmoceras. Paleobiology*, **7**: 200–215.

RAUP, D. M. and GOULD, S. J. 1974. Stochastic simulation and evolution of morphology towards a nomothetic paleontology. *Systematic Zoology*, **23**: 305–322.

RAUP, D. M. and STANLEY, S. M. 1971. *Principles of Paleontology.* Freeman and Co., San Francisco, x + 388 pp.

READ, D. W. and LESTREL, P. E. 1986. Comment on uses of homologous point measures in systematics: a reply to BOOKSTEIN *et al. Systematic Zoology*, **35**: 241–253.

RENDEL, J. M. 1967. *Canalisation and Gene Control.* Academic Press, London, 166 pp.

REYMENT, R. A. 1958. Some factors in the distribution of fossil cephalopods. *Stockholm Contributions in Geology*, **1**: 97–184.

REYMENT, R. A. 1961. A note on geographical variation in European *Rana. Growth*, **25**: 219–227.

REYMENT, R. A. 1963a. Studies on Nigerian Upper Cretaceous and Lower Tertiary Ostracoda. Part 2, Danian, Paleocene and Eocene Ostracoda. *Stockholm Contributions in Geology*, **10**: 1–286 pp.

REYMENT, R. A. 1963b. Multivariate analytical treatment of quantitative species associations: an example from Palaeoecology. *Journal of Animal Ecology*, **32**: 535–547.

REYMENT, R. A. 1966a. Preliminary observations on gastropod predation in the western Niger Delta. *Palaeogeography, Palaeoclimatology, Palaeoecology*, **2**: 81–102.

REYMENT, R. A. 1966b. Studies on Nigerian Upper Cretaceous and Lower Tertiary Ostracoda. Part 3, Stratigraphical, Palaeoecological and Biometrical conclusions. *Stockholm Contributions in Geology*, **14**: 1–151.

REYMENT, R. A. 1969a. A multivariate paleontological growth problem. *Biometrics*, **22**: 1–8.

REYMENT, R. A. 1969b. Covariance structure and morphometric analysis—a contribution to paleogenetics. *Mathematical Geology*, **1**: 185–197.

REYMENT, R. A. 1970. Spectral breakdown of morphometric chronoclines. *Mathematical Geology*, **2**: 365–376.

REYMENT, R. A. 1971a. *Introduction to Quantitative Paleoecology.* Elsevier, Amsterdam, xii + 226 pp.

REYMENT, R. A. 1971b. Multivariate normality in morphometric analysis. *Mathematical Geology*, **3**: 357–368.

REYMENT, R. A. 1972. Models for studying the occurrence of lead and zinc in a deltaic environment. In *Mathematical Models of Sedimentary Processes* Ed. T. W. MERRIAM, Plenum, New York, 233–245.

REYMENT, R. A. 1973. Factors in the distribution of fossil cephalopods. 3, experiments with exact models of certain shell types. *Bulletin of the Geological Institutions of the University of Uppsala*, NS 4: 7–41.

REYMENT, R. A. 1975. Canonical correlation analysis of hemicytherinid and trachyleberinid ostracods in the Niger Delta. *Bulletin of American Paleontology*, **65**: 141–145.

REYMENT, R. A. 1976. Chemical components of the environment and Late Campanian microfossil frequencies. *Geologiska Föreningens Förhandlingar*, **98**: 322–328.

REYMENT, R. A. 1979. On the interpretation of the smallest principal component. *Bulletin of the Geological Institutions of the University of Uppsala*, NS 8: 1–4. (Translation of original in publications of Akademia Nauk, USSR.)

REYMENT, R. A. 1980. *Morphometric Methods in Biostratigraphy*. Academic Press, London 175 pp.

REYMENT, R. A. 1982a. Threshold characters in a Cretaceous foraminifer. *Palaeogeography, Palaeoclimatology, Palaeoecology*, **38:** 1–7.

REYMENT, R. A. 1982b. Phenotypic evolution in a Cretaceous foraminifer. *Evolution*, **36:** 1182–1199.

REYMENT, R. A. 1982c. Speciation in a Late Cretaceous lineage of *Veenia* (Ostracoda). *Journal of Micropalaeontology*, **1:** 37–44.

REYMENT, R. A. 1982d. Analysis of trans-specific evolution in Cretaceous ostracods. *Paleobiology*, **8:** 292–305.

REYMENT, R. A. 1982e. Morphological variation in time of a Paleocene species of *Cytherella*. In *Fossil and Recent Ostracoda*, Ed. R. H. BATE, E. ROBINSON and L. M. SHEPPARD, London, 165–168.

REYMENT, R. A. 1983. Phenotypic evolution in microfossils. *Evolutionary Biology*, **16:** 209–254.

REYMENT, R. A. 1985a. Multivariate morphometrics and analysis of shape. *Mathematical Geology*, **17:** 591–609.

REYMENT, R. A. 1985b. Phenotypic evolution in a lineage of the Eocene ostracod *Echinocythereis*. *Paleobiology*, **11:** 74–194.

REYMENT, R. A. 1987. Multivariate analysis in geoscience: fads, fallacies and future. *Chemometrics and Intelligent Laboratory Systems*, **2:** 79–91.

REYMENT, R. A. 1988. Does sexual dimorphism occur in Upper Cretaceous ammonites. *Senckenbergiana Lethaea*, **69:** 109–119.

REYMENT, R. A. 1989. Compositional data analysis. *Terra Nova*, **1:** 29–34.

REYMENT, R. A. 1990. Reification of classical multivariate statistical analysis in morphometry. In F. J. ROHLF and F. J. BOOKSTEIN (Eds.) (1990) *ibid.* 122–144.

REYMENT, R. A. and BANFIELD, C. F. 1976. Growth-free canonical variates applied to fossil foraminifers. *Bulletin of the Geological Institutions of the University of Uppsala*, NS 7: 11–21.

REYMENT, R. A. and BANFIELD, C. F. 1981. Analysis of asymmetric relationships in geological data. In *Modern Advances in Geomathematics*, Eds. R. CRAIG and M. LABOVITZ, Pion Ltd, London, 236–253.

REYMENT, R. A. and BRÄNNSTRÖM, B. 1962. Certain aspects of the physiology of *Cypridopsis* (Ostracoda, Crustacea). *Stockholm Contributions in Geology*, **9:** 207–242.

REYMENT, R. A. and KENNEDY, W. J. 1991. Phenotypic plasticity in a Cretaceous ammonite analyzed by multivariate statistical methods. *Evolutionary Biology*, **25:** 411–426.

REYMENT, R. A. and REYMENT, E. R. 1989. A note on spinosity in *Afrobolivina afra* (Foraminifera). *Bulletin of the Geological Institutions of the University of Uppsala*, NS 15: 1–6.

REYMENT, R. A. and STURESSON, U. P. A. 1987. Correlation of chemical and physical environmental fluctuations in a Late Cretaceous borehole sequence—a multivariate study. *Sedimentary Geology*, **53:** 311–325.

REYMENT, R. A. and VAN VALEN, L. 1969. *Buntonia olokundudui* sp nov. (Ostracoda, Crustacea): a study of meristic variation in Paleocene and Recent ostracods. *Bulletin of the Geological Institutions of the University of Uppsala*, NS 1: 83–94.

REYMENT, R. A., BLACKITH, R. E. and CAMPBELL, N. A. 1984. *Multivariate Morphometrics*, Second Edition, Academic Press, vii + 233 pp.

REYMENT, R. A., BOOKSTEIN, F. L., MCKENZIE, K. G., MAJORAN, S. 1988. Ecophenotypic variation in *Mutilus pumilus* (Ostracoda) from Australia studied by canonical variate analysis and tensor biometrics. *Journal of Micropalaeontology*, **7:** 11–20.

REYMENT, R. A., HAYAMI, I. and CARBONNEL, G. 1977. Variation of discrete morphological characters in *Cytheridea* (Crustacea, Ostracoda). *Bulletin of the Geological Institutions of the University of Uppsala*, NS 7: 23–36.

REYMENT, R. A., REYMENT, E. R. and HONIGSTEIN, A. 1987. Predation by boring gastropods on Late Cretaceous and Early Palaeocene ostracods. *Cretaceous Research*, **8:** 189–209.

RISKA, B. 1981. Morphological variation in the horseshoe crab *Limulus polyphemus*. *Evolution*, **35:** 647–658.

RISKA, B. 1985. Group size factors and geographic variation of morphometric correlation. *Evolution*, **39:** 792–803.

ROHLF, F. J. 1986. Relationships among eigenshape analysis, Fourier Analysis and analysis of coordinates. *Mathematical Geology*, **18:** 845–854.

ROHLF, F. J. and ARCHIE, J. W. 1984. The comparison of Fourier methods for the description of wing shape in mosquitoes (Diptera: Culicidae). *Systematic Zoology*, **33**: 302–317.

ROHLF, F. J. and BOOKSTEIN, F. L. 1987. A comment on shearing as a method for size correction. *Systematic Zoology*, **36**: 356–367.

ROHLF, F. J. and BOOKSTEIN, F. L. 1990. *Proceedings of the Michigan Morphometrics Workshop.* Special Publication No. 2, University of Michigan Museum of Zoology, viii + 380 pp.

ROHLF, F. J. and FERSON, S. 1983. Image analysis. In J. FELSENSTEIN (Ed.). *Numerical Taxonomy*, NATO ISI Series 6, Ecological Science No. 1, Springer Verlag, New York, pp. 583–599.

ROHLF, F. J. and SLICE, D. 1990. Extensions of the Procrustes method for the optimal superimposition of landmarks. *Systematic Zoology*, **39**: 40–59.

ROHLF, F. J. and SOKAL, R. R. 1969. *Statistical Tables.* Freeman and Co., San Francisco, 253 pp.

RÖHRS, M. 1959. Neue Ergebnisse und Probleme der Allometrieforschung. *Zeitschrift für wissenschaftliche Zoologie.*, **162**: 1–95.

ROUGHGARDEN, J. 1979. *Theory of Population Genetics and Evolutionary Ecology: an Introduction.* Macmillan Publishing Co., New York, 634 pp.

ROY, S. N., GNANADESIKAN, R. and SRIVASTAVA, J. N. 1971. *Analysis and Design of Certain Multiresponse Experiments.* Pergamon Press, Oxford.

RUDMAN, A. J. and LANKSTON, R. W. 1973. Stratigraphic correlation of well logs by computer techniques. *Bulletin of the American Association of Petroleum Geologists*, **57**: 577–588.

SAMPSON, P. D. and SIEGEL, A. F. 1984. The measure of "size" independent of "shape" for multivariate lognormal populations: definition and applications. *Proceedings of the International Biometrics Conference, Tokyo*, August, 1984, 10 pp.

SAVAZZI, E. and REYMENT, R. A. 1989. Subaerial hunting behaviour in *Natica gualteriana* (naticid gastropod). *Palaeogeogr., Palaeoclimat., Palaeoecol.*, **74**: 355–364.

SCHEINER, S. M. and LYMAN, R. F. 1989. The genetics of phenotype plasticity. *J. evol. Biol.*, **2**: 95–107.

SCHMALHAUSEN, I. I. 1949. *Factors of Evolution: the Theory of Stabilizing Selection.* Blakiston, Philadelphia, xiv + 327 pp.

SCHOENBERG, P. 1970. On metric multidimension unfolding. *Psychometrika*, **35**: 349–366.

SCHÖNEMAN, P. H. and CARROLL, R. M. 1970. Fitting one matrix to another under choice of a central dilation and a rigid motion. *Psychometrika*, **35**: 245–256.

SCHWARTZ, M., GREEN, S. and RUTLEDGE, W. A. 1960. *Vector Analysis*, Harper and Brothers, New York, xii + 556 pp.

SEARLE, S. R. 1961. Phenotypic, genetic and environmental correlations. *Biometrics*, **17**: 474–480.

SEARLE, S. R. 1966. *Matrix Algebra for the Biological Sciences*, Wiley and Sons, New York, xii + 296 pp.

SEBER, G. A. F. 1984. *Multivariate Observations.* Wiley and Sons, New York, xx + 686 pp.

SEN GUPTA, B. K. and STRICKERT, D. P. 1982. Living benthic Foraminifera of the Florida–Hatteras slope: distribution, trends and anomalies. *Bulletin of the Geological Society of America*, **93**: 218–224.

SHEA, B. J. 1985. Bivariate and multivariate growth allometry: statistical and biological considerations. *Journal of the Geological Society of London*, **206**: 367–380.

SIEGEL, A. F. and BENSON, R. H. 1982. A robust comparison of biological shapes. *Biometrics*, **38**: 341–350.

SIMPSON, G. G. 1944. *Tempo and Mode in Evolution.* Columbia University Press, New York.

SIMPSON, G. G. 1953. *The Major Features of Evolution.* Columbia University Press, New York, xx + 434 pp.

SIMPSON, G. G., ROE, A. and LEWONTIN, R. C. 1960. *Quantitative Zoology.* Harcourt Brace, New York, xii + 440 pp.

SLATKIN, M. 1982. Pleiotropy and parapatric speciation. *Evolution*, **36**: 263–270.

SLICE, D. E. 1989. *Fractal-D.* Exeter Publishing Co., Ltd, New York.

SMITH, C. A. B. 1969. *Biomathematics.* Volume 2, Griffin and Co., x + 682 pp.

SMITH, M. F. and PATTON, J. L. 1988. Subspecies of pocket gopher: causal bases for geographic differentiation in *Thomomys bottae. Systematic Zoology*, **37**: 163–178.

SMITH, T. B. 1987. Bill size polymorphism and intraspecific niche utilization in an African finch. *Nature*, **329**: 717–719.

SOMERS, K. M. 1986. Multivariate allometry and removal of size with principal components analysis. *Systematic Zoology*, **35:** 359–368.

SPRENT, P. 1972. The mathematics of size and shape. *Biometrics*, **28:** 23–37.

STANLEY, S. M. and YANG, X. 1987. Approximate evolutionary stasis for bivalve morphology over millions of years: a multivariate multilineage study. *Paleobiology*, **13:** 113–139.

STONE, M. 1974. Cross validatory choice and assessment of statistical procedures. *J. roy. Statist. Soc.*, **B 36:** 111–148.

STONE, M. and BROOKS, R. J. 1990. Continuum regression: cross-validated sequentially constructed prediction embracing ordinary least squares, partial least squares and principal components regression. *J. roy Statist. Soc.*, **52:** 237–269.

STRAUSS, R. E. and BOOKSTEIN, F. L. 1982. The truss: body form reconstructions in morphometrics. *Systematic Zoology*, **31:** 113–135.

STRICKBERGER, M. W. 1976. *Genetics*. Second Edition, Macmillan, New York.

STÜTZLE, W., GASSER, T., MOLINARI, L., LARGO, R. H., PRADER, A. and HUBER, P. J. 1980. Shape-invariant modelling of human growth. *Annals of Human Biology*, **7:** 507–528.

SWAN, A. R. H. and SAUNDERS, W. B. 1987. Function and shape in late Paleozoic (mid-Carboniferous) ammonoids. *Paleobiology*, **13:** 297–311.

TER BRAAK, C. F. J. 1986. Canonical correspondence analysis: a new eigenvector technique for multivariate direct gradient analysis. *Ecology*, **67:** 1167–1179.

TER BRAAK, C. F. J. 1987. The analysis of vegetation-environment relationships by canonical correspondence analysis. *Vegetation*, **69:** 69–77.

THOMPSON, D. W. 1917. *On Growth and Form*. Cambridge University Press, 1116 pp.

THURSTONE, L. L. 1947. *Multiple Factor Analysis*. University of Chicago Press, Chicago, xix + 535 pp.

TISSOT, B. 1988. Geographic variation and heterochrony in two species of cowries (genus *Cypraea*). *Evolution*, **42:** 103–117.

TOBLER, W. 1978. Comparison of plane forms. *Geographical Analysis*, **10:** 154–162.

TORGERSON, W. S. 1958. *Theory and Methods of Scaling*. Wiley and Sons, New York, 460 pp.

TURELLI, M. 1988. Phenotypic evolution, constant covariances, and the maintenance of additive variance. *Evolution*, **42:** 1342–1347.

TURELLI, M., GILLESPIE, J. H. and LANDE, R. 1988. Rate tests for selection on quantitative characters during macroevolution and microevolution. *Evolution*, **42:** 1085–1089.

TURNER, J. R. G. 1986. The genetics of adaptive radiation and Neo-Darwinian theory of punctuational evolution. pp. 183–202 In *Patterns and Processes in the History of Life*, D. M. RAUP and D. JABLONSKI Eds., Springer Verlag, Berlin.

VAN NORDWIJK, A. J., VAN BALEN, J. H. and SCHARLOO, W. 1980. Heritability of ecologically important traits in the Great Tit. *Ardea*, **68:** 193–203.

VAN DE GEER, J. P. 1971. *Introduction to Multivariate Analysis for the Social Sciences*. Freeman and Co., xi + 293 pp.

VAN VALEN, L. 1962. A study in fluctuating asymmetry. *Evolution*, **16:** 124–142.

VAN VALEN, L. 1969. The variation genetics of extinct animals. *American Naturalist*, **103:** 193–224.

VAN VARK, G. N. and HOWELLS, W. W. (Eds.). 1984 *Multivariate Statistical Methods in Physical Anthropology*. D. Reidel Publ. Co., Dordrecht, x + 433 pp.

VEITCH, L. G. 1978. Size, shape and allometry in *Uca*, a multivariate approach. *Mathematical Scientist*, **3:** 35–45.

VERMEIJ, G. J. 1987. *Evolution and Escalation*. Princeton University Press, xv + 527 pp.

VIA, S. and LANDE, R. 1985. Genotype-environment interaction and the evolution of phenotypic plasticity. *Evolution*, **39:** 505–522.

WAGNER, G. P. 1984. On the eigenvalue distribution of genetic and phenotypic dispersion matrices: evidence for non-random organization of quantitative character variation. *Journal of Mathematical Biology*, **21:** 77–95.

WAKE, D. B., ROTH, G. and WAKE, M. H. 1983. On the problem of stasis in organismal evolution. *Journal of Theoretical Biology*, **101:** 211–224.

WALLACE, B. 1981. *Basic Population Genetics*. Columbia University Press, New York, xii + 688 pp.

WARTENBERG, D., FERSON, S. and ROHLF, F. J. 1987. Putting things right: a critique of detrended correspondence analysis. *American Naturalist*, **129:** 434–448.

WATSON, G. S. 1970. Orientation statistics in the Earth Sciences. *Bulletin of the Geological Institutions of the University of Uppsala*, NS 2: 73–89.

WATSON, G. S. 1981. The interaction of statistics and geology—finite deformations. In D. F. MERRIAM Ed. *Down to Earth Statistics: Solutions looking for Geological Problems*, Syracuse University, Geology Contribution **8**, 17–27.

WATSON, G. S. 1989. Comment on D. G. KENDALL "A survey of the statistical theory of shape" *Statistical Science*, **4**: 113–115.

WILTSE, W. I. 1980. Predation by juvenile *Polinices duplicatus* (Say) on *Gemma gemma* (Toth). *Jl Experimental Marine Biology and Ecology*, **42**: 187–199.

WOLD, S. 1978. Cross-validatory estimation of the number of components in factor and principal component models. *Technometrics*, **20**: 397–405.

WOOD, A. M., LANDE, R. and FRYXELL, G. A. 1987. Quantitative genetic analysis of morphological variation in an Antarctic diatom grown at two light intensities. *J. Phycol.*, **23**: 42–54.

WRIGHT, S. 1931. Evolution in Mendelian populations. *Genetics*, **16**: 97–159.

WRIGHT, S. 1932. The roles of mutation, inbreeding, crossbreeding and selection in evolution. *Proceedings of the Sixth International Congress on Genetics*, **1**: 356–366.

WRIGHT, S. 1968. *Evolution and the Genetics of Populations. Vol. 1. Genetic and Biometric Foundations.* University of Chicago Press, Chicago, vii + 469 pp.

YOUNG, G. and HOUSEHOLDER, A. S. 1938. Discussion of a set of points in terms of mutual distances. *Psychometrika*, **3**: 19–22.

ZAHN, C. R. and ROSKIES, R. Z. 1972. Fourier description for plane closed curves. *IEEE Transactions on Computers*, *c*-21: 269–281.

ZERA, A. J. 1984. Differences in survivorship, development rate and fertility between the longwinged and wingless morphs of the waterstrider, *Limnoporus canaliculatus. Evolution*, **38**: 1023–1032.

ZIEGELMEIER, E. 1954. Beobachtungen über den Nahrungserwerb bei der Naticide *Lunatia nitida* Donovan (Gastropoda Prosobranchia). *Helgoländer wissenschaftliche Meeresforschung*, **5**: 1–33.

Glossary

Informal explanations of some of the terms used in the text. Definitions of many other terms may be obtained by reference to the Index.

Abscissa: The horizontal coordinates in a two-dimensional system, such as a two-dimensional graph.

Acquired Character: One that appears during the life of an individual due to an environmental and/or functional cause.

Adaptation: The fitness of a structure, function or entire organism for a particular environment.

Adductor: A muscle that draws a part towards the median axis; fossil ostracods and bivalves often display the impressions of these muscles (adductor muscle scars).

Algorithm: Some special process of solving a certain kind of mathematical problem.

Allele: The alternative forms of a gene, having the same locus in homologous chromosomes and capable of segregating as a unit Mendelian factor.

Allopatric: An allopatric relationship between two populations or species means that they occupy different geographical regions.

Ammonite: An extinct group of the molluscan class Cephalopoda, having a chambered shell.

Analogy: Similarity of external features or function, but not of origin.

Anova: Acronym for the Analysis Of Variance. An important branch of statistics due to Sir Ronald Fisher.

Apomixis: Parthenogenetic reproduction in which the individual develops from an unfertilized egg or somatic cell.

Asexual: Not related to sex, i.e. not involving gametes or fusion of their nuclei.

Association Matrix: A symmetric, Q-mode matrix of associations or similarities between all paired comparisons of specimens of a sample. The off-diagonal elements are usually made to run from nought to one. The diagonal elements are usually ones.

Behrens–Fisher Problem: The problem of determining the probability of drawing two random samples, the means of which differ by k (which can be nought) from normal populations with known mean differences but unknown variances.

Bipolar latent vector: A vector containing both positive and negative components, which are interpreted in principal component analysis as indicative of shape changes.

Canonical Form of a Matrix: The simplest form to which a square matrix can be reduced by a certain kind of transformation. The canonical form of a matrix has non-zero elements only in the main diagonal.

Canonical axes: A reference axis in canonical variate or canonical correlation space.

Cartesian coordinates: In the plane, a point can be located by its distances from two intersecting straight lines, the axes. The coordinate measured from the y-axis parallel to the x-axis is called the abscissa and the other coordinate, the ordinate.

Caudal: Pertaining to the tail (L. *cauda* = tail), or posterior part of the body (in ostracods, the carapace, which may be provided with a caudal process).

Centroid: A vector, each element of which is the mean value for each variable of a multivariate sample (or population). Synonym; Mean Vector.

Chi-square:

$$\chi^2 = \sum_{i=1}^{k} x_i^2$$

where the x_i are independently and normally distributed with mean of zero and variance of 1. This distribution was discovered by Helmert in 1876. For a sample size in excess of 30, $\sqrt{2\chi^2}$ is distributed approximately normally with mean $(2n - 1)^{\frac{1}{2}}$ and with unit variance.

Circumcentre: The centre of the circumscribed circle (i.e. the circle passing through the three vertices of a triangle; it is the point of intersection of the perpendicular bisectors of the sides.

Component scores: The scores of original data vectors after transformation to principal component space.

Compressed: Reduced in breadth; flattened laterally. Many ammonites are referred to as being compressed in shape.

Concomitant variable: A dependent variable; a covarying variable of a different class from the other set of variables with which it is being examined.

Cretaceous: The Mesozoic Era is divided into three Systems, the Cretaceous (youngest), Jurassic, and the Triassic (oldest).

Deformation (*Elastic*): The change in the position of the points of a body accompanied by a change in the distances between them.

Deme: Local population of closely related plants and animals (from Grk *demos* = the people).

Depressed: Flattened vertically, from above. Some ammonite shells are depressed in shape.

Dimorphism: Existing under two distinct forms, as in sexual dimorphism.

Dirac Function: The Dirac Function vanishes when its arguments are different from zero, becomes infinite for zero arguments, and has the property that its integral over the whole space has the value 1.

Distal: Away from the point of attachment or site of reference (L. *distare* = stand apart).

Entropy: Definitions of entropy vary greatly from one context to another. Examples of relevance in this book are

1. The definition in ecology. When two or more species inhabit adjacent areas, the boundary where they mix, has high entropy.
2. The definition of entropy as a measure of uncertainty in biological systems.

Epistasis: Term applied to the phenotype or fitness of two or more gene loci when their joint effect differs from the sum of the effects of the loci taken separately.

Fitness: The average contribution of one allele or genotype to the next generation, or to succeeding generations, compared with that of other alleles or genotypes. A fundamental concept in quantitative genetics.

Foraminifera: A class of protozoans, the members of which secrete a calcarous shell or construct a shell (test) by agglutinating minute grains of sediment.

Freedom: Degrees of freedom in Statistics are the number of free (that is, unrestricted in the sense of random sampling) variables entering into a statistic.

Gaussian: Multivariate Gaussian (multivariate normal) in reference to the generalized Distribution of Gauss (Johan Carl Friedrich Gauss (1777–1855), German mathematician).

Genotype: The internal genetic or hereditary constitution of an organism without regard to its external appearance. Compare Phenotype.

Growth trajectory: A graphical representation, or function, describing some aspect of the size of an organism over a time.

Growth vector: In principal component analysis, a vector (usually the first) with components all positive and consistent with an hypothesis of size increase in a sample of organisms.

Habitat: The natural or usual dwelling place of an individual or group of organisms.

Heteroscedasticity: Inequality of population variances in the univariate case and of dispersions in the multivariate case.

Homologous: Of like source in structure and embryonic development from primitive origin.

Homology: Fundamental similarity; structural likeness of an organ or part in one kind of animal with the comparable unit in another, resulting from a common descent. Compare Analogy.

Hooke's Law: The basic law of proportionality of stress and strain, discovered by Robert Hooke in 1678. The generalized statement of the law says that for sufficiently small strains, each component of the stress tensor is a linear function of the other components of this tensor.

Ichthyosaurs: A group of extinct marine reptiles considered now to be most closely related to the diapsids and placed in that sub-class by recent workers.

Information Theory: The branch of probability theory founded by C. E. Shannon and which deals with the likelihood of transmission of messages, accurate within specified limits, when the "bits" of information constituting the messages are subject to probabilities of failure in transmission.

Kurtosis: Pertaining to the peakedness of a unimodal distribution. Curves flatter than normal are called **platykurtic** and those sharper than normal, **leptokurtic**.

Macroevolution: A blurry term used for evolution manifested in big changes in the phenotype and which usually are great enough to justify a taxonomic assignation of high order to the evolutionary product and its descendants.

Mandible: Lower jaw of a vertebrate; either jaw of an arthropod.

MANOVA: Acronym for the Multivariate Analysis of Variance.

Mean Vector: See Centroid.

Microevolution: A conveniently blurry term applied to slight evolutionary changes within a species. Meant to be opposite of macroevolution.

Migration: A term used in theoretical population genetics as a synonym for gene flow among populations. In other connexions, the term is applied to directed movement of organisms that is not necessarily accompanied by gene flow.

Multiple correlation coefficient: The correlation between the observed and the predicted values of the dependent variable in multiple regression.

Multiple regression: A method of relating *p* independent variates to one dependent variate. Can be viewed as a special case of canonical correlation.

Neogene: The upper subdivision of the Tertiary Sub-Era comprises two Systems, the Pliocene (younger) and the Miocene (older).

Norm of Reaction: The total phenotypic expression of a genotype under a set of environmental conditions. Compare with Phenotypic Plasticity.

Ordinate: The vertical axis of a graph in two-dimensional Cartesian space. See Abscissa.

Ordination: A graphical exploratory technique for studying multivariate properties by placing the individuals, observed on *p* variates, in some natural ordering.

Organism: A single plant or animal; one that functions as a unit.

Ostracoda: A group of the bivalved Crustacea having, usually, a calcareous shell. Both marine and non-marine. Of exceptional importance in Micropalaeontology.

Paedomorphism: The maintenance of the body form commonly found in immature stages when the animal is adult, or substantially fully grown.

Paleogene: The Cenozoic is divided into two major parts, the younger Neogene, and the older Paleogene. The Paleogene comprises the Paleocene, Eocene and Oligocene.

Parthenogenesis: Development of a new individual from an unfertilized egg, as in some freshwater ostracods. See Apomixis.

Partial correlation coefficient: The correlation between pairs of variables while one, or more, other variables are held fixed.

Phenotype: The external appearance of an individual without regard to its genetic or hereditary constitution. Compare Genotype.

Phenotypic plasticity: The expression of the same genotype as different phenotypes as a response to different ecological conditions.

Pleiotropy: The phenotypic effect of a single gene on more than one characteristic.

Polymorphism: Existence of individuals of more than one form in a species.

Proximal: Toward or nearer the place of attachment or reference of the centre of the body. Opposite of Distal.

Pseudo-landmark: A point with a reliable operational definition but which is not a **landmark** homologous from form to form.

Reification: Examination of the morphometric implications of a latent vector by identifying its biological or physical implications.

Robustness: As applied to Statistics, means the capability of a statistical procedure to produce a fair result despite deviations in the data from the theoretical premises upon which the procedure is based.

Shape: The geometry of the organism after information about scale, position and orientation has been removed.

Skewness: Pertaining to the asymmetry of a frequency function as a departure from the normal distribution. For unimodal distribution, left (negative) skewness indicates an extension towards lower values of the variable, and positive (right) skewness towards higher values.

Spectrum: Of a square symmetric matrix is the set of all its latent roots.

Sphericity: Refers to axes defining a spheroid; when two or more derived vectors (e.g. principal component axes) are of equal length, hence arbitrarily defined.

Spurious correlations: Apparently meaningful correlation but owing to extraneous factors such as data recording, etc. is a nonsense correlation. The classical reference is the spurious correlation between the price of wheat and the appearance of sunspots. Spurious correlations arise when methods appropriate to Cartesian space are applied to compositions, properly considered in respect of simplex space.

Stepwise discriminant analysis: The analysis proceeds by the addition or subtraction of a variable at a time in a search for the optimal discriminator. The method has its weaknesses.

Sympatric: Term applied to two species or populations that occupy the same geographical locality.

Trend: Secular trend in a time series is the part of the variation resulting from slowly changing, long-lasting forces. Usually characterized by a monotonically increasing or decreasing function of time.

Zapping: Colloquialism (U.S.A., from the "space-comics") for applying an ordinating procedure, such as principal components, to a poorly known set of multivariate observations to see whether the data are structured and/or contain multivariate outliers. A more serious data-analytical procedure than implied by the name.

Name Index

359

Subject Index

SUPPLEMENT

Computer Programs for Multidimensional Palaeobiology

L. F. Marcus

Introduction

A set of programs is provided to support most of the methods discussed in the text. They have been written using a very special matrix-algebra-like language, available within the statistical package Statistical Analysis System (SAS). SAS is available on main frames, many mini-computers, work stations and IBM PC computers and clones. One additional stand alone program TPSPLINE, provided as an executable module, was kindly contributed by F. James Rohlf, State University of New York, Stonybrook (see Appendix III).

SAS is organized into procedures (PROC) and one called PROC IML (Interactive Matrix Language) is a very powerful tool for handling data arrays and doing multivariate and multidimensional computations, as well as graphics.

A short list of IML matrix operations are compared to those from Chapter 2, together with an example. A full list of all relevant operations and functions is given in Appendix II. Code for all of the programs is printed here, with a short explanation and cross-reference to the main text. The data used are all to be found associated with the problems, in the book, or on diskette available from the Supplement author on $5\frac{1}{4}$ inch or $3\frac{1}{2}$ inch media. The programs along with their output are also on the diskette.

The SAS package, including PROC IML, must be available in order to use these procedures. IML is a separately licensed module and not all computer centres or site-licences for the PC software may include it.

The PC SAS version 6.04 requires 640K RAM. A math co-processor is advised but not essential, and SAS can make use of Expanded Memory. When one runs SAS on a PC, very little memory should be devoted to memory resident files. I have found it advisable to keep at least 540K RAM free. The package is big, and the programs will run far faster on an 80286 CPU (IBM AT and clones) or 80386 CPU machine. Very few routines are slow. If they are, it will be pointed out in their description. Newer algorithms are being explored to increase speed. The programs were not tested on the main frame or other computers but SAS claims that the version 6.06, already released, for an IBM main frame should be compatible with the PC version. Minor modifications may have to be made.

A version of "PC SAS on a Page" is included as Appendix I. It is also available in Rohlf and Bookstein (1990) and Marcus and Corti (1989). Both of these references contain additional useful SAS and IML programs.

The main purposes of this supplement are to make the matrix algebra and algorithms explicit for the methodology. This will reinforce the discussions and examples in the text. The more important matrix and other IML statements are highlighted in boldface. It is recommended that the reader compare the code to the prose and formulae in the main text.

Example of IML Code and Results

The following complete program will do a principal component analysis on a small data array imbedded in the code. As will be shown later, data may be accessed from external files in a variety of ways. This complete example will work directly and it is suggested that readers try it. Comments are enclosed in slash asterisk, /* comment */, asterisk slash as you will see below. It is especially important to note that all SAS statements end with a ; (semi-colon). SAS pays no attention to lower and upper case in the programs and code itself and translates all array names to upper case in the output. Strings defined as character variables and used as labels for output stay in the case in which they were entered.

PCAONE.SAS

```
proc iml;          /* this invokes the IML procedure in sas  */
reset fuzz nolog;  /* fuzz - rounds near 0 to 0 in printing */
                   /* nolog - results to OUTPUT window */
X={11.10 10.38 9.78,   /* matrix X consists of first 5 rows and 3 */
   11.02 10.42 9.84,   /* columns of Brizalina data from Chpt. 3 */
   11.02 10.41 9.82,
   10.98 10.42 9.94,
   11.09 10.49 10.08};
N=nrow(X);         /* number of rows in data arraY        */
p=ncol(X);         /* number of columns in data arraY     */
mn=j(N,1)`*X/N;    /* mean vector using j vector of column of 1's */
Y=X-j(N,1)*mn;     /* compute deviations from mean */
S=Y`*Y/(N-1);      /* produces variance-covariance matrix */
call eigen(a,E,S); /* a for eigen-values; E for eigen-vectors */
print N p mn, S, a E;      /* print results to date  */
Pcscores=Y*E;      /* compute mean centered Pcscores         */
call pgraf(Pcscores[,1:2]);  /* plot PC2 against PC1 */
Check=Pcscores`*Pcscores/(N-1); /* test for orthogonality */
print Check;
```

```
/* check should have diagonal elements equal to eigenvalues a */;
quit;  /* correct way to leave IML though not necessary */
/* if you want to do additional steps without reinvoking IML
   then leave out the quit; statement          */
```

You see that in very few lines we have a Principal Component (PC) analysis on a covariance matrix, using code that looks like ordinary matrix algebra—with a CALL to a function to return the eigenvalues and eigenvectors, and another CALL to a function to plot the resulting PC scores.

The Log and Output from this example run on a PC clone using the Queen's College site-licence is given below. The lines of original code are numbered. This example was run using a colour monitor—where results and SAS notes appear in different colours. This is simulated using a bold font for "NOTE" and result lines. If there had been errors, error pointers would appear just below the faulty code in error in yet another colour. The NOLOG option on the RESET line in IML causes the results to be written in the OUTPUT window. These results are reproduced below the Log Window.

Log Window

NOTE: Copyright(c) 1985,86,87 SAS Institute Inc., Cary, NC 27512-8000, U.S.A.
NOTE: SAS (r) Proprietary Software Release 6.03
          Licensed to QUEENS COLLEGE ACADEMIC COMPUTER CENTER, Site 11047001.
NOTE: AUTOEXEC processing completed.

```
    1    proc iml;          /* this invokes the IML procedure in sas  */
IML Ready
    2    reset fuzz nolog;  /* fuzz - rounds near 0 to 0 in printing */
    3                       /* nolog - results to OUTPUT window */
    4    X={11.10 10.38 9.78,   /* matrix X consists of first 5 rows and 3 */
    5       11.02 10.42 9.84,   /* columns of Brizalina data from Chpt. 3 */
    6       11.02 10.41 9.82,
    7       10.98 10.42 9.94,
    8       11.09 10.49 10.08};
    9    N=nrow(X);         /* number of rows in data arrayY          */
   10    p=ncol(X);         /* number of columns in data arrayY       */
   11    mn=j(N,1)'*X/N;    /* mean vector using j vector of column of 1's */
   12    Y=X-j(N,1)*mn;  /* compute deviations from mean */
   13    S=Y'*Y/(N-1);  /* produces variance-covariance matrix */
   14    call eigen(a,E,S);  /* a for eigen-values; E for eigen-vectors */
   15    print N p mn, S, a E;   /* print results to date  */
   16    Pcscores=Y*E;    /* compute mean centered Pcscores       */
   17    call pgraf(Pcscores[,1:2]);  /* plot PC2 against PC1 */
   18    Check=Pcscores'*Pcscores/(N-1); /* test for orthogonality */
   19    print Check;
   20    /* check should have diagonal elements equal to eigenvalues a */;
   21    quit;  /* correct way to leave IML though not necessary */
NOTE: The PROCEDURE IML used 7.00 seconds.
   22    /* if you want to do additional steps without reinvoking IML
   23       then leave out the quit; statement          */
```

Output Window

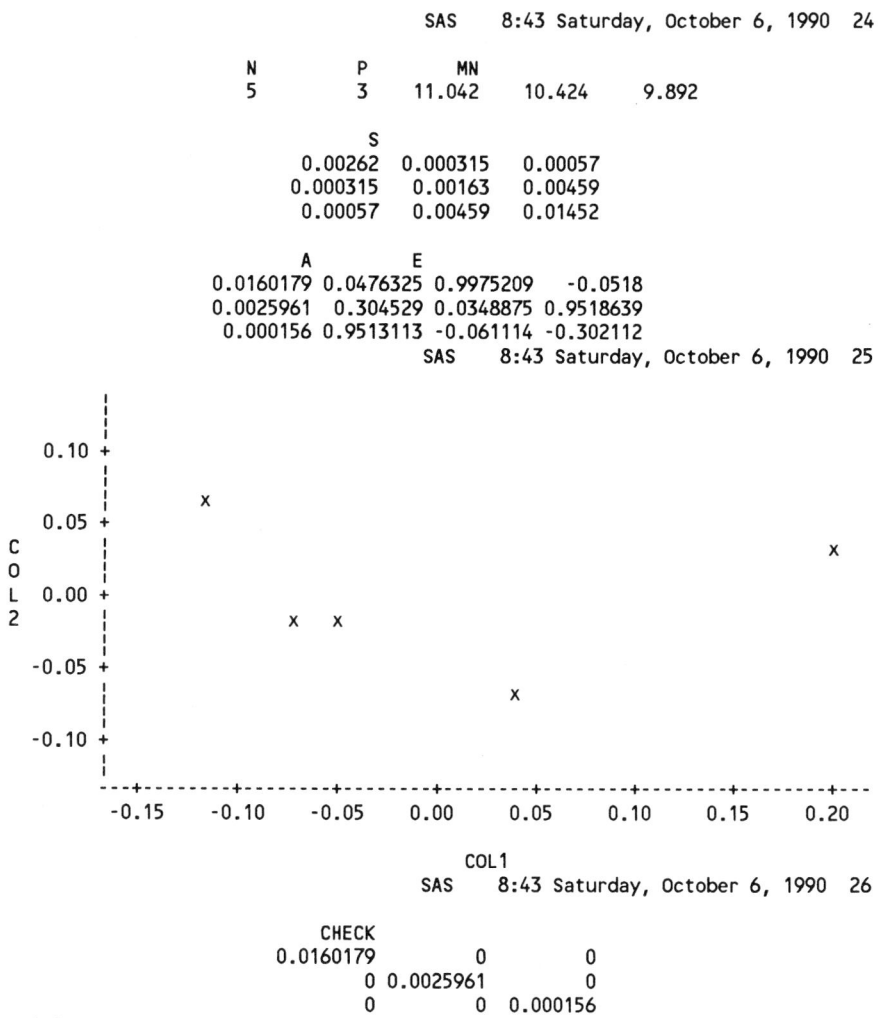

```
                              SAS     8:43 Saturday, October 6, 1990  24

              N         P        MN
              5         3      11.042    10.424     9.892

                        S
                0.00262  0.000315   0.00057
                0.000315  0.00163   0.00459
                0.00057   0.00459   0.01452

                A           E
        0.0160179 0.0476325 0.9975209   -0.0518
        0.0025961  0.304529 0.0348875 0.9518639
        0.000156 0.9513113 -0.061114 -0.302112
                              SAS     8:43 Saturday, October 6, 1990  25
```

```
        0.10 +
             |
             |     x
        0.05 +
  C          |
  O          |                                                    x
  L   0.00 +
  2          |
             |          x    x
       -0.05 +
             |                        x
       -0.10 +
             |
        ---+--------+--------+--------+--------+--------+--------+--------+---
         -0.15    -0.10    -0.05     0.00     0.05     0.10     0.15     0.20

                              COL1
                              SAS     8:43 Saturday, October 6, 1990  26
```

```
                 CHECK
        0.0160179         0         0
                0 0.0025961         0
                0         0  0.000156
```

Exiting IML.

Comparison of IML Code to Matrix Algebra

In Chapter 2 of the text is a short summary of Matrix Algebra. Below I have given Reyment's matrix notation, and the corresponding IML code for some common symbols and operators. I have left the required semi-colons (;) for SAS out of this table.

Definition	Reyment Notation	SAS IML
Symbol for a matrix	A	A
Transpose	A^T	$A`$
Symbol for a vector	$\mathbf{a}$	a
Transpose of vector	$\mathbf{a}^T$	$a`$
Addition of 2 vectors	$a + b$	$a + b$
Minor product of 2 vectors	$a^T \cdot b$	$a` * b$
Major product of 2 vectors	$M = a \cdot b^T$	$M = a * b`$
Matrix multiplication	$C = AB$	$C = A * B$
Major product moment	$C = XX^T$	$C = X * X`$
Minor product moment	$E = X^T X$	$E = X` * X$
Vector of ones (length n)	$\mathbf{1}$	$j(n, 1)$
Individual x element	x_{ij}	$x[i, j]$

Some simple statistics shown in the Matrix Algebra section of the text are similarly shown using both systems.

Statistic	Reyment Notation	SAS IML
Mean vector	$\bar{x} = 1^T X / 1^T 1$	$x\,\text{hat} = j(N, 1)` * X / j(N, 1)` * j(N, 1)$
Deviate score	$y_{ij} = x_{ij} - \bar{x}_j$	$y[i, j] = x[i, j] - x\,\text{hat}[, j];$
Variance j	$s_j^2 = y_j^T y / 1^T 1$	$s[j, j] = y[, j]` * y[, j] / j(1, N) * j(N, 1)$
Standard score	$z_{ij} = y_{ij}/s_j$	$z[i, j] = y[i, j]/s[j, j] \# \# .5$
Covariance	$s_{ij} = y_i^T y_j/(N - 1)$	$s[i, j] = y[, i]` * y[, j]/(N - 1)$
Covariance matrix	$S = Y^T Y/(N - 1)$	$S = Y` * Y/(N - 1)$
Correlation	$r_{ij} = S_{ij}/S_i S_j$	$r[i, j] = s[i, j]/(s[i, j] * s[j, j]) \# \# .5$
Correlation matrix	$R = Z^T Z/N$	$R = Z` * Z/N$

By now you can see some obvious relations between Reyment's matrix notation and IML notation. Subtraction, division by a scalar and other simple algebraic operations are very similar in SAS's PROC IML to ordinary matrix algebra. I have tried to be consistent using N (upper case) following Reyment for sample size; and a capital letter, e.g. X, or a word starting with a capital letter, e.g. Score, to indicate matrices. Scalars and vectors are given in lower case.

When SAS is run interactively on a PC, the screen is divided into three windows: the PROGRAM EDITOR window, the LOG window and the OUTPUT window. A brief introduction to managing these is given in Appendix I to the Supplement. All three windows are available on the screen at one time or any one may be enlarged (zoomed) to fill the screen. PROC IML only makes use of the OUTPUT window, if the option NOLOG is invoked on the RESET line. This option is used here. An example for a small program presenting print-outs of the contents of all three windows is reproduced below. SAS may also be run from DOS in batch mode (see SAS manuals).

Example of Getting Data into IML

Program in File DATAIN.SAS

```
data briz; /* name I assign SAS data set for Brizalina data */
infile 'brizdat.one'; /* pointer to file containing data */
input x1-x3;  /* defines variables and free format in this case */
proc print; /* to see that data is now SAS data set call briz */
proc iml;
use briz;  /* Open the Briz data set for reading */
read all into X; /* put all 15 values in matrix X */
print X;    /* to see if data really in X */
quit;
```

Contents of LOG Window in File DATAIN.LOG

```
   201    data briz; /* name I assign SAS data set for Brizalina data */
   202    infile 'brizdat.one'; /* pointer to file containing data in current
directory */
   203    input x1-x3;  /* defines variables and free format in this case */
   204    proc iml;
NOTE: The infile 'brizdat.one' is file D:\SUPPLEMT\BRIZDAT.ONE.
NOTE: 5 records were read from the infile D:\SUPPLEMT\BRIZDAT.ONE.
      The minimum record length was 17.
      The maximum record length was 17.
NOTE: The data set WORK.BRIZ has 5 observations and 3 variables.
NOTE: The DATA statement used 3.00 seconds.
IML Ready
   205    reset nolog;
   206    use briz; /* Open the Briz data set for reading */
   207    read all into X; /* put all 15 values in matrix X */
   208    print X;          /* to see if data really in X */
   209    quit;
NOTE: The PROCEDURE IML used 1.00 seconds.
```

Contents of OUTPUT Window in File DATAIN.OUT

```
                          SAS    8:43 Saturday, October 6, 1990   27

                   X
                11.1     10.38     9.78
                11.02    10.42     9.84
                11.02    10.41     9.82
                10.98    10.42     9.94
                11.09    10.49    10.08
Exiting IML.
```

Chapter 3 Programs

These programs do a great deal of multivariate morphometrics and produce the same results as in the text, with a few exceptions. They can be used for computations with other data sets with minor modifications.

Principal Component Analysis

This program computes PCAs for both the correlation and covariance matrix. It uses the first 50 crabs in the CRAB.DAT data file (explained in the text). When you run it you should get the same results as in Tables 3.1 and 3.2.

Program in File CRAB1.SAS

```
/*      Program to Compute PCA for Covariance and Correlations
 This example is described in chapter 3 and produces the same
 results as are found in Tables 3.1 and 3.2                    */
data crab;infile 'crab.dat' obs=50; /* use first 50 crabs */
    input FL RW CL CW BD;
    sex="female" ;  /* assigns character variables for sex and species */
    morph="orange"; /* for labelling. */
proc iml;
reset fuzz nolog noname  /* noname leaves matrix name off */
pagesize=66;             /*  pagesize for 8.5"x11" paper */
use _last_; /* Note below vars are variable name; morph is species */
read all into X[colname=vars rowname=morph];
/* X=log(X); if wanted to use logs would include this line */
N=nrow(X);
p=ncol(X);
print "N= " n  " P= " p, ; /* final comma skips a line for neater output */
I=j(n,1);   /* makes notation look like Reyment's */
mean=I'*X/(I'*I);
print "Means",mean[colname=vars],; /* print prints a blank line */
Y=X-I*mean;   /* deviations from mean */
S=(Y'*Y)/(N-1);    /* covariance matrix */
call eigen(l, E, S);
D=diag(vecdiag(S)##-.5); /* produces diagonal matrix of standard devs.*/
A=D*E*diag(l##.5); /* matrix of loadings or correlations between X
                       and pc scores */
ltot=sum(l); /* three lines to get eigenvalues as %'s */
lpercent=100*l/ltot;
l=l||lpercent;  /* horizonal concatenation of l and lpercent */
print "Covariance Matrix", S[format=7.4 rowname=vars colname=vars],;
lab1={Value Percent}; /* this contains a vector of string values */
print "Eigenvalues of Covariance Matrix", l[format=7.4 colname=lab1],;
lab2={E1 E2 E3 E4 E5};
print "Eigenvectors of Covariance Matrix", E[format=7.4 rowname=vars
         colname=lab2],;
print "Correlations of X and PC Scores",A[format=7.4 rowname=vars],;
R=D*S*D; /* Correlation matrix */
print "Correlation Matrix",r[format=6.4 rowname=vars colname=vars],;
call eigen(l, E, R);
ltot=sum(l);
```

```
lpercent=100*l/ltot;
l=l||lpercent;
print "Eigenvalues of Correlation Matrix", l[format=7.4 colname=lab1],;
print "Eigenvectors of Correlation Matrix", E[format=7.4 rowname=vars
      colname=lab2];
```

Programs for Principal Coordinate Analysis

Two versions of this program are given. One reproduces the results given in the text; while the other will produce exactly the same plots as an R mode principal components run on the same data using the covariance matrix.

a. Principal Coordinates using Gower's (1971) Association Matrix

This program will reproduce the scores given in Table 3.3 except for a difference in the signs, which is not unexpected as scores are only defined to be orthogonal. Note the strong horseshoe effect in the plot of the first two principal coordinates and contrast this result to the plot in the b. version below.

Program in File GOWCRAB.SAS

```
/*   Illustration of Principal Coordinates on 1st 10 female and
     10 male orange crabs; using Gower's Association Measure
     (See and run similar next program that uses distance squared*/
data crab;infile 'crab.dat';
   i+1;                      /* this logic in next 7 steps selects */
      input FL RW CL CW BD; /* 1st 10 female and male orange crabs */
      if i<=10|
       (i>50 & i<=60);
      color="orange";
      if i<=10 then sex="f";
      if i>50 & i<=60 then sex="m";
   drop i;
proc iml;
reset fuzz nolog noname pagesize=66;
use crab;
read all into X[rowname=sex colname=vars];
N=nrow(X);
p=ncol(X);
range=x[<>,]-x[><,]; /* Required for Gower's association measure */
print range[colname=vars];
A=I(N); /* set aside space for association matrix a */
do i=1 to N-1;  /* these next 10 steps computes Gower's measure */
   do j=i+1 to N;   /* between all crabs */
      a[i,j]=(p-(abs(x[i,]-x[j,])*(range##-1)'))/p;
      a[j,i]=a[i,j];
   end;
end;
rmean=a[:,];   /*  these three steps centers matrix - see Gower */
gmean=rmean[,:];
Q=A-j(n,1,1)*rmean-rmean'*j(n,1,1)'+j(n,n,1)*gmean;
```

```
call eigen(a,E,Q);
E=e[,1:3];
a=a[1:3,];
print "Eigenvalues ", a;
Coord=E*diag(a##.5); /* find principal coordinates */
lab1={PCOORD1 PCOORD2 PCOORD3};
/* Results below same as Table 3.3 in Text - note sign reversals! */
print Coord[format=5.2 rowname=sex colname=lab1];
XY=Coord[,1:2]; /* copies 1st 2 columns of Coords to XY matrix */
/* Note the form of graphs and contrast to results using distance**2 */
call pgraf(XY,sex,"COORD1","COORD2"); /* plots 2nd prin. coord against 1st */
XY=Coord[,{1 3}]; /* like above but 1st and 3rd column */
call pgraf(XY,sex,"COORD1","COORD3");
XY=coord[,2:3];  /* like above but 2nd and 3rd column */
call pgraf(XY,sex,"COORD2","COORD3");
```

b. Principal Coordinates using Euclidean Distance Square between Crabs

Program in File CRABCORD.SAS

```
/*    Illustration of Principal Coordinates on 1st 10 female and
10 male orange crabs; using Euclidean distance squared
(See and run similar program using Gower's Association)*/
data crab;infile 'crab.dat';
.
. NOTE - LEFT OUT LINES exactly the same as in CRABPCA.SAS above.
.          They are included in the File on Disk.
.
XSS=X*X'; /* major product moment of data for crabs */
ss=vecdiag(XSS); /* elements are sums of squares for each crab */
Dist=SS*j(n,1,1)'+j(n,1,1)*ss'-2*XSS; /* computes distances */
rmean=dist[:,]; /* form mean centered association matrix */
gmean=rmean[,:];
Q=-.5*(Dist-j(n,1,1)*rmean-rmean'*j(n,1,1)'+j(n,n,1)*gmean);
call eigen(a,E,Q);
E=E[,1:3];
a=a[1:3,];
print "Eigenvalues ", a,;
Coord=E*diag(a##.5); /* find principal Coordinates */
lab1={PCOORD1 PCOORD2 PCOORD3};
/* Results below are very different from Table 3.3 in Text since
   the Association Measure is based on Euclidean Distance Squared */
print Coord[format=5.2 rowname=sex colname=lab1],;
reset pagesize=24;
XY=Coord[,1:2]; /* copies 1st 2 columns of Coords to XY array */
/* Note the form of graphs and contrast to Gower's Association */
call pgraf(XY,sex,"I","II"); /* plots 2nd prin. Coord against 1st */
XY=Coord[,{1 3}]; /* like above but 1st and 3rd column */
call pgraf(XY,sex,"I","III");
XY=Coord[,2:3];  /* like above but 2nd and 3rd column */
call pgraf(XY,sex,"I","III");
```

Correspondence Analysis

This program uses the same crab data as in the principal coordinates example above. The method is designed for frequency data, but is used here only to compare the ordinations produced by the various methods. Note that the patterns are similar except when the Gower association is used for the Principal Coordinates. The program will work on any data matrix.

Program in File CRABCORR.SAS

```
data crab;infile 'crab.dat';
  i+1;                      /* this logic in next 7 steps selects */
    input FL RW CL CW BD; /* 1st 10 female and male orange crabs */
    if i<=10|              /* from data set with 200 crabs       */
    (i>50 & i<=60);        /* same as in GOWCRAB.SAS             */
    color="orange";
    if i<=10 then sex="f";
    if i>50 & i<=60 then sex="m";
  drop i;
proc iml;
reset fuzz nolog noname spaces=3 pagesize=66;
use _last_;
read all into K[ROWNAME=SEX COLNAME=VARS]; /* K will contain data
for the crabs */
tot=sum(K); /* finds sum of elements in matrix K */
K=K/tot;   /* standardizes all elements by dividing through by total */
N=nrow(K);
p=ncol(K);
print "N = " N " p = " p,;
CD=diag(K[+,]##-.5); /* These two produce diagonal matrices*/
RD=diag(K[,+]##-.5); /* of reciprocal sq. root of row and col.totals */
X=RD*K*CD;         /* X is re-scaled matrix */
Sim=X'*X;   /* Association matrix for re-scaled x */
print "Similarity Matrix",Sim [Format=6.4 ROWNAME=VARS COLNAME=VARS],;
call svd(U,q,V,X); /* Singular Value Decomposition */
lat_root=q[2:p,]##2; /* Re-scales singular values to roots */
total=sum(lat_root);
inertia=100*lat_root/total;
percent=cusum(inertia);
Out2=inertia||percent;
lab1={ INERTIA PERCENT};
print "LATENT ROOT  % INERTIA  CUM %",lat_root[FORMAT=9.7] out2[FORMAT=8.2],;
XSC=RD*U[,2:3]*diag(q[2:3,]); /* Correspondence axial loadings */
YSC=CD*V[,2:3]*diag(q[2:3,]);
print "QMODE" XSC[FORMAT=6.3] "RMODE" YSC[FORMAT=6.3],;
reset pagesize=24;
call pgraf(XSC,sex,,,"Q-Mode");
call pgraf(YSC,vars',,,"R Mode");
quit;
```

Program for Discrimination of Two Groups and Hotelling's T^2

This program uses the same crab data as above, but it finds a discriminant function for the two sexes. The code will produce results comparable to those given in the text of the book; note that these programs can be used for almost any data set. The program also tests the equality of the mean vectors for the two sexes using Hotelling's T^2 statistic. Quadratic discrimination errors are compared to those for linear discriminant analysis.

Program in File CRABDSQ.SAS

```
/* This program is a modification of the CRABCVA program, for only two
   groups. It computes classification error results for linear and quadratic
   discriminators */
data crab;infile 'crab.dat' obs=100; /* orange crabs only */
do grp=1 to 2;      /* Data input and identified with sex */
   do j=1 to 50;
      input FL RW CL CW BD;
      if grp=1  then sex="f";
      if grp=2 then sex="m";
      ind=grp;  /* index variable for group membership - see below */
      output;
end; end;
drop j grp;
 proc iml worksize=80;
reset fuzz nolog noname pagesize=66;
use _last_;
read all into X [colname=var rowname=sex];
p1=ncol(X); /* number of variables plus IND */
p=p1-1;
   VAR=VAR[1:p]; rowvar=var';
   NIND=NROW(X);
   NGRP=X[NIND,p1]; /* This is the number of groups */
print "No. Inds. ="NIND "p="p"No. Groups="NGRP, "Variables" rvar ,;
   GROUP=X[,p1];     /* Vector containing group membership */
   D=DESIGN(group);  /* Produces an NINDxP1 matrix for group indices*/
   NI=D[+,];         /* Row Vector of number of inds. in each group */
   X=X[,1:p];
 Mni=(diag(NI)**-1)*D'*X;
 mn_dif=Mni[1,]-Mni[2,];
  lab1={female male};
  print "Group Means",Mni[FORMAT=6.2 COLNAME=VAR ROWNAME=LAB1],;
print "Mean Differences",mn_dif[FORMAT=5.3 COLNAME=VAR],;
DEV=X-D*MNI; /* deviations about group means */
SW=DEV'*DEV/(NIND-NGRP);
print "Pooled Within Covariance Matrix",SW[FORMAT=7.2 ROWNAME=VAR COLNAME=VAR],;
DIGW=DIAG(VECDIAG(SW)##-.5);
RW=DIGW*SW*DIGW;
print "Within Correlation Matrix",RW[FORMAT=6.4 COLNAME=VAR ROWNAME=VAR],;
Coeff=mn_dif*INV(SW);
PRINT "Discriminant Coefficients",Coeff[FORMAT=5.2 COLNAME=VAR],;
Stcoeff=Coeff*diag(sqrt(vecdiag(SW)));
print "Standardized Coefficients", Stcoeff[FORMAT=5.2 COLNAME=VAR],;
dsq=Coeff*mn_dif';
```

```
PRINT "Mahalanobis D**2=" dsq[FORMAT=6.2],;
tsq=(1/NI[,1]+1/NI[,2])##-1*dsq/(NI[,1]+NI[,2]-2);
PRINT "Hotelling TSQ = " tsq[FORMAT=6.2],;
den_df=nind-p-1;
f=den_df*tsq/p;
PRINT " F = " F[FORMAT=5.2] "with " p " AND " den_df "df",;
p=1-PROBF(f,p,den_df);
PRINT "Probability of a larger F = "p,;
Score=(X-j(nind,1)*x[:,])*Coeff';
ierror=j(1,2,0); /* These 6 lines find error of classification */
do i=1 to nind; ;
   if i<=NI[,1] & Score[i,]<=0 then ierror[,1]=ierror[,1]+1;
   if i>NI[,1] & Score[i,]>0 then ierror[,2]=ierror[,2]+1;
end;
print"Linear discriminator", "Errors in Group 1 & 2"ierror,;
Dgrp=j(nind,2);
sw1=dev[1:NI[,1],]'*dev[1:NI[,1],]/(NI[,1]-1);
sw2=dev[NI[,1]+1:NIND,]'*dev[NI[,1]+1:NIND,]/(NI[,2]-1);
ldet1=log(det(sw1));
isw1=inv(sw1);
isw2=inv(sw2);
ldet2=log(det(sw2));
do i=1 to 2;
 if i=1 then do;si=isw1;ldet=ldet1;end;
 if i=2 then do;si=isw2;ldet=ldet2;end;
 do j=1 to NIND;
    dgrp[j,i]=(x[j,]-mni[i,])*si*(x[j,]-mni[i,])'+ldet;
end;end;
ierror=j(1,2,0);
do I=1 to  nind;
   if i<=ni[,1] & dgrp[i,1]>dgrp[i,2] then ierror[,1]=ierror[,1]+1;
   if i>ni[,1] & dgrp[i,1]<dgrp[i,2] then ierror[,2]=ierror[,2]+1;
end;
print"Quadratic discriminator", "Errors in Group 1 & 2"ierror,;
```

Canonical Variate Analysis

This program is illustrated with the crab data again, but this time for all four groups—2 species with each of the 2 sexes. Tests are not done here, but one can follow the matrix algebra required. It is suggested that most analyses be done using PROC CANDISC or DISCRIM in SAS.

Program in File CRABCVA.SAS

```
/* CVA on crab data  using all 200 specimens*/
data crab;infile 'crab.dat';
do grp=1 to 4;       /* Data input and identified with sps-sex */
   do j=1 to 50;
      input FL RW CL CW BD;
      if grp=1 then sps_sex="of";
      if grp=2 then sps_sex="om";
```

```
      if grp=3 then sps_sex="bf";
      if grp=4 then sps_sex="bm";
      ind=grp;  /* index variable for group membership - see below */
      output;
end; end;
drop j grp;
/* proc discrim canon all;class sps_sex; */
 proc iml worksize=80;
reset fuzz noname nolog pagesize=66;
use _last_;
read all into X [colname=var rowname=sps_sex];
p1=ncol(X); /* number of variables plus IND */
p=p1-1;
    var=var[1:p]; rvar=var';
    NIND=nrow(X);
    Ngrp=X[NIND,p1]; /* This is the number of groups */
print "No. Inds.="NIND"p=" p "No. Groups="Ngrp,"Variables" rvar,;
    GROUP= X[,p1];      /* Vector containing group membership */
    D=DESIGN( group);   /* Produces an NINDxP1 matrix for group indices*/
    NI=D[+,];           /* Row Vector of number of inds. in each group */
    X=X[,1:p];
    tmn= x[:,]; /* Grand mean of data array */
print "Grand Means",tmn[FORMAT=6.2 COLNAME=VAR],;
    T=(X'*X-tmn'*tmn*NIND);
print"Total Sum of Squares and Cross Product Matrix",
      T [FORMAT=8.2 ROWNAME=VAR COLNAME=VAR],;
 mni=(diag(NI)**-1)*D'*X;
  LAB1={of om bf bm};
  print "Group Means",mni[FORMAT=6.2 COLNAME=VAR ROWNAME=LAB1],;
DEV=X-D*mni; /* deviations about group means */
W=DEV'*DEV;
print "Within Sum of Squares Matrix",W[FORMAT=7.2 ROWNAME=VAR COLNAME=VAR],;
DIGW=diag(vecdiag(W)##-.5);
RW=DIGW*W*DIGW;
print "Within Correlation Matrix",RW[FORMAT=6.4 COLNAME=VAR ROWNAME=VAR],;
B=T-W;
free T D RW ;
print "Between Groups Matrix of Sums of Squares and Cross Products",
       B [format=7.2 colname=var rowname=var],;
call geneig(a,E,B,W); /* Eigenvalues and vectors of W**-1*B */
nroots=min(p,Ngrp-{1});
a=a[1:nroots,]; /* nroots=lesser of (Ngrp-1) and p variables */
print "Latent Roots",a [FORMAT=5.3],;
E=E[,1:nroots];
Stde=Diag(vecdiag(W)##.5)*E;
print "Standardized Eigen Vectors ",Stde[FORMAT=8.4 ROWNAME=VAR],;
Canmean=(mni-j(ngrp,1)*tmn)*E*(nind-ngrp)##.5;
print, "Canonical Means",Canmean[FORMAT=6.3  ROWNAME=LAB1],;
Dtemp=Canmean*Canmean';
ddiag=vecdiag(Dtemp);
Dss=ddiag*j(1,ngrp);
Dsq=Dss+Dss'-2*Dtemp;
print "Mahalanobis D**2", Dsq[FORMAT=6.2 ROWNAME=LAB1 COLNAME=LAB1],;
Canscr=X*E;
call pgraf(Canscr[,1:2],sps_sex,"CANVAR1","CANVAR2");
quit;
```

Shrunken Estimators for Canonical Variate Analysis based on Campbell (1980)

This procedure adjusts for the contribution of very small or relatively small latent roots in the pooled within covariance matrix and produces a new ordination based on these adjusted estimates. The procedure is very similar to the kind of estimates used in ridge regression. Campbell's criterion, based on contribution of the pooled within (W) latent roots, was used rather than deleting variables as in Reyment.

Program in File CAMPRIDG.SAS

```
data crab;infile 'crab.dat';
do i=1 to 4;              /* alternate method of reading data and */
   if i=1 then name='of';  /* associating group names */
   if i=2 then name='om';
   if i=3 then name='bf';
   if i=4 then name='bm';
   do j=1 to 50;
      input fl rd cl cw bd;
      output;
   end;end;
drop i j ;
proc freq;  /* let program find number in each group */
   table name/noprint out=nlist;
proc iml worksize=80;
reset fuzz nolog noname pagesize=66;
use nlist;  /* these 4 steps get into iml and print counts for groups */
vname={count};
read all var vname into N[rowname=name];
N=N';
use crab;  /* bring in crab data  - can be any other data set */
Ntot=N[,+]; /* determines total observations and prints */
print "Total number of specimens = " Ntot,;
Nloc=ncol(N); /* determines number of groups (localities here) */
print "Number of groups = " Nloc,;
print "Number of specimens in each group",N,;
ipoint=1;        /* next ~19 lines read from file and compute */
do l=1 to Nloc; /* pooled within covariance matrix  W  */
  do i=1 to N[1,l];      /* unlike earlier routines this one does not
    store the data, but reads and computes so it consumes
    less array space.  A much larger data set can be handled this way */
      read point ipoint into x[colname=vari];
        if i*l=1 then do;
            m=ncol(x);
            print "No. Variables =" m,; /* automatically determined */
            W=j(m,m,0);
            end;
        if i=1 then y=j(1,m);
        ipoint=ipoint+1;
        if i=1 then y=x;
          else Y=Y//x;
  end;
  mn=y[:,];                    /* mean vector for group i */
```

```
  if l=1 then Mns=mn;        /* sets up array to store group means */
  if l>1 then Mns=Mns//mn;
  W=W+Y`*Y-mn`*mn*n[1,l];
end;
  S=diag(vecdiag(w)##-.5);   /* diagonal with reciprocal st. devs. */
  Wstar=S*W*S;/* within correlation matrix */
print "Within Correlation matrix",Wstar[format=5.3 colname=vari rowname=vari],;
call eigen(e,U,Wstar);   /* find and print eigenvalues; ecs of W* */
print "Eigenvalues of Wstar",e[format=5.3 ],;
print "Eigenvectors of Wstar",U[format=6.3 rowname=vari],;
print "Group Means",Mns[format=5.3 rowname=name colname=vari],;
tmn=N*Mns/Ntot;   /* grand mean */
print "Grand Mean", tmn[format=5.3 colname=vari],;
B=(Mns`*diag(N`)*Mns-Ntot*tmn`*tmn); /* Among sum squares matrix */
Bstar=S*B*S; /* computes among matrix standardized by w/in st. devs. */
Nact=min(m,Nloc-{1}); /* number of canonical variates */
 call geneig(a,V,Bstar,Wstar); /* solution to INV(Wstar)*Bstar */
 a=a[1:nact,];
V=V[,1:nact];
 print "Canonical Eigenvalues",a[format=6.4],;
 print "Canonical variate coefficients", V[format=7.3 rowname=vari],;
tra=sum(a);
print "Sum of Eigenvalues" tra[FORMAT=6.3],;
bsq=vecdiag(diag(e##-1)*U`*Bstar*U); /* this line corresponds to
     last column of Table 2 (Campbell, 1980) and gives contribution
     of within eigen structure to among */
print "Between groups SQ",bsq [format=6.4],;
bsqtot=BSQ[+,];
print "Total for Between groups SQ",bsqtot[format=6.4],;
bsq=100*bsq/bsqtot;
print "BSQ as percentages ",bsq[format=4.1],;
do i=1 to m;     /* This part computes ridge constants for
                       within using rule suggested in discussion of
                       Campbell (1980)        */
  if bsq[i,1]<10  then do;  /* change 10% to other value if want */
    ind=bsq[i,1]*e[i,1]*100/sum(e); /* ridge adjustment to Wstar
                                        eigenvalues */

    e[i,1]=e[i,1]+.5/(1+2*ind);
  end;
end;
print "Ridge adjusted eigenvalues of Wstar",e[format=5.3],;
/* Note main adjustment is to 3rd eigenvalue in this example */
Other=diag(e##-.5)*U`*Bstar*U*diag(e##-.5); /*  new coeffs. */
call eigen(a1,E1,Other);
a1=a1[1:nact,];
E1=E1[,1:nact];
New=U*diag(e##-.5)*E1;  /* Shrunken estimators of coefficients */
print "Shrunken canonical eigenvalues ",a1[format=6.4],;
print "Shrunken canonical eigenvectors ",E1[format=6.3 rowname=vari],;
print "Shrunken canonical coefficients ",New[format=6.3 rowname=vari],;
tra1=sum(a1); print tra1;
Canmns=(Mns-j(nloc,1)*tmn)*S*V*(ntot-nloc)##.5;
print "Canonical Variate Means before Shrinking",Canmns[FORMAT=6.3 ROWNAme=name],;
reset pagesize=24;
call pgraf(canMns[,1:2],name); /* plot of Un-shrunk Canonical Means */
Canmns=(Mns-j(nloc,1)*tmn)*S*New*(ntot-nloc)##.5;
print "Shrunken canonical means",canMns[format=6.3 rowname=name];
call pgraf(Canmns[,1:2],name); /* plot of Shrunken Canonical Means */
```

Canonical Correlation and Redundancy Analysis

Program in File CANCOR.SAS

```
/* Example of canonical correlation with ostracod Veenia fawwarensis
   Note $$$$$ where have to replace values for other data sets */
data ostracod; infile 'veenia2.dat';
input x1-x7;
proc iml worksize=70;
reset fuzz fw=5 noname nolog /* fw sets field widths for output */
pagesize=66;
use _last_;
read all into X[colname=var];
Nu=nrow(X);
p=ncol(X);
print "Number of inds. = " Nu " Number of variables " p,;
xvar=4; /* $$$$$$$$$ Input Number of variables in X set
        number of variables in Y set computed by subtraction */
m=xvar;
m1=m+1;
n=p-m;
varx=var[,1:m];
vary=var[,m+1:p];
S=(X'*X-nu*x[:,]'*x[:,])/(nu-1); /* next three lines total R */
Di=diag(vecdiag(S)##-.5);
R=Di*S*Di;
R11=R[1:m,1:m];    /* partition R to X, XY, and Y part */
R22=R[m1:p,m1:p];
R12=R[1:m,m1:p];
print "Correlation matrix ",R[format=5.3 colname=var rowname=var],;
print "Correlation matrices",R11[format=5.3 COLNAME=varx ROWNAME=varx],
R12[format=5.3 colname=vary rowname=varx],
R22[format=5.3 colname=vary rowname=vary],;
Result=R11**-1*R12*R22**-1*R12';
U=root(R22);
C=inv(U')*R12'*INV(R11)*R12*INV(U); /* symmetrical matrix for
                                        roots and vectors */
/* print C; */
call eigen(a,E,C);
nroots=min(m,n);
cc=sqrt(a[1:nroots,]);
print "Canonical Correlations",CC [FORMAT=6.3],;
Vec=inv(U)*E;
B=(INV(R11)*R12)*VEC*DIAG(a##-.5);
print "Left-hand coefficients ",B[FORMAT=5.2 rowname=varx],;
print "Right-hand coefficients ",Vec[FORMAT=5.2 rowname=vary],;
RUX=R22*VEC;
RUY=R11*B;
print "Left-hand structure",RUY[format=6.3 rowname=varx],;
print "Right-hand structure",RUX[format=6.3 rowname=vary],;
/* Z=(X-j(nu,1)*x[:,])*di; These lines are check for computations
Ui=Z[,1:m]*B;
Vi=Z[,m1:p]*VEC;
R1=Ui'*Ui/(Nu-1);
R2=Vi'*Vi/(Nu-1);
R3=Ui'*Vi/(Nu-1);
r4=a##.5;
print r1,r2,r3,r4;     */
```

Programs Motivated by Integrated Morphological Analysis on Brizalina

Create a SAS data Set

This program takes the original *x,y* coordinates digitized by Bookstein (1990) from the file BRIZALIN.DAT and converts them to the logs of the euclidean distances squared between landmarks; and then stores them in BRIZLOG.SSD. The variable "mag" is a magnification factor that needs to be used as a multiplicative correction.

Program in File BRIZLOG.SAS

```
libname new '';
data new.brizlog;
infile "brizalin.dat";
input sample $ mag xB yB xC yC xD yD xA yA xE yE xF yF ;
log_ab=log(mag*sqrt((xb-xa)**2+(yb-ya)**2));
log_ac=log(mag*sqrt((xc-xa)**2+(yc-ya)**2));
log_ad=log(mag*sqrt((xa-xd)**2+(ya-yd)**2));
log_bc=log(mag*sqrt((xb-xc)**2+(yb-yc)**2));
log_bd=log(mag*sqrt((xb-xd)**2+(yb-yd)**2));
log_cd=log(mag*sqrt((xd-xc)**2+(yd-yc)**2));
drop mag xB yB xC yC xD yD xA yA xE yE xF yF ;
proc print;
```

Covariance and D^2 in a Module

Since several of the programs to follow will use the same data on *Brizalina*, and the covariance matrix as well as Mahalanobis D^2 of the observations to the Centroid; these have been put in a module called INDATA which will be stored in a disk file called by default IMLSTORE.SCT. This saves space. Otherwise this subroutine would have to be included in each program.

Program in File MODSTORE.SAS

```
/* this program creates a module for the input steps and
computation of covariance matrix and Mahalanobis Dsq
for each observation to be used in other modules
The module is stored in the current SASUSER directory */
proc iml worksize=64;
start indata;
  use _last_;
  read all into X[colname=name];
  N=nrow(X);
  p1=ncol(X);
  p=p1-1;
  X=X[,1:p];
  mn=X[:,];
```

```
Y=X-j(N,1)*mn;
S=Y'*Y/(N-1);
SI=inv(S);
dsq=j(N,1);
do i=1 to N;
   Dsq[i,]=Y[i,]*SI*Y[i,]'; /* Mahalanobis Dsq to each observation */
end;
finish;
store module=(indata);
```

Multivariate Skew, Kurtosis and Quantile-Quantile Plots of D²

These programs will use the *Brizalina* data which has been stored in the SAS data set BRIZLOG.SSD. This speeds up input and access to the data which is NOT stored in ASCII. The original BRIZLOG.DAT as printed in Table 3.8 is still available as a data set.

Program in File BRIZMAR2.SAS

```
/*  This program is written in modules and includes:
       1. Q-Q plots for Mahalanobis Dsquared and Chisquared Quantiles;
       2. Mardia multivariate skew;  3. Mardia multivariate kurtosis.
    Skew is very slow and it is deactivated below with comments.
    Any of the three parts can be ignored by commenting out the
    corresponding 'run' statement.  Note: Answers differ from Reyment */
 libname new '';
data new2;set new.brizlog;samp=1*sample;
proc iml worksize=64;
reset fuzz fw=7 nolog noname pagesize=24;
load module=(indata); /* Note uses module created in MODSTORE.SAS */
start quantile;  /****** part unique to Q-Q plot ******/
   create new from Dsq; /* 6 lines to sort using IML sort routine */
   append from Dsq;
   close new;
   sort new by col1;
   use new;
   read all into Dsq;
   chisq=cinv(do(.5,N-.5,1)/N,p);
   XY=Chisq'||Dsq;
   call pgraf(XY,,"Chi-Sq","D-Square","Q-Q Plot" );
finish;
start kurtose; /****** part unique to kurtosis test ******/
/*  Mardia's kurtosis statistic from formula in Gnanadesikan */
   b2p=sum(Dsq##2)/n;
   z=(b2p-p*(p+2))/sqrt(8*p*(p+2)/n); /* treated as N(0,1 */
   prob=1-probnorm(z);
   print "Kurtosis B2 =" b2p[FORMat=6.2],
       "Z ="z[format=6.3]"with Probability="prob[format=6.4],;
finish;
start skew; /***** Mardia's skew statistic test *****/
   m111=j(p*p,p,0);
   do i=1 to p;
     do j=1 to p;
       ind=(i-1)*p+j;
       do k=1 to p;
```

```
        do l=1 to N;
          m111[ind,k]=m111[ind,k]+y[l,i]*y[l,j]*y[l,k]/n;
        end;
      end;
    end;
  end;
b1p=0;
do i=1 to p;
  do j=1 to p;
    ind1=(i-1)*p+j;
    do k=1 to p;
      do i2=1 to p;
        do j2=1 to p;
          ind2=(i2-1)*p+j2;
          temp=si[i,i2]*si[j,j2]*m111[ind1,k];
          do k2=1 to p;
            b1p=b1p+temp*si[k,k2]*m111[ind2,k2];
          end;
        end;
      end;
    end;
  end;
end;
Chisq=N*b1p/6;
df=p*(p+1)*(p+2)/6;
prob=1-probchi(Chisq,df);
print "Skew b1 =" b1p,"Chisquared=" Chisq"with" df"df and prob" prob,;
finish;
run indata;
run kurtose;
run quantile;
/* run skew;  Commented out as very slow */
```

Comparison of Latent Roots and Latent Vectors for Samples

Program in File BRIZANPC.SAS

```
/************************************************************************
Program to compare eigenvalues and eigenvectors among samples;
finding only the first few eigenvectors; and printing them out
for each sample.
Must SORT the data BY group to make sure the order is correct.
We make use of the PROC SORT and PROC PRINCOMP here to save space
and IML code for Principal Components was completely given earlier
*************************************************************************/
libname new '';
proc sort data=new.brizlog; by sample;
PROC PRINCOMP COV NOPRINT OUTSTAT=PCAOUT ;BY SAMPLE;
   /*PCA for each Locality saving 3 components */
proc sort;by _TYPE_;  /* to get means etc. together */
proc iml;
reset nolog noname fuzz pagesize=66;
reset fuzz fw=7;
```

```
use pcaout where (_TYPE_="MEAN");
read all into mean [rowname=sample colname=var];
ngrp=nrow(mean);
p=ncol(mean);
nr=2; /*$$$$ choose number of eigenvalues and eigenvectors to retain */
print "Level Sample Means","Level"mean[FORMAT=6.2 COLNAME=VAR ROWNAME=SAMPLE],;
use pcaout where (_TYPE_="EIGENVAL");
read all into eigval;
percent=(eigval[,1:nr]'*diag(eigval[,+]##-1))*100;
eigval=eigval[,1:nr]';
print "First three EigenValues",eigval[FORMAT=6.3 COLNAME=SAMPLE],;
print "as Percentages",percent[format=6.1 colname=sample],;
use pcaout where (_TYPE_="SCORE");
read all into Eigvec [rowname=_NAME_];
tname=_NAME_[1:nr];
 do i=1 to ngrp;
  print nr "Eigenvectors for Level" i,;
    Temp=eigvec[(i-1)*p+1:(i-1)*p+nr,];
    print Temp[FORMAT=7.4 COLNAME=VAR rowname=tname],;
 end;
grp1=3; grp2=5; /* $$$$$ choose groups you want angles between */
eig1=eigvec[(grp1-1)*p+1:(grp1-1)*p+nr,];
eig2=eigvec[(grp2-1)*p+1:(grp2-1)*p+nr,];
x=eig1*eig2';
ang=j(nr,nr);
print "Cosines of angles ",x[ROWNAME=tname colname=tname],;
    ang=arcos(x)*180/3.14159;
print "Angles for levels " grp1 grp2,ang[FORMAT=5.1 colname=tname rowname=tname]
```

Krzanowski W Criterion for Number of Principal Components

Note this routine runs very slowly because of the number of SVDs required. It will not take large data sets. A larger version is available from L. F. Marcus for larger data sets, and has been tested against Krzanowski (1986).

Program in BRIZKRAS.SAS

```
libname new '';
proc iml worksize=64;
reset fuzz noname nolog pagesize=66;
use new.brizlog;
read all into Y[rowname=sample colname=var];
N=nrow(Y);
p=ncol(Y);
mn=y[:,];
S=(Y'*Y-mn'*mn*N)/(N-1);
DSD=diag(vecdiag(s)##-.5);
Y=(Y-j(N,1)*mn)*DSD;
R=Y'*Y/(N-1);
call eigen(a,E,R);
```

```
print "Number of individUals=" N "Number of Variables = " p,;
print "Means", mn[FORMAT=6.2 colname=var],;
print "Covariance Matrix", S[format=6.3 colname=var rowname=Var],;
print "Correlation Matrix", R[format=6.4 colname=var rowname=Var],;
free mn S DSD R;
print "Eigenvalues", a[format=6.4],;
free a E;
Yest=j(N,p,0);
call svd(U,d,V,Y);
 Ye=U*diag(d)*V';
press=j(p-1,1,0); /* to store Krzanowski's Press Values */
Ujall=j(p*n,p-1);  /* temporary storage for results of svd */
Djall=j(p*(p-1),1); /* loop creates some values needed later */
 do j=1 to p;
   uindl=(j-1)*n+1;
   uindu=j*n;
   vindl=(j-1)*(p-1)+1;
   vindu=j*(p-1);
    if j=1 then Z=Y[,2:p]; /* these 3 lines delete one column of Y */
    if j=p then Z=Y[,1:p-1];
    if j>1 & j<p then  Z=Y[,1:j-1]||Y[,j+1:p];
       call svd(Uj,dj,Vj,Z);
     ujall[uindl:uindu,]=uj;
     djall[vindl:vindu,]=dj;
 end;
do t=1 to p-1;  /* actUal compUtation of Press ValUes */
 do i=1 to n;
  if i=1 then Z=Y[2:N,];   /* these 3 lines delete one row of Y */
  if i=N then Z=Y[1:N-1,];
  if i>1 & i<N then   z=Y[1:i-1,]//Y[i+1:n,];
      zmn=z[:,];
      Z=Z-j(n-1,1)*zmn;
    call svd(Ui,di,Vi,Z);
    free Ui;
  do j=1 to p;
   uindl=(j-1)*n+1;
   uindu=j*n;
   vindl=(j-1)*(p-1)+1;
   vindU=j*(p-1);
   Uj=Ujall[uindl:uindu,];
   Dj=Djall[vindl:vindu,];
     ytemp=uj[i,t]*dj[t,1]##.5*di[t,1]##.5*vi[j,t];
     ytemp0=u[i,t]*d[t,1]*v[j,t];
     if ytemp*ytemp0<0 then ytemp=-ytemp;
     yest[i,j]=yest[i,j]+ytemp;
       press[t,1]=press[t,1]+(yest[i,j]-ye[i,j])##2;
   end;
 end;
  press[t,1]=press[t,1]/(N-1);
end;
/* Ydif=Ye-Yest;  if want differences between values and estimates
   print Ydif;  */
print "Krzanowski Press Values",press[format=6.3],;
ss0=ssq(Ye); /* sum of squares of deviations from svd) */
dr=(N-1)*p; /* Values required to compute W values */
dm=N+p-2;
dr=dr-dm;
```

```
w=j(p-1,1);   /* to store W ValUes - below computations of wi*/
w[1,1]=((ss0-press[1,1])/dm)/(press[1,1]/dr);
do j=2 to p-1;
   dm=N+p-2*j;
   dr=dr-dm;
   w[j,1]=((press[j-1,1]-press[j,1])/dm)/(press[j,1]/dr);
end;
print " W statistics ",w[FORMAT=6.3];
quit;
```

Krzanowski Procedure for Choosing Important Variables

Program in BRIZVRSL.SAS

```
libname new '';  /* refers to current directory */
/* Krzanowski module to select variables */
data new2;set new.brizlog;samp=1*sample;/* samp is now numeric */
proc iml worksize=64;
reset nolog noname;
load module=(indata);
run indata;
wselect=3;  /* number of variables selected using W criterion */
DSD=diag(vecdiag(S)##-.5);
Y=Y*DSD;
/* R=Y'*Y/(N-1); If you want correlations*/
call svd(U,d,V,Y);
ZSC=U[,1:wselect]*diag(d[1:wselect,]);
mean=zsc[:,];
ZSC=ZSC-j(n,1)*mean;
free U V d;
 do j=1 to p;
  if j=1 then Z=Y[,2:p];
  if j=p then Z=Y[,1:p-1];
  if j>1 & j<p then    Z=Y[,1:j-1]||Y[,j+1:p];
     ztemp=z[1,];
     call svd(Uj,dj,Vj,Z);
     free Vj;
     ZSCj=Uj[,1:wselect]*diag(dj[1:wselect,]);
     mean=zscj[:,];
     ZSCj=ZSCj-j(n,1)*mean;
     Temp=ZSCj'*ZSC;
     call svd(Us,ds,Vs,Temp);
     Q=Vs*Us';
     free Us Vs ;
     msq=sum(zsc[,##])+sum(zscj[,##])-2*trace(ZSCj*Q'*ZSC');
print"SS for variable "j" = " msq;
end;
```

Krzanowski Influence for Observations

Program in File BRIZANGL.SAS

```
 libname new '';
/* Krzanowski module for angles */
data new2;set new.brizlog;samp=1*sample;
proc iml worksize=64;
reset nolog noname;
load module=(indata);
run indata;
wselect=3; /* $$$$$$ number of variables selected using W criterion */
DSD=diag(vecdiag(s)##-.5);
Y=Y*DSD;
/* R=Y'*Y/(N-1); */
call svd(U,d,V,Y);
angle=j(N,1);
spec=j(N,1);
 do i=1 to N;
  spec[i,]=i;
  if i=1 then Z=Y[2:N,];
  if i=n then Z=Y[1:N-1,];
  if i>1 & i<N then    Z=Y[1:i-1,]//Y[i+1:N,];
     zmn=z[:,];
     Z=Z-j(n-1,1)*zmn;
    call svd(Ui,di,Vi,Z);
    Temp=V[,1:wselect]'*Vi[,1:wselect];
    call svd(Ut,dt,Vt,Temp);
    angle[i,1]=arcos(min(dt))*180/3.14159;
 end;
angle=spec||angle;
lab={"Specimen" "Minangle"};
print "Influence angles ",angle[FORMAT=5.2 colname=lab];
```

Canonical Variate Analysis for Brizalina

Note that the Canonical Variate Analysis in the text can be done using the Canonical Variate Program illustrated with the crab data. Changes required in the program to accommodate the *Brizalina* data are shown highlighted below with # # # . Of course, different statements before PROC IML are needed to access the appropriate data. A statement is added to print the unstandardized Canonical Vectors as well.

Program in File BRIZCVA.SAS

```
libname new '';
data new2;set new.brizlog;
ind=1*sample; /* defines numerical grp. index in same place as in crabs*/
 proc iml worksize=80;
reset fuzz noname nolog pagesize=66;
use _last_;
```

```
read all into X [colname=var rowname=sample];  /* ### */
.
.
. Note: Program lines same as CRABCVA.SAS.  They are in the file.
.       Lines that are different are marked with ###.
.
.
 mni=(diag(NI)**-1)*D'*X;
   lab1={One Two Three Four Five}; /* ### */
DEV=X-D*mni; /* deviations about group means */
.
E=E[,1:nroots];
print "Eigenvectors (Canon. Var. Vectors",e[format=8.4 rowname=VAR],; /* ### */
.
Canscr=X*E;
reset pagesize=24;
call pgraf(Canscr[,1:2],sample,"CANVAR1","CANVAR2"); /* ### */
quit;
```

Program for Robust Estimation of Means and Covariances

This program is based on Campbell (1980) and uses the VEENIA data. It offers two options for weighting the data both based on the Mahalanobis distance of an observation from the centroid of the data. It also produces Q–Q plots for data to look for outliers.

CAMPROB1.SAS

```
data veenia;infile 'veenia';
/* Robust estimation of Means and Variance Covariance Matrix
   This program follows Campbell(1980).
   Also includes Q-Q plot before and after weighting.
   The b0 and b1 values are those Campbell suggested for the
   "re-descending" weights.  For the non-descending form you
   can use a very large value of b2 and the weight will then
   be proportional to Mahalanobis Dsquared */
input y1-y7;
proc iml worksize=64;
reset fuzz noname nolog ;
start quantile; /****** Q-Q plot subroutine ******/
.
. Note: Same as in BRIZMAR2.SAS
.
finish;
use veenia;
read all into Y [colname=var];
N=nrow(Y);
p=ncol(Y);
nu=p;
b1=2;
```

```
b2=1.25;  /* Make this large for non-descending form */
d0=sqrt(nu)+b1/sqrt((2)); /* Distance Criterion for weight */
print "Campbell Criteria", "b1=" b1 "b2=" b2 "d0=" d0[FORMAT=7.4],;
wm=j(n,1);
test=1;k=0;
testcrit=.001;  /* convergent criterion based on max weight change */
do until (test<testcrit);
k=k+1;
 mn=wm'*Y/sum(wm);
 Ydev=Y-j(n,1)*mn;
 S=j(p,p,0);
 S=Ydev'*diag(wm##2)*Ydev/(wm'*wm-1);
 Sinv=inv(S);
 om=j(n,1);
 old=wm;
 /* Next few lines Compute New Weights */
   do i=1 to n;
      om[i,]=sqrt(ydev[i,]*Sinv*ydev[i,]');
      if om[i,]<=d0 then wm[i,]=om[i,];
        else wm[i,]=d0*exp(-.5*(om[i,]-d0)##2/(b2*b2));
   end;
   if k=1 then om1=om;
    if k=1 then do;
     print "Unweighted Mean Vector",mn[FORMAT=7.4 COLNAME=VAR],;
     print "Unweighted Covariance Matrix",S[format=7.4 COLNAME=VAR ROWNAME=VAR],;
     print "Q-Q plot for Original Data",;
      dsq=om##2;
      call quantile;
     end;
    wm=wm/om;
    test=abs(max(old-wm));
   outer=wm||om;
/* print outer; */
end;
PRINT "Number of iterations = " k;
print "Final Weighted Means",mn[format=7.4 COLNAME=VAR],;
print "Final Weighted Covariance Matrix",S[format=7.4 ROWNAME=VAR COLNAME=VAR],;
print "Q-Q plot for Weighted Data";
dsq=om##2;
call quantile;
ind=cusum(j(n,1));
print "Final Weights, Final Distances and Original Distances",
ind wm[FORMAT=5.3] om[FORMAT=6.3] om1[FORMAT=6.3];
```

Chapter 4

A program is included to compute Bookstein's shape coordinates as well as baseline size and centroid size. The uniform factor scores are also computed. This is a slight modification of the UNICOORD.SAS routine in Rohlf and Bookstein (1990). That example uses another data set. A small additional IML routine takes the coordinates produced and plots them for each landmark; and for the mean landmarks for a sample.

Thin-plate Splines discussed in Section 4.6 is available in the program TPSPLINE. Directions for its use are given in Appendix III using the

Agrenocythere data listed in Problem 6 of this Chapter. Also see the N.B. accompanying that problem. The data is supplied on the program disk available from the author of the Supplement.

```
/*              Shape Coordinates Analysis
                      UNICOORD.SAS
   Data Set is Assumed to consist of alternating x and y
   landmarks, like data here for samples of Brizalina.
   All other numerical data is stripped from the data set in
   the DATA statements. One Alphanumeric variable will be identified
   as a ROW label in IML. Other variables may be put back later with
   a subsequent DATA   ; SET ; or MERGE  ; series of statements.
   The output consists of Base Size=distance between basal landmarks,
   Centroid Size (in square root form) and the x and y uniform components.
   These may be used in Tsquared, MANOVA, regression, plotting etc.
   The data is also put out to a disk file called DIGROT.OUT.
   Shape coordinates are computed by rotating the data to a base
   line, and then dividing by the length of the base line.  SIZE
   is in original units of base line length; as is Centroid Size.
   The formats must be correct.  Base line subscripts, and whether data
   is x,y or y,x must be entered on the lines with a series of $$$$$.
   Also the name of ROWNAME variable must be changed.
   */
data new;
infile "brizalin.dat";   /* give path if necessary */
input sample $  mag xb yb xc yc xd yd xa ya xe ye xf yf;
          /* statements above and below must reflect actual form
             of data in file; extraneous variable mag dropped below.
             x,y variable names not used, but labels preserve sense
             of original data - important in this example; but usually not */
drop mag;
proc iml worksize=64; /* may have to increase for larger arrays */
reset fuzz;
base1=1; /* $$$$$ index of the 0,0 variable here */
base2=4; /* $$$$$ index of the 1,0  or 0,1 variable here */
switch=0; /* $$$$$  put 0 if x,y data; 1 if a y,x data   */
use _last_;
read all  into R[colname=coords rowname=sample]; /* $$$$$  replace sample */
/* print r[format=4.0 colname=coords rowname=sample];*/
N=nrow(R);
m=ncol(R);
p=m/2;   /* actual number of coordinate pairs */
 X=j(n,p); /* saves space for necessary arrays */
 Y=j(n,p);
 W=j(n,p);
 Z=j(n,p);
 do i=1 to p;  /* separates out coordinates in x and y arrays */
     x[,i]=r[,(i-1)*2+1+switch];
     y[,i]=r[,(i-1)*2+2-switch];
 end;
xmn=X-x[,:]*j(1,p);  /* three steps to compute centroid size=c_size */
ymn=Y-y[,:]*j(1,p);
c_size=sqrt(xmn[,##]+ymn[,##]);
free xmn ymn;
/* print r[format=4.0 colname=coords rowname=sample];     */
 X=X-x[,base1]^J(1,p);   /* translate all specimens to base1 as */
 Y=Y-y[,base1]*j(1,p);          /* the 0,0 point */
```

```
theta=j(n,1);
theta=atan(y[,base2]/x[,base2]);    /* angle to rotate coordinates */
W=diag(cos(theta))*X+diag(sin(theta))*Y;  /* actual rotation to base line */
Z=-diag(sin(theta))*X+diag(cos(theta))*Y;
/* print w[format=6.2] z[format=6.2];  take out comment if want to print here */
do j=1 to p;          /* put coords back in r, but not size adjusted */
 r[,(j-1)*2+1+switch]=w[,j];
 r[,(j-1)*2+2-switch]=z[,j];
end;
size=w[,base2];    /* set aside baseline length */
Uni=j(n,2);        /* next 5 lines compute uniform components */
wmn=w[:,];
zmn=z[:,];
denom=zmn[,##];
uni[,1]=(W*zmn')/denom;
uni[,2]=(Z*zmn')/denom;
coords={'size' 'c_size' 'uni_x' 'uni_y'}||coords; /* adjust output labels */
R=diag(size##-1)*R;      /* divide all coordinates by baseline length */
R=size||c_size||uni||R;   /* create array with all variables */
   create new3 from R [colname=coords rowname=sample];/* $$$$$ change sample*/
   append from R [rowname=sample];    /* $$$$$ to row variable name above */
data one;set new3;
file 'digrot.out'; /* $$$$$ you will have to adjust number of x,y pairs
                        according to the number of landmarks in your data */
put sample size c_size uni_x uni_y xb yb xc yc xd yd xa ya xe ye xf yf;
/* proc print; if you want to look at what was produced */
/*  you would plot shape coordinates, do regression of c_size on
      coordinates to look for allometry, Hotelling's Tsquared,
      Manova or what have you:
PROC PLOT;PLOT y2*x2=sample x2*c_size=sample uni_x*uni_y=sample;  etc.
PROC REG;
   MODEL C_SIZE=x2-x6 y2-y6; Note that x1,y1,x7,y7 left out as
                             these were chosen as the ends of the
                             base line. */
/* Below is an example of MANOVA for the three sample run
     through PROC DISCRIM - one of many ways - others may use
     PROC CANDISC, PROC GLM or PROC ANOVA  - MANOVA is the
     multivariate generalization of ANOVA; as Hotelling T**2
     is to the t test */
/*PROC DISCRIM  anova simple manova; CLASS SPECIES;
   Var x2-x6 y2-y6;  */
     run;
```

Plot Shape Coordinates

Program in PLOTCOOR.SAS

```
/* Program to take shape coordinates and plot them using IML
    Separate plots for all data and sample means.         */
DATA coords; infile 'digrot.out';
input sample $ size csize uni_x uni_y xb yb xc yc xd yd xa ya
xe ye xf yf;
proc iml;
reset fuzz nolog;
use coords; /* data set created in unicoord.sas */
read all into X[rowname=sample colname=coords];
```

```
p=ncol(x)-4; /* pairs of coordinates - sizes and x,y uniform */
X=X[,5:p+4];
coords=coords[,5:p+4];
/* print x[rowname=sample format=5.2 colname=coords]; */
/*  Must know columns in which coords for C, D, E and F are found
    if you don't know - a preliminary run with PRINT COORDS;
    would indicate this, or PROC CONTENTS DATA=NEW3; */
call pgraf(X[,3:4],sample,"XC","YC","Triangle ABC");
call pgraf(X[,5:6],sample,"XD","YD","Triangle ABD");
call pgraf(X[,9:10],sample,"XE","YE","Triangle ABE");
call pgraf(X[,11:12],sample,"XF","YF","Triangle ABF");
sampmn={"1" "2" "3" "4" "5"};   /* new labels for means */
D=design(num(sample));
Mean=diag(d[+,]##-1)*D'*X;
print mean[format=5.2 rowname=sampmn colname=coords];
call pgraf(Mean[,3:4],sampmn',"XC","YC","Mean Triangle ABC");
call pgraf(Mean[,5:6],sampmn',"XD","YD","Mean Triangle ABD");
call pgraf(Mean[,9:10],sampmn',"XE","YE","Mean Triangle ABE");
call pgraf(Mean[,11:12],sampmn',"XF","YF","Mean Triangle ABF");
quit;
```

Chapter 6

A program illustrates the use of Aitchison's principal component analysis on data which is constrained to sum to a constant. Results are given for an example in Reyment and some additional output is given corresponding to the various forms of association matrices given in Aitchison (1986).

Program in File ECHIAITC.SAS

```
/* Aitchison analysis of constrained frequency data
   uses data from Table 1 in Chapter 6 for Echinocythereis
   Data is a small data set so is accessed directly rather
   than from a file */
data echino;
input spec $ morph1 morph2 morph3;
cards;
1 .73 .21 .06
2 .33 .66 .01
3 .16 .79 .05
4 .26 .58 .16
5 .13 .74 .13
6 .42 .58 0
7 .21 .71 .07
8 .08 .88 .04
proc iml;
reset fuzz nolog noname pagesize=66;
use _last_;
read all into X[colname=varnm rowname=spec];
N=nrow(X);
d=ncol(X);
print "N= " N  "Number of variables =" d,;
```

```
  print "Original Data",X[colname=varnm rowname=spec],;
 delta=.005; /* Round off value for 0 replacement */
 do i=1 to N; /* 12 lines do Aitchison 0 replacement */
   cnt=0;
   do j=1 to D;
      if x[i,j]=0 then cnt=cnt+1;
   end;
   zero_rpl=delta*(cnt+1)*(D-cnt)/(D*D);
   correc=delta*cnt*(cnt+1)/(D*D);
   if cnt>0 then do j=1 to D;
   if x[i,j]=0 then x[i,j]=zero_rpl;
   else x[i,j]=x[i,j]-correc;
  end;
 end;
 print "Adjusted Data",X[colname=varnm rowname=spec],;
 LogX=log(X);
 dsmall=d-1;                 /* Dimension of simplex  */
 X=diag(x[,+]##-1)*X; /* Standardizes to Col. Sum=1 if not already done */
 mnx=x[:,];
 Xdev=X-j(N,1)*mnx;
 Var=Xdev'*Xdev/(N-1);    /* Covariance Matrix of Proportions */
 Dig=diag(vecdiag(var)##-.5);
 Rx=Dig*Var*Dig;       /* Correlation Matrix of Proportions */
 call eigen(a,E,Var);
  a=a[1:d,];  /*  Reduced Roots and Vectors since Simplex */
  E=E[,1:d];
 Xscore=Xdev*E;
 print "Data Matrix Rows Add to 1",X[format=5.2 colname=varnm],;
 print "Correlation Matrix", Rx[format=7.4 rowname=varnm colname=varnm],;
 rmnlogx=logx[:,];
 print "Eigenvalues",a[FORMAT=8.5],;
 print "Eigenvectors",E[FORMAT=6.3 ROWNAME=varnm],;
 Z=Logx-rmnlogx*j(1,d); /* produces a log ratio matrix */
 /* sumz=z[,+]; */  /* check that rows sum to 0 */
 print sumz;   */  /* check that rows sum to 0 */
 mnz=z[:,];
 Zdev=Z-j(N,1)*mnz;
 Gamma=Zdev'*Zdev/(N-1); /* Simplex covariance matrix */
 Dg=diag(vecdiag(gamma)##-.5);
 Rg=Dg*Gamma*Dg;          /* Simplex correlation matrix */
 print "Gamma Correlation",Rg[FORMAT=7.4 rowname=varnm colname=varnm],;
 Colgam=vecdiag(Gamma)*j(1,D);
 Tau=Colgam+Colgam'-2*Gamma; /* Tau form of simplex */
 Dcoltau=tau[,d]*j(1,d);
 Sigma=.5*(Dcoltau+Dcoltau'-Tau);
 Sigma=Sigma[1:dsmall,1:dsmall];   /* Sigma form of Simplex */
 print "Gamma Covariance",Gamma[FORMAT=6.4 rowname=varnm colname=varnm],;
 print "Tau Matrix",Tau[FORMAT=6.4 rowname=varnm colname=varnm],;
 print "Sigma Matrix",Sigma[FORMAT=6.4 rowname=varnm colname=varnm],;
 call eigen(a,E,Gamma);
 E=E[,1:dsmall];
 a=a[1:dsmall,];
 print "Eigenvalues of Gamma",a [FORMAT=7.4],;
 print "Eigenvectors of Gamma",E[FORMAT=6.3 ROWNAME=VARNM],;
 Zscore=Zdev*E;
 reset pagesize=24;
 call pgraf(Zscore[,1:2],spec,"PC1","PC2","Simplex Scores");
 call pgraf(Xscore[,1:2],spec,"PC1","PC2","Raw Data Scores");
```

Chapter 7

Analysis of Asymmetrical Matrices

This program uses as an example the data presented in Table 7.15. The data are transformed into proportions, and then partitioned into a symmetrical matrix and skew-symmetrical matrix following Gower (1977). The symmetrical part is analysed using Principal Coordinate Analysis using the same techniques used earlier; and the skew symmetric is analysed using the SVD following Gower. Each pair of coordinates are rotated before being plotted using a PCA analysis.

SKEWMC.SAS

```
/* Analysis of Asymmetrical comparison matrix  - see Chapt 7*/
data dickmc;input env $ x1-x7;  /* single letter codes for environ */
cards;
n 0 117 286 148 0 2 0
p 38 0 5 2 0 0 1
w 301 10 0 175 1 138 12
m 538 3 168 0 29 320 281
b 0 0 0 9 0 0 8
l 2 0 168 292 0 0 20
s 0 1 147 617 161 25 0
proc iml;
reset fuzz noname nolog pagesize=66;
use dickmc;
read all into X[rowname=env];
p=ncol(X);
pt=int(p/2);
/* Next three rows turn McCammon data into proportions - would
   usually leave these rows out */
Z=diag(x[,+]);
Prop=inv(Z)*x+I(p);
print "Data as Proportions",Prop[format=7.4 rowname=env],;
/* Next steps compute symmetrical part and do a Principal
   Coordinates Analysis on the M matrix    */
M=(PROP+PROP')/2;
print "Symmetrical Partition ", M[FORMAT=7.4],;
mr=m[:,];
mg=mr[,:];
Mcent=M-j(p,1)*mr-(j(p,1)*mr)'+mg*j(p,p);
call eigen(a,E,Mcent);
Es=E[,1:p-1]*diag(a[1:p-1,]##.5);
print "Principal Coordinate Scores",Es[format=7.4],;
reset pagesize=24;
call pgraf(Es[,1:2],env,"PCOORD1","PCOORD2"); /* could do 3:4 also */
/* Rest of Program does Skew Symmetric Part */
N=(PROP-PROP')/2;
print "Skew Symmetric Partition ",N[format=7.4 rowname=env],;
call svd(U,sing,V,N);
/* print U[format=7.4],sing[format=7.4], V[format=7.4]; */
/* The U matrix is treated as pairs of columns, and a PCA analysis
```

```
    is done on each pair to rotate the plots */
  do i=1 to pt;
    ind=(i-1)*2+1;
    Ut=U[,ind:ind+1];
    umn=ut[+,];
    Devu=Ut-j(p,1)*umn;
    Su=Devu'*Devu/(p-1);
    call eigen(a,E,Su);
    Sct=Ut*E*sqrt(sing[ind,1]);
    PRINT "Plots for Pair of Axes for Plane" i;
    call pgraf(Sct,env);
    if i=1 then Scu=Sct;
    if i>1 then Scu=Scu||Sct;
  end;
  print "Scores for SVD of Skew Symmetric Matrix",Scu[format=7.4];
quit;
```

Mantel Test

Test of correlation between two similarity or distance matrices. This procedure is discussed in the text, but no direct example is given. Two small test data matrices are used in this example.

Program in File MANTELT.SAS

```
/*******************************************************************
MANTEL TEST for similarity of two square symmetrical matrices.
Matrices one and two are square, or lower diagonal (in the form
of a vector). May be distance, association, or what have you
matrices.  Result gives Asymptotic probabilities.  If you don't want
Random permutation - then add comment indicators to last
10 lines.  You may choose number of permutations - nrep=100 now.
For large matrices the asymptotic test can be quite slow.
Assumes distance matrices have 0 on diagonal, otherwise will
consider as association matrix.
*******************************************************************/
    data one;input x @@;cards;   /* note is lower triangular distance */
    0 1 0 1.4 1.1 0.0 .9 1.6 .7 0
    data two;input y @@;cards;   /* also lower tri. dist., but could  be square */
    0 .5 0 .8 .5 0 .6 .9 .4  0
    proc iml worksize=64;
    reset noname nolog fuzz;
    z=uniform(0);   /* seed random number generator */
    use one;        /* read in first matrix */
    read all into x;
    ncx=ncol(X);
    if ncx=1 then X=sqrsym(x); /* lower triangle vs square */
    if sum(diag(X))>0 then X=X-diag(X); /* distance vs all else */
    ax=sum(X);
    bx=ssq(X);
    dx=ssq(x[+,]);
    gx=ax##2;
    hx=dx-bx;
    kx=gx+2*bx-4*dx;
```

```
 use two;
 read all into y;
 ncy=ncol(y);
 if ncy=1 then Y=sqrsym(y);
 if sum(diag(Y))>0 then y=y-diag(Y);
 ay=SUM(Y);
 by=ssq(Y);
 dy=ssq(y[+,]);
 gy=ay##2;
 hy=dy-by;
 ky=gy+2*by-4*dy;
 N=nrow(X);
 xrand=j(N,1);
 df=N*(N-1);
 xvar=(BX-GX/df);
 yvar=(BY-GY/df);
 den=(xvar*yvar)##.5;
 ez=sum(x)*sum(y)/df;
 varz=(2*bx*by+4*hx*hy/(N-2)+kx*ky/((N-2)*(N-3))-gx*gy/df)/df;
 zobs=sum(X#Y);
 cov=(zobs-ez);
 r=cov/den;
  z=cov/sqrt(varz);
prob=1-probnorm(z);
PRINT "Pearson correlation coefficient r=" r[format=6.4],;
PRINT "z=" z[format=6.4] " Asymptotic probability for z=" prob[format=6.4],;
/*Add slash-star and star-slash to lines below to remove permutation
  test & alternative probability. Set nrep to whatever you want */
 nrep={100};
 ncnt=0;
 do i=1 to nrep;
       Mmat=design(rank(uniform(xrand)));
 New=Mmat*X*Mmat';
 if sum(New#Y)>zobs then ncnt=ncnt+1;
 end;
 prob= ncnt/nrep;
 print ncnt " out of " nrep, "Empirical prob=" prob[format=6.4];
```

Thin-plate Splines

The data for the ichthyosaur comparison discussed in Section 8.12 is supplied on the Program disk together with the TPSPLINE program. The use of this program is discussed in Appendix III.

Appendix I

Pc SAS Guide on a Sheet

L. F. Marcus

A semi-colon ;;;;;;;;;;;;;;; terminates each SAS statement.

Function keys:

F1—help—brings up a help screen—**[PgDn]** to see more.

F2—keys—gives you a longer version with more keys defined.

F3—log—takes you to log window.

F4—output—takes you to output window.

F5—next—takes you to "next" window—rotates through windows.

F6—pgm—takes you to the program window.

F7—zoom—windows to whole screen; or toggles back to part.

F8—subtop—submits first statement in the program window.

F9—recall—recalls last program and so on for all programs.

F10—submits statements in program window for execution.

[Alt] F9—returns last command to command line (and so on).

Other combinations are explained by **F2** and more **[PgDn]** after F2.

[Home] key takes you to the command line in a window.

[End] key deletes to the end of a line.

Important COMMANDS on the Command line of the Program Window

include 'a:fn.ft'—takes file fn.ft from a: into program window.

file 'a:this.put'—stores any window contents in file a:this.put.

bye—terminates SAS session.

x—temporarily exits to DOS; **EXIT** in DOS returns to SAS.

clear—clears almost any window contents.

end—closes and removes window for KEYS, OPTIONS or HELP etc.

file 'prn'—sends contents of window to printer.

BASIC SAS JOB (almost) ALWAYS INCLUDES FOLLOWING

DATA whatever; /∗ Data name optional up to 8 characters ∗/
INPUT X @@; /∗ @@ means continue to read data from same line ∗/
CARDS; /∗data follows—comments can be included this way ∗/
7 14 15 18 19
PROC PRINT; /∗ do something to the data using a proc ∗/
RUN; /∗ execute the last proc—another proc does the same thing

Note—comments do not work on a data line ∗/

Some Edit procedures in the Program Window

Arrow keys, [PgDn], [PgUp], [Ins], [Del]—right of keyboard.
[**D**]elete, [**I**]nsert, [**C**]opy and [**M**]ove—in number area [**Enter**].
DD, CC, MM—in number area for line range; must appear twice.
[**A**]fter and [**B**]efore where to copy or move lines to.

See other KEY combinations for more or you can program them.

HELP or **[F1]** is extensive. Explore it. Explore **OPTIONS** and **KEYS** also, eg.
options page = 66; will set lines on a page to 66 so that printing Log or Output
Window will not make the printer advance paper so often. You can **FILE**
output or the log to a file and **INCLUDE** it to the Program Window to edit
before printing.

Appendix II

Summary of IML

A number of special functions are available. At this point I will review and summarize most of the matrix algebra operations and functions that will be needed in the supplement. For a full discussion of the many other functions and applications available in IML, which include high resolution graphics, data base applications, linear programming and many functions not used here, it is recommended that one get the SAS/IML guide available from SAS. We will only use a very small portion of the power of IML in this supplement. A few special non-standard operations are convenient and indicated below; their use will be minimized. In all that follows, unless otherwise indicated, the word matrix will stand for matrix, vector or scalar as the latter two can be considered as special (simple) matrices.

Operators

Function performed	SAS operator	Example
Reverse sign of array	–	A = –A
Addition of matrices!	+	C = A + B
Subtraction of matrices!	–	C = A – B
Matrix multiplication!	*	C = A*B
Matrix power	**	C = A**2
Division! (and element Division)	/	C = A/3
Transpose (NOTE Left ')	`	C = B`
Subscript	[ ]	aij = A[i,j]
Element power	# #	St_Dev = S # # .5
Inverse (square matrix)	** – 1 or INV	B = A** – 1 or B = INV(A)
Enclose multiple values	{ }	A = {1 2, 3 4}
Vertical concatenation!	//	A = B//C
Horizontal concatenation!	‖	A = B‖C

Element maximum	$< >$	$max = A[< >,]$
Element minimum	$> <$	$min = A[> <,]$
Logical and	**&**	$(a < 3) \& (C > 7)$
Logical or	¦	$(a = 3) ¦ (b = 0)$
Less than	$<$	$a < b$
Greater than	$>$	$b > a$
Less or equal	$< =$	$a < = b$
Greater or equal	$> =$	$a > = b$
Not	$\wedge$	$\wedge b$
Not equal	$\wedge =$	if $a \wedge = 0$
Addition (subscript reduction)	$+$	$Sum\ x = x[+,]$
Mean (subscript reduction)	$:$	$Mean\ x = x[:,]$
Sum of squares (subscript reduction)	$\# \#$	$SSX = x[\# \#,]$
Element product!	$\#$	$A = B \# C$

! Matrices must conform for the operation to execute.

Some special matrices are of considerable value and are briefly explained below with the functions.

Trigonometric, Transformation and Probability Functions. These operate on all the elements of a single argument matrix. A vector is always assumed to be a column vector unless otherwise indicated.

Single Matrix Argument Functions. These appear on the right side of an equation with the matrix argument in parentheses, eg. $B = SIN(A)$ will take all of the angles in radians in A and put the sines in B.

ARCOS—arc cosine	**MIN**—minimum value
ARSIN—arc sine	**NCOL**—number of columns
ABS—absolute value	**NROW**—number of rows
ATAN—arc tangent	**SIN**—sine
COS—cosine	**SQRT**—square root
DET—determinant	**SSQ**—sum of squares of elements
EXP—exponential	**SUM**—sum of all elements
INV—inverse	**TAN**—tangent
LOG—natural logarithms	**TRACE**—trace
MAX—maximum value	

More complicated functions or subroutines and some special matrices; all appear on the right of an = signs except the CALL functions, e.g. to plot a scattergram you say CALL PGRAF(XY) where the XY array contains the two columns to plot against each other.

Function and use

APPEND—adds data to the end of a SAS data set

CLOSE—close an open data set (for rereading e.g. see CAMPRIDG.SAS)

CONTENTS—returns variable names of a SAS data set

CREATE—sets up a new SAS data set

CUSUM—sums elements in row major order

DESIGN—creates a design matrix, e.g. $A = \{1,1,3\}$; $D = DESIGN(A)$; creates a 3×3 matrix with 1's in first col of row 1 and 2, and a 1 in row 3, col 3.

DET—returns the determinant of a non-singular matrix, e.g. **DET(A)**;

DIAG—forms a matrix of diagonal elements of the matrix **A** and 0 elsewhere, e.g. $D = DIAG(A)$;

—forms a matrix of diagonal elements from the vector **a**, and 0 elsewhere, e.g. $D = DIAG(a)$;.

DO—with **END** for loops.

—fills a matrix with a sequence of values in row major order, e.g. $A = DO(.5,20.5,1)$; generates the sequence .5, 1.5, 2.5, 3.5, 4.5, . . . 20.5 in A.

EIGEN—**CALL EIGEN(A,E,R)** for symmetrical **R**: **A** is a column vector of latent roots; and **E** the set of column latent vectors. $R = E*DIAG(A)*E'$.

FINISH—last statement in subroutine module

FREE—free memory space by removing matrices, e.g. **FREE A B C;**

GENEIG—generalized eigen solution: **CALL EIGEN(M,E,A,B);** where $A*E = B*E*DIAG(M)$

—A and B are symmetric, and B is positive definite.

GINV—$G = GINV(A)$; produces, Moore-Penrose generalized inverse

I—identity matrix, e.g. **IDEN** = **I(n)** defines IDEN as an $n \times n$ identity matrix

J—matrix of identical values

—$n \times n$ matrix of 1's—**j(n,n,1)**

—$n \times 1$ matrix of 1's—**j(n,1)**

—$n \times n$ null matrix—**j(n,n,0)**

—$1 \times n$ row vector of 3's—**j(1,n,3)**

LOAD—load from disk storage; used here for modules

NORMAL—pseudo-random normal deviate, e.g. **XRAN = NORMAL(seed);**

PGRAF—to produce a scatter plot. **CALL PGRAF(XY);** where XY has x's in column 1 and y's in column 2, also labelling etc. available (see programs here for examples)

PRINT—prints arrays, e.g. **PRINT A,B,C;** labels and formats are possible as shown in examples

PROBCHI—chi-squared probabilities, e.g. **PCHI = PROBCHI(A,nu);** with nu degrees of freedom, left tail; **PNCHI = PROBCHI(A,nu,nc);** with nc non-centrality parameter

PROBF—F distribution probabilities: **PF = PROBF(A,nd,dd);** nn is numerator; dd = denominator degrees of freedom **PNF = PROBNF(A,nd,dd,nc);** see PROBCHI above

PROBIT—inverse normal(0,1), e.g. **Z = PROBIT(.5);**

PROBNORM—normal probability of z variable

PROBT—student's t: **Pt = PROBT(A,df);** left tail; **PNt = PROBT(A,df,nc);**

READ—read data from data set to a matrix, e.g. **READ ALL INTO X;**

RESET—sets many options such as **FW** = field width for numbers; **FUZZ** to print near 0 values as 0; **NOLOG** sends output to OUTPUT window instead of default **LOG** window; **NONAME** prints matrices without names instead of default **NAME.**

ROOT—Cholesky decomposition

RUN—run a sub-program module

SORT—sort a SAS data set

SQRSYM—takes lower triangular matrix in vector form and makes into a square symmetric matrix

START—first line in subroutine module, e.g. **START QUANTILE;**

SVD—singular value decomposition: **CALL SVD(U,Q,V,A);** returns U,Q,V for A where $A = U*DIAG(Q)*V'$.

SYMSQR—square symmetrical to vector of lower triangle

UNIFORM—random number 0 to 1

USE—opens an external data set to be read, e.g. **USE_LAST_;**

VECDIAG—vector from diagonal of square matrix vd = VECDIAG(A)

WORKSIZE—actually an option in PROC IML, e.g. **PROC IML WORKSIZE = 80;** used to allocate additional memory to IML.

There are a number of programming statements not given above that are illustrated and used in the programs, e.g. **IF ... THEN ... , ELSE, GOTO,** and **DO** loops. They are used in much the same manner as in other programming languages such as BASIC, PASCAL, FORTRAN and C and will not be explained here.

Appendix III

Thin-plate Spline

The TPSPLINE program is a stand alone program which can be run on almost any IBM PC (or clone) with graphics capability (CGA, EGA, VGA, Hercules, 8514 etc.). A mouse makes the program more versatile, but is not necessary. It is useful, but not required, to have a math co-processor (8087 chip). 640K of RAM is necessary for realistic analyses. The program is small enough to run on floppy disks, but a hard disk is recommended.

The program is in its own directory together with a README.TPS file that gives full instructions for installation, data formats, running and interpretation of results. Each set of homologous coordinates is in its own data file, and any pair of patterns to be compared must have the same number of points. At present the maximum number of points is 25. The current upper limit is given by pressing the **Help** key after executing the program.

After consulting the README.TPS file you can execute the program by typing TPSPLINE [Return] when in the TPSPLINE directory.

One can then interactively enter data points ·with a mouse, or two files containing the homologous data sets to compare, following the example in README.TPS.

In addition to the examples supplied by Rohlf, there are four data files containing two examples from the Reyment text. The files are in the TPSPLINE directory on the Supplement disk.

1. Comparison of *Agrenocythere hazelae* and *radula*. Files are printed in Chapter 4, problem 6; and are named as AGRENO.HAZ and AGRENO.RAD on the disk. Compare the bending energy and warping energies to the values in the discussion of Example 4.8 in the text.

2. A discussion of the comparison of the ichthyosaurs *Grippia* and *Cymbospondylus* is given in Chapter 8, Section 8.12. The data for the two forms are in the files GRIPPIA.DAT and CYMBOSPO.DAT. Compare the output to values given in the Reyment text.

ORDER FORM

MULTIDIMENSIONAL PALAEOBIOLOGY PROGRAMS

Programs are described and listed in the Supplement. Requirements: IBM compatible microcomputer with SAS (Statistical Analysis System) installed including PROC IML. (TPSPLINE requires graphics, but does not require SAS to be present.) Programs will be available not described in the Supplement.

Programs written for SAS IML version 6.04. Will run with slight modification on main frame computers (IBM, DEC, UNIX suporting, etc. that support SAS). All relevant data sets and IML source code are supplied in ASCII form.

$5\frac{1}{4}''$ or $3\frac{1}{2}''$ floppy disk. $20.00

Postage and packing. $4.00

(1991–92 prices) Institutional orders or prepaid US$ checks. NY residents please add $8\frac{1}{4}\%$ sales tax.

NAME. . . . .

TEL . . . .

STREET . . . .

CITY . . . .

STATE. . . . . ZIP. . . . .

COUNTRY. . . . .

P.O. Box 524,
Planetarium Station,
New York, NY 10024